T0344872

Matrix Metalloproteinase Biology

Matrix Metalloproteinase Biology

Edited by

Irit Sagi
Professor of Biological Chemistry and Biophysics,
Department of Biological Regulation,
Weizmann Institute of Science,
Rehovot, Israel

Jean P. Gaffney
Assistant Professor of Chemistry at Baruch College,
Department of Natural Sciences,
City University of New York,
New York, NY, USA

WILEY Blackwell

Published by John Wiley & Sons, Inc., Hoboken, New Jersey

Published simultaneously in Canada

Library of Congress Cataloging-in-Publication Data:

Matrix metalloproteinase biology / edited by Irit Sagi and Jean P. Gaffney.
p. ; cm.
Includes bibliographical references and index.
ISBN 978-1-118-77232-4 (cloth)
I. Sagi, Irit, editor. II. Gaffney, Jean P., editor.
[DNLM: 1. Matrix Metalloproteinases. QU 136]
QP552.M47
572′.696–dc23

2015000037

Typeset in 11/13pt Times by SPi Global, Chennai, India.

Printed in Singapore by C.O.S. Printers Pte Ltd

10 9 8 7 6 5 4 3 2 1

1 2015

Contents

List of Contributors

Christoph Becker-Pauly
Biochemisches Institut
Medizinische Fakultät
Christian-Albrechts-Universität zu Kiel
Kiel, Germany

Jian Cao
Department of Medicine
State University of New York at Stony Brook
Stony Brook, NY, USA

Jillian Cathcart
Department of Medicine
State University of New York at Stony Brook
Stony Brook, NY, USA

Ivan E. Collier
Departments of Medicine
Division of Dermatology
Washington University School of Medicine
St. Louis, MO, USA

Howard C. Crawford
Department of Cancer Biology
Mayo Clinic
Jacksonville, FL, USA

Antoine Dufour
Department of Oral Biological & Medical Sciences and Department of Biochemistry and Molecular Biology
Centre for Blood Research
University of British Columbia
Vancouver, BC, Canada

Gregg B. Fields
Torrey Pines Research Institute for Molecular Studies
Port St. Lucie, FL, USA

Marco Fragai
Magnetic Resonance Center and Department of Chemistry
University of Florence
Florence, Italy

Gregory I. Goldberg
Departments of Medicine
Biochemistry and Molecular Biophysics
Washington University School of Medicine
St. Louis, MO, USA

Barbara Grünwald
Institute for Experimental Oncology and Therapy Research
Klinikum rechts der Isar
Technische Universität München
Munich, Germany

Ulrich auf dem Keller
Department of Biology
Institute of Molecular Health Sciences
ETH Zurich,
Zurich, Switzerland

Achim Krüger
Institute for Experimental Oncology and Therapy Research
Klinikum rechts der Isar
Technische Universität München
Munich, Germany

Claudio Luchinat
Magnetic Resonance Center and Department of Chemistry
University of Florence
Florence, Italy

Dmitry Minond
Cancer Research
Torrey Pines Research Institute for Molecular Studies
Port St. Lucie, FL, USA

Christopher M. Overall
Department of Oral Biological & Medical Sciences and Department of Biochemistry and Molecular Biology
Centre for Blood Research
University of British Columbia
Vancouver, BC, Canada

Ashleigh Pulkoski-Gross
Department of Medicine
State University of New York at Stony Brook
Stony Brook, NY, USA

Stefan Rose-John
Biochemisches Institut
Medizinische Fakultät
Christian-Albrechts-Universität zu Kiel
Kiel, Germany

Pascal Schlage
Department of Biology
Institute of Molecular Health Sciences
ETH Zurich,
Zurich, Switzerland

M. Sharon Stack
Harper Cancer Research Institute
University of Notre Dame
South Bend, IN, USA

Maciej J. Stawikowski
Torrey Pines Research Institute
Torrey Pines, FL, USA

Stanley Zucker
VA Medical Center
Northport, NY, USA

1 Matrix Metalloproteinases: From Structure to Function

Maciej J. Stawikowski[1] and Gregg B. Fields[1]

Departments of Chemistry and Biology, Torrey Pines Institute for Molecular Studies, Port St. Lucie, USA

1.1 Introduction

Members of the matrix metalloproteinase (MMP) family are known to catalyze the hydrolysis of a great variety of biological macromolecules. Proteomic approaches have significantly expanded the number of known MMP substrates. However, the mechanisms by which macromolecular substrates are processed have often proved elusive. X-ray crystallography and NMR spectroscopy have yielded detailed information on structures of MMP domains and, in a few cases, full-length MMPs. As structures of MMPs and their substrates have been reported, examination of MMP•substrate complexes has provided insight into mechanisms of action. We examine the structures of MMPs and their substrates and consider how the various structural elements of MMPs contribute to the hydrolysis of biological macromolecules.

1.2 Structures of MMPs

1.2.1 General MMP structure and domain organization

MMPs belong to the M10 zinc metalloproteinase family [1]. All MMPs have the characteristic zinc binding motif HExxHxxGxxH in their catalytic domain. MMPs possess similar domain organizations. Most MMPs consist of a signal peptide followed by four distinct domains, the N-terminal prodomain (propeptide), catalytic (CAT) domain, linker (hinge) region, and C-terminal hemopexin-like (HPX) domain (Fig. 1.1). The membrane-type (MT) MMPs contain an additional transmembrane (TM) domain that anchors them to the cell membrane. Following the TM domain is a small cytoplasmic "tail".

There are several exceptions to this general domain organization. MMP-7 and MMP-26 (matrilysins) lack the linker region and HPX domain and thus are referred to as "minimal MMPs". MMP-2 and MMP-9 possess three repeats of fibronectin type

Matrix Metalloproteinase Biology, First Edition. Edited by Irit Sagi and Jean P. Gaffney.

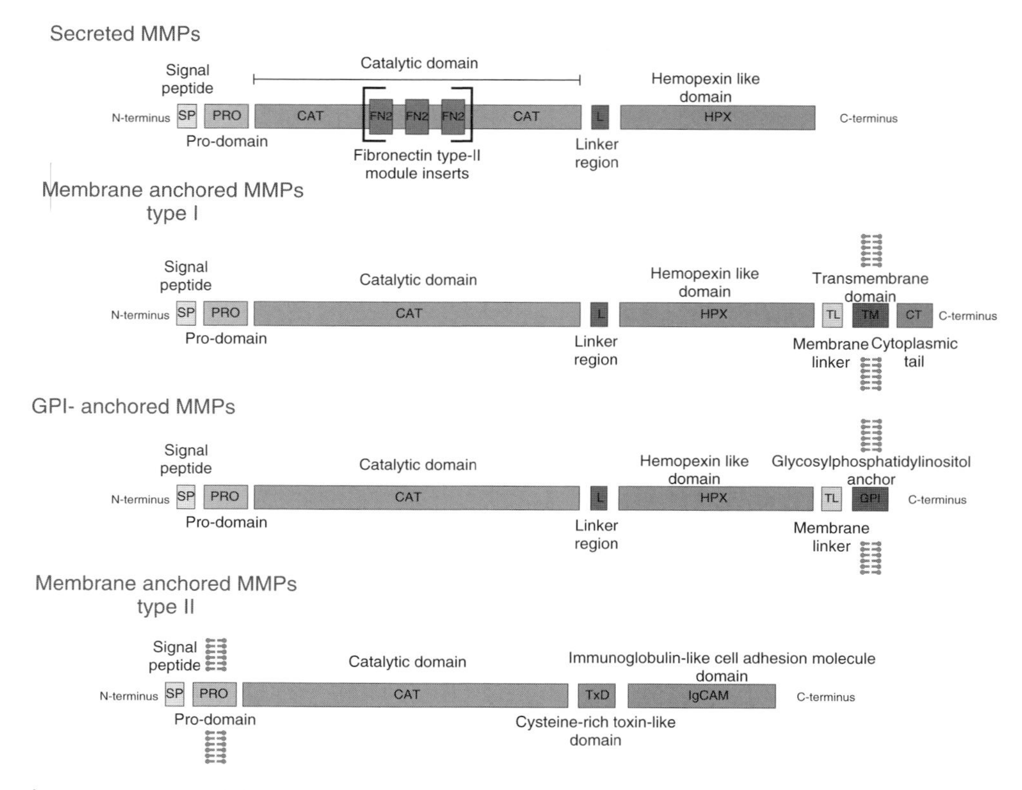

Figure 1.1 General domain organization of MMPs. (*See insert for color representation of this figure.*)

II-like motifs within the CAT domain. MMP-17 and MMP-25 are type I TM enzymes anchored to membranes through a C-terminal glycosylphosphatidylinositol (GPI) residue [2]. The N-terminal MMP-23 pro-domain contains a type II TM domain that anchors the protein to the plasma membrane. Instead of the C-terminal HPX domain common to other MMPs, MMP-23 contains a small toxin-like domain (TxD) and an immunoglobulin-like cell adhesion molecule (IgCAM) domain.

1.2.2 Catalytic domain

The topology of the CAT domain is similar among all MMPs. The CAT domain is composed of a five- stranded β-sheet which is interrupted by three α-helices (Fig. 1.2). Four of the five β-strands are aligned in a parallel fashion, while only the smallest "edge" strand runs in the opposite direction. Between strands III and IV there is an S-loop fixed by a structural Zn atom. The center of the catalytic site is located at helix B and the loop connecting it with helix C. This center helix provides the first and second His residues of the Zn-binding motif along with "catalytic" Glu residue. The loop behind this helix provides the third zinc binding His residue. Further down along this loop there is a 1,4 β-turn forming Met residue. This residue is highly conserved among metzincins and is believed essential for the structural integrity of the zinc-binding site. However, MMP-2 mutants where the conserved Met was replaced

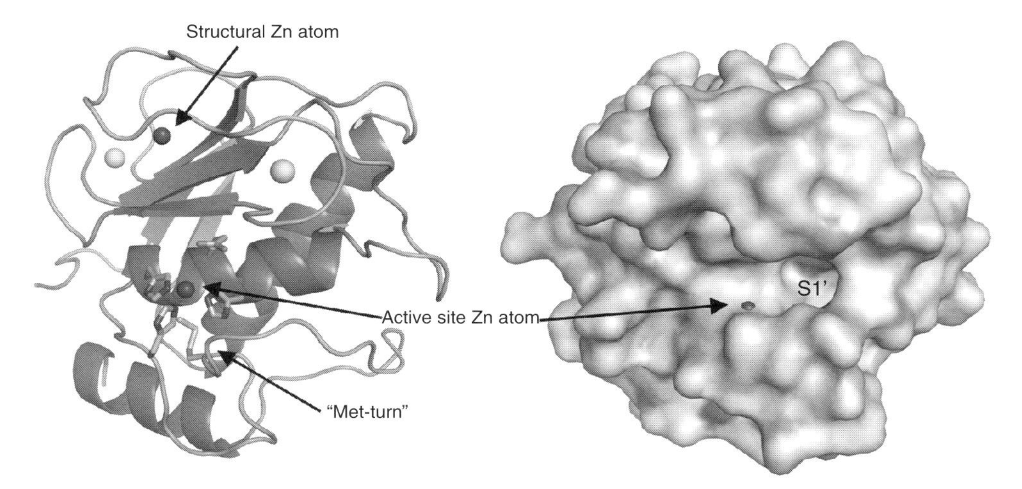

Figure 1.2 Typical structure of the CAT domain of MMPs. Characteristic structural elements are highlighted with arrows. Figure generated using MMP-8 structure (PDB 2OY2) [4]. (*See insert for color representation of this figure.*)

with Leu or Ser were able to cleave gelatin, type I collagen, and chemokine monocyte chemoattractant protein-3 with similar efficiency as wild-type MMP-2 [3].

1.2.3 Catalytic mechanism

On the basis of early structural information, a catalytic mechanism for MMPs was proposed (Fig. 1.3) [5, 6]. The carbonyl group of the scissile bond coordinates to the active site zinc (II) ion. A water molecule is hydrogen bonded to a conserved Glu residue and coordinated to the zinc (II) ion. The water molecule donates a proton to the Glu residue, allowing the generated hydroxide ion to attack the carbonyl at the scissile bond. This attack results in a tetrahedral intermediate, which is stabilized by the zinc (II) ion. The Glu residue transfers a proton to the nitrogen of the scissile amide, the tetrahedral intermediate rearranges, and amide bond hydrolysis occurs. During this catalytic process, the carbonyl from a conserved Ala residue helps to stabilize the positive charge at the nitrogen of the scissile amide.

1.2.4 Fibronectin type II-like inserts

Gelatinases (MMP-2 and MMP-9) bind to gelatin and collagen with significant contribution from their three fibronectin type II-like (FN2) repeats. MMP-2 and MMP-9 are unique among the MMPs in that the three FN2 modules (Col-1, Col-2, and Col-3) are inserted in their CAT domain in the vicinity of the active site [7]. More specifically, the FN2 modules of MMP-2 and MMP-9 are inserted between the fifth β-strand and helix B in the CAT domain (according to active enzyme domain organization). The basic fold of the FN2 module comprises a pair of β-sheets, each made from two antiparallel strands, connected by a short α-helix (Fig. 1.4). The two β-sheets form a hydrophobic pocket that is part of a hairpin turn, which orients the surrounding

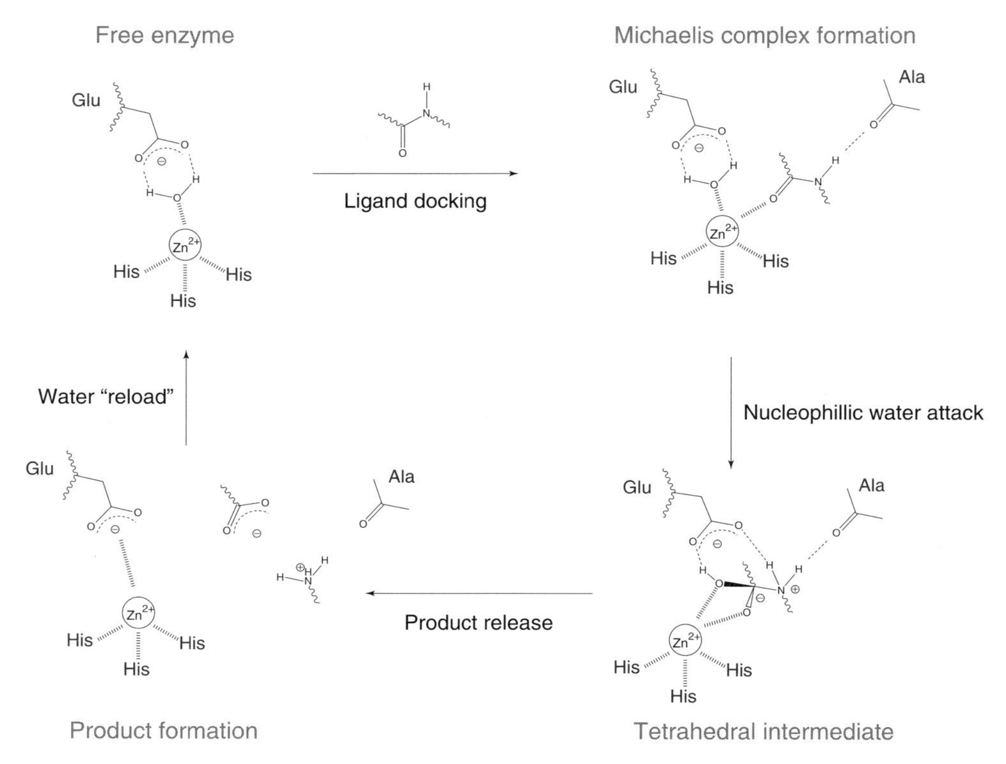

Figure 1.3 Mechanism of proteolysis catalyzed by MMPs. (Figure prepared based on mechanism proposed by Lovejoy et al. [5]). (*See insert for color representation of this figure.*)

aromatic side chains into the hydrophobic pocket. These pockets are the structural hallmark of the FN2 modules and contribute to substrate binding (see below) [8].

1.2.5 Linker region

The CAT domain is connected to the HPX domain via a linker (hinge) region. The length of this linker varies from 8 to 72 amino acids, depending on the enzyme (Fig. 1.5). The linker regions may be posttranslationally modified with sugar moieties. The conformational flexibility of the linker region contributes to MMP function. For example, in the case of MMP-9, it has been suggested that the long (72 residue), glycosylated, and flexible linker region mediates protein-substrate interactions by allowing the independent movement of the enzyme CAT and HPX domains [9]. Independent domain movements were also proposed to mediate enzyme translocation on collagen fibrils [10–12]. Domain flexibility may contribute to MMP activation via promoting long-range conformational transitions induced by the binding of activator proteins or ligand [13–15]. Finally, the linker region may help to re-orient the CAT domain with respect to the HPX domain during catalysis of collagen [16]. Domain flexibility may be rationalized for most MMPs by considering

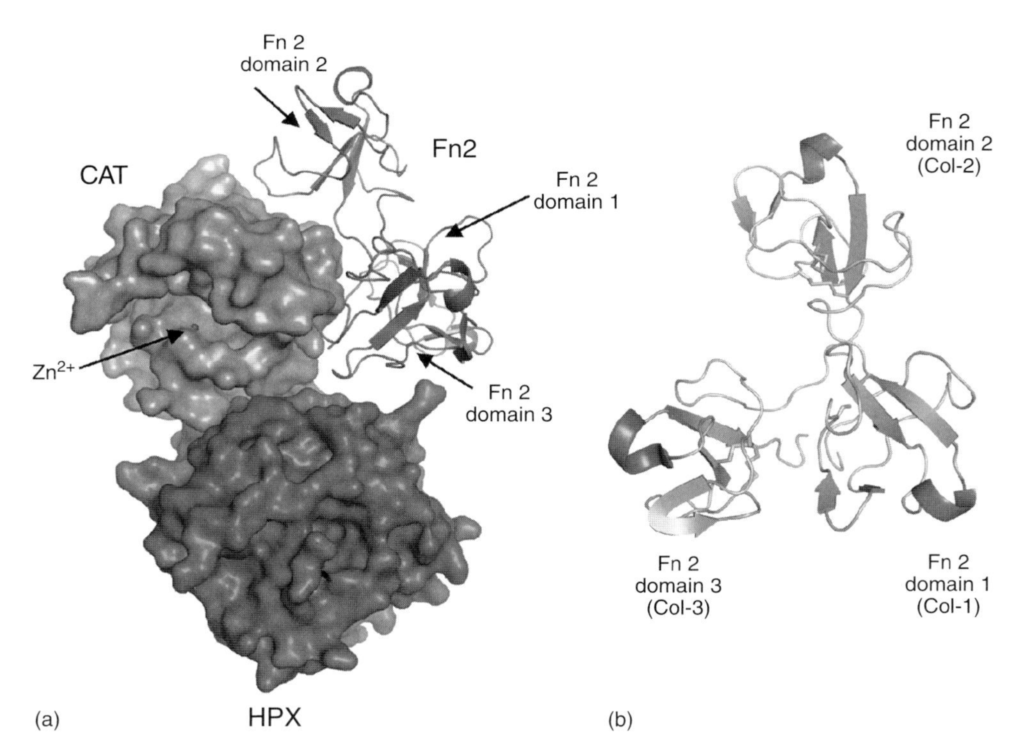

Figure 1.4 Fibronectin type II-like module structure and organization. (a) General orientation of FN2 modules of MMP-2. (b) Top view of FN2 modules. Figure prepared using MMP-2 structure (PDB 1CK7) [8]. (*See insert for color representation of this figure.*)

MMP	Linker region sequence	Length
MMP-26	GEK	3
MMP-23	GCLDRLFV	8
MMP-7	GKRSNSRKK	9
MMP-21	GSCEGSFDTAFDWI	14
MMP-1	GRSQNPVQPIGPQTPKA	17
MMP-13	GPGDEDPNPKHPKTPDK	17
MMP-8	GLSSNPIQPTGPSTPKP	17
MMP-27	GGLPKEPAKPKEPTIPHA	18
MMP-12	GDPKENQRLPNPDNSEPAL	19
MMP-2	GASPDIDLGTGPTPTLGPVTPEI	23
MMP-20	GPRKVFLGKPTLPHAPHHKPSIPDL	25
MMP-10	GPPPASTEEPLVPTKSVPSGSEMPAK	26
MMP-3	GPPPDSPETPLVPTEPVPPEPGTPAN	26
MMP-19	GKKSPVIRDEEEEETELPTVPPVPTEPSPMPDP	33
MMP-14	GGESGFPTKMPPQPRTTSRPSVPDKPKNPTYGPNI	35
MMP-11	GQPWPTVTSRTPALGPQAGIDTNEIAPLEPDAPPDA	36
MMP-17	GVRESVSPTAQPEEPPLLPEPPDNRSSAPPRKDVPHR	37
MMP-25	GKAPQTPYDKPTRKPLAPPPQPPASPTHSPSFPIPDR	37
MMP-28	GKPLGGSVAVQLPGKLFTDFETWDSYSPQGRRPETQGPKY	40
MMP-16	GPPDKIPPPTRPLPTVPPHRSIPPADPRKNDRPKPPRPPTGRPSYPGAKPNI	52
MMP-24	GPPAEPLEPTRPLPTLPVRRIHSPSERKHERQPRPPRPPLGDRPSTPGTKPNI	53
MMP-15	GTPDGQPQPTQPLPTVTPRRPGRPDHRPPRPPQPPPPGGKPERPPKPGPPVQPRATERPDQYGPNI	66
MMP-9	GPRPEPEPRPPTTTTPQPTAPPTVCPTGPPTVHPSERPTAGPTGPPSAGPTGPPTAGPSTATTVPLSPVDDA	72

Figure 1.5 Comparison of MMP linker lengths and sequences. Table was generated after alignment of human MMPs using sequences from the Uniprot database [19] and SeaView 4 [20] and Jalview [21] programs.

the amino acid composition (i.e., Gly and Pro residues) and the various lengths of linker regions (Fig. 1.5). The linker region and HPX domain of MT1-MMP and MMP-9 are proposed to offer allosteric control of enzyme dimer formation, which in turn modulates biological function [17, 18].

Glycosylation of MT1-MMP, which occurs in the linker region (residues 291, 299, 300, and 301), is required for the recruitment of tissue inhibitor of metalloproteinase 2 (TIMP-2) on the cell surface and subsequent formation of the MT1-MMP/TIMP-2/proMMP-2 trimeric complex and activation of proMMP-2 [22]. Glycosylation does not affect MT1-MMP collagen hydrolysis or autolytic processing [22].

1.2.6 Hemopexin-like domain

Except for MMP-7 and MMP-26, all vertebrate and human MMPs are expressed with a C-terminal HPX domain. The HPX domain is organized in four β-sheets (I to IV), arranged almost symmetrically around a central axis in a consecutive order (Fig. 1.6). The end result is a four-bladed propeller of pseudo-fourfold symmetry. Each propeller blade is formed by four antiparallel β-strands connected in a W-like topology, and is strongly twisted. The small C-terminal helix of the blade IV is tethered to the entering strand of blade I via a single disulfide bridge, stabilizing the whole domain. Within the central tunnel, up to four ions ($2Ca^{2+}$, $2Cl^{-}$) have been identified although their function is not clear [23].

The HPX domain mediates binding of MMP-1, MMP-8, MMP-13, MT1-MMP, and MMP-3 to collagen [24–28]. The HPX domain of MMP-2 was shown to possess critical secondary binding sites (exosites) required for the interactions of MMP-2 with fibronectin, and fibronectin was cleaved at a significantly reduced rate by an

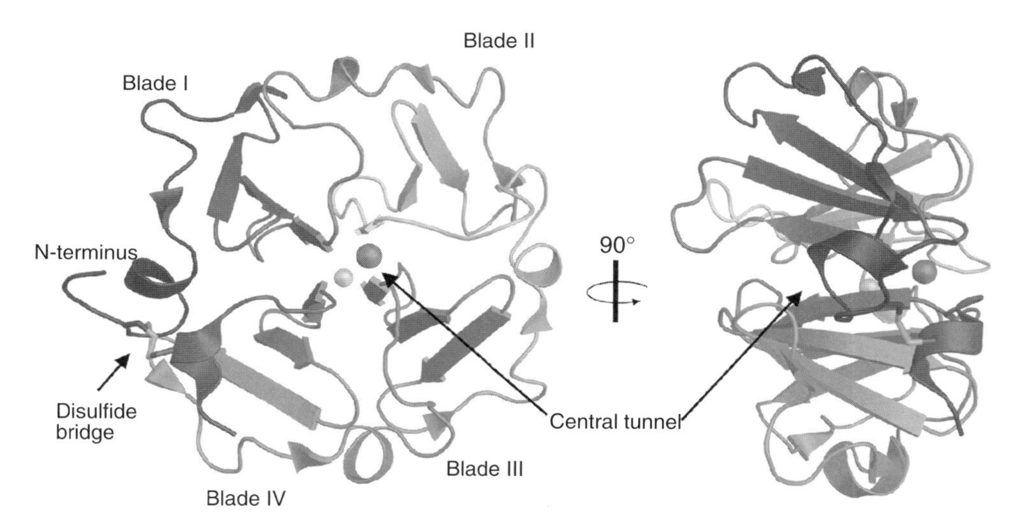

Figure 1.6 Typical structure of the HPX domain. The propeller-like structure is composed of four blades (I-IV) and stabilized by a single disulfide bridge, designated with an arrow. In the central tunnel, up to four different ions have been identified (here Ca^{2+} is orange and Cl^{-} is yellow). This figure was generated using the HPX domain of MT1-MMP (PDB 3C7X) [23]. (*See insert for color representation of this figure.*)

MMP-2 variant where the HPX domain was deleted [29]. In the case of MMP-2 and MMP-9, the HPX domain is important for interactions with TIMPs. The HPX domain of MMP-2 has also been shown to play a role in zymogen activation by MT1-MMP [30].

HPX domains modulate interaction of MMPs with cell-surface biomolecules. For example, the HPX of MMP-2 plays a role in the binding of the enzyme to the αvβ3 integrin [31, 32]. MT1-MMP has numerous cell surface binding partners, including tetraspanins (CD9, CD63, CD81, CD151, and/or TSPAN12), the α2β1 and αvβ3 integrins, and CD44 [33–39]. The HPX domain of MT1-MMP binds to CD63 and CD151 [35, 40]. Tetraspanins protect newly synthesized MT1-MMP from lysosomal degradation and support delivery to the cell surface [36].

CD44 also binds to MT1-MMP via the HPX domain of the enzyme, specifically blade I of the HPX domain [34, 41]. The association with CD44 leads to MT1-MMP localization to lamellipodia [34] [40]. The MT1-MMP/CD44 interaction promotes signaling through EGFR activation to the MAPK and PI3K pathways, enhancing cell migration [41]. CD44 also binds to MMP-9 via the HPX domain [40].

Highly efficient collagenolysis requires homodimerization of MT1-MMP, where association includes interactions of the HPX domain [42]. Homodimerization is symmetrical, involving residues Asp385, Lys386, Thr412, and Tyr436 in blades II and III of the HPX domain [43].

1.2.7 Transmembrane domain and cytoplasmic tail

On the basis of their method of attachment to the cell membrane, MT-MMPs may be classified into two groups, TM-type and glycosylphosphatidyl-inositol (GPI)-type. MT1-MMP (MMP-14), MT2-MMP (MMP-15), MT3-MMP (MMP-16), and MT5-MMP (MMP-24) are type I TM proteins with a short cytoplasmic tail that is involved in the regulation of intracellular trafficking and activity of these proteases [44–46]. MT4-MMP (MMP-17) and MT6-MMP (MMP-25) are bound to the cell surface by a GPI-mediated mechanism [2, 47].

Although the structure of the TM domain has not been solved experimentally, a model has been generated (Fig. 1.7). Besides facilitating cellular localization, the TM domain allows MT-MMPs to process a unique set of substrates, interact uniquely with TIMPs, and participate in a non-conventional mechanism of regulation involving enzyme internalization, processing, and ectodomain shedding [48, 49].

The cytoplasmic tail of MT1-MMP is distinct from those of MT2-MMP, MT3-MMP, and MT5-MMP, and is well characterized. The cytoplasmic tail of MT1-MMP is important in the ERK activation cascade [52], S1P-dependent G_i protein signaling [53], and VEGF upregulation through Src tyrosine kinase pathways [54]. The multifunctional gC1qR proteins can bind to the cytoplasmic tail of MT1-MMP in a similar manner to the cytoplasmic portion of adrenergic receptor [55]. More recently, Uekita et al. [56] have identified a new 19 kDa MT1-MMP cytoplasmic tail binding protein-1 (MTCBP-1). MTCBP-1 is localized between three subcellular compartments (membrane, cytoplasm, and nucleus) that can regulate gene expression and may suppress the invasion and migration-promoting activity of MT1-MMP [56]. The cytoplasmic tail of MT1-MMP increases the expression of

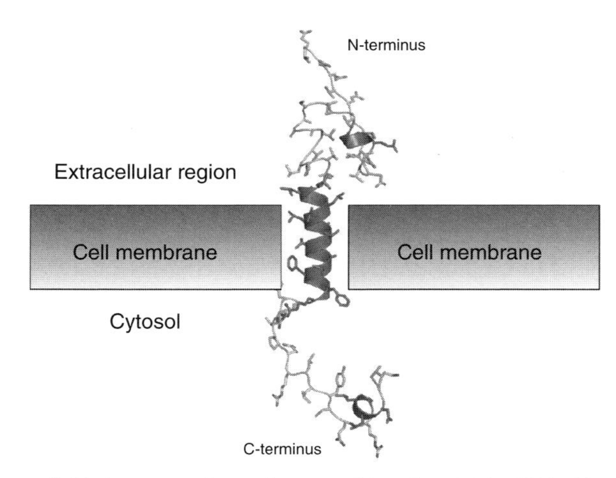

Figure 1.7 Structure of TM domain and cytoplasmic tail (residues 518–582) of human MT1-MMP generated by homology modeling [50, 51]. (*See insert for color representation of this figure.*)

hypoxia inducible factor-1 (HIF-1) target genes, which in turn stimulates aerobic glycolysis [57].

Phosphorylation of cytoplasmic Tyr573 of MT1-MMP is imperative for tumor cell migration and proliferation in three-dimensional collagen matrices and tumor growth in mice [58, 59], while phosphorylation of cytoplasmic Thr567 enhances tumor invasion of and growth within collagen matrices [60]. Interestingly, Tyr573 phosphorylation does not affect proteolytic activity, but may act by inducing relocalization of the enzyme and increasing the proportion of quiescent cells [58].

MT1-MMP undergoes both clathrin-mediated and caveolae-dependent endocytosis [61,62] [42]. The cytoplasmic tail has been implicated as necessary for endocytosis [61].

1.3 Overview of MMP substrate specificity

Extensive sequence specificity studies of many MMPs provided a number of important insights into the differences and similarities in subsite preferences among these enzymes. Substrate specificity studies have been performed with proteins and synthetic peptides.

Significant interactions between the MMPs and their substrates or inhibitors occur between the S_1' subsite and P_1' residue. MMPs may be classified as falling into two broad structural classes dependent on the depth of the S_1' pocket. This "selectivity pocket" is relatively deep for the majority of the enzymes (e.g., MMP-2, MMP-3, MMP-8, MMP-12, and MMP-13) but shallow in the case of MMP-1, MMP-7, and MMP-11.

The substrate-binding groove is relatively open at the S_3-S_1 and S_3' subsites and narrows at the S_1' and S_2' subsites. The S_1' subsite is a well-defined pocket that penetrates the surface of the enzyme. Differences between the various MMPs in the S_3–S_1

subsite region are relatively subtle. Interestingly, Pro is a preferred P_3 subsite moiety for many MMP substrates.

The S_2' subsite is a solvent-exposed cleft with a general preference for hydrophobic P_2' residues in both substrates and inhibitors. The S_3' subsite is a relatively poorly defined solvent-exposed region. While there are some variations in residues for this subsite for the various MMPs, the introduction of different P_3' substituents in general tends to have only a modest effect on inhibitor selectivity.

In addition to active site subsites, the specificity of MMPs is modulated by discrete binding sites outside of the catalytic center (exosites). Substrate interaction with exosites can influence the behavior of a proteinase in a number of ways. Exosites modulate and broaden the substrate specificity profile of MMPs by providing an additional contact area not influenced by the primary specificity subsites. In this way, the function of the proteinase is refined and can be made, in general, more specific or efficient. In addition to bringing substrates to the enzyme for potential hydrolysis, exosites may be involved in essential "substrate preparation" prior to cleavage. For example, the localized "unwinding" of native collagen substrates by MMPs is facilitated by exosites [16]. Exosites can also target the enzyme to substrates in tissues or to cell-associated substrates.

In collagenolytic MMPs (MMP-1, MMP-8, MMP-13, and MT1-MMP), exosites are found in the HPX domain, and in gelatinases (MMP-2 and MMP-9), on the three FN2 modules. In MMP-1, MMP-8, MMP-13, MT1-MMP, and MMP-3, the HPX domain binds native collagen. The FN2 modules in gelatinases form a collagen binding domain (CBD) which lies proximal to the S_3' subsite. The matrix binding properties of the CBD also have the potential to localize the enzyme to collagen, either in the extracellular matrix (ECM) or on the cell surface linked to β1 integrins.

1.3.1 ECM substrates

The repertoire of MMP substrates is extremely rich. To study proteolytic processes in detail (referred to as the protease web), a broad approach including gene deletion, transgenic mouse models, and genomic and proteomic profiling techniques is necessary. Degradomics, the characterization of all proteases, inhibitors, and protease substrates present in an organism using genomic and proteomic techniques, is a well-established method for MMP substrate identification [63].

The ECM is composed of two main classes of macromolecules: proteoglycans (PGs) and fibrous proteins [64, 65]. The main fibrous ECM proteins are collagens, elastin, fibronectin, and laminins [66]. PGs fill the majority of the extracellular interstitial space within the tissue in the form of a hydrated gel [64].

Collagen is the most abundant fibrous protein within the interstitial ECM and constitutes up to 30% of the total protein mass of a multicellular animal. Collagens provide tensile strength, regulate cell adhesion, support chemotaxis and migration, and direct tissue development [67]. Collagen associates with elastin, another major ECM fiber. Elastin fibers provide recoil to tissues that undergo repeated stretch. A third fibrous protein, fibronectin, is intimately involved in directing the organization of the interstitial ECM and has a crucial role in mediating cell attachment and function.

Collagenases (MMP-1, MMP-8, MMP-13, and MT1-MMP) catalyze the degradation of fibrillar collagens in their native triple-helical supersecondary structure. The physiological role of collagenases has been proposed to be the remodeling of the collagenous component of the ECM, including involvement in the wound healing process. Furthermore, since collagen is the predominant ECM deposit in fibrotic organs, collagenases are believed to be the main proteases responsible for the resolution of fibrosis and restoration of the normal ECM environment. Numerous ECM components, including types I, II, and III collagen, fibronectin, vitronectin, laminins 111 and 332, fibrin, and proteoglycans are substrates for MT1-MMP [68].

Gelatinases (MMP-2 and MMP-9) have been proposed to be involved in inflammatory processes and in tumor progression [69, 70]. However, gelatinases have also been found to have protective roles against cancer [71–74]. Gelatinases have been more recently recognized as participating in cardiovascular and auto-immune diseases. In the case of cardiovascular diseases, gelatinases participate in both the genesis of atherosclerotic lesions and to the acute event (i.e., stroke or myocardial infarction). In the case of auto-immune diseases, gelatinases are involved in the generation of remnant epitopes and in the modulation of cross-talk between immune system compartments.

Stromelysins (MMP-3, MMP-10, and MMP-11) share the ability to degrade types IV and IX collagen, laminin, fibronectin, elastin, and proteoglycans, although with significantly different affinities among them. Additional substrates include cytokines, growth factors, and soluble regulatory molecules [75]. Each stromelysin has a different physiological distribution in human tissues, hence the types of processes which are modulated are largely variable.

Among the matrilysins (MMP-7 and MMP-26), MMP-7 is widely expressed in human tissues and mainly in epithelial-derived ones. MMP-7 catalyzes the hydrolysis of cytokines, growth factors, and receptors [76]. MMP-7 biological functions mainly concern ECM remodeling and immune system modulation. The biological aspects of MMP-26 are so far restricted to ECM turnover and remodeling in a limited cohort of tissues both in physiological and pathological conditions [77, 78].

The matrix metalloproteinase term initially related to enzymes processing ECM proteins, but recent findings prove that the role of MMPs is much more sophisticated. MMPs contribute to processing of cytokines, chemokines, hormones, adhesion molecules, and membrane-bound proteins, resulting in modulation of normal cellular behavior, cell-cell communication, and tumor progression. The reader is referred to several excellent reviews on MMPs that have compiled an extensive list of substrates [79–81].

1.3.2 Cell surface substrates

Proteolytic events at the cell surface are of interest because of their potential to affect cellular functions. Cell surface-associated MMP-2, MMP-9, and MMP-13 can activate latent transforming growth factor-beta (TGFβ) [82]. MT1-MMP modulates the bioavailability of TGFβ (i) by activating MMP-13 and MMP-2 [83], (ii) by releasing active TGFβ from cell surface complexes involving the αvβ3 integrin [84], and/or (iii) by releasing a membrane-anchored proteoglycan, betaglycan, that binds TGFβ [85].

MT-MMPs can cleave and shed a variety of cell surface adhesion receptors and proteoglycans. CD44 (a multifunctional adhesion molecule) [86] and syndecan-1 [87] can be directly shed by MT1-MMP and MT3-MMP. The αv chain of the αvβ3 integrin, which is reported to play a crucial role in tumor angiogenesis, invasion, and metastasis, is processed by MT1-MMP into a functional form [88]. The multifunctional receptor of complement component 1q (gC1qr) is also susceptible to MT1-MMP proteolysis [89]. Low-density lipoprotein receptor related protein (LRP1/CD91) is a cell surface-associated endocytic receptor, implicated in the internalization and degradation of multiple ligands such as thrombospondins (1 and 2), α2-macroglobulin-protease complexes, urokinase- and tissue-type plasminogen activators, MMP-2, MMP-9, and MMP-13 [90, 91]. The cleavage of LRP1 by MT1-MMP in breast cancer and fibrosarcoma cells may thus lead to the control of the bioavailability and fate of many ligands and soluble MMPs in cancer progression [91]. MT1-MMP also sheds transglutaminase (Belkin et al., 2001), death receptor-6, MHC class I chain-related molecule A, E-cadherin, and ECM metalloproteinase inducer [92–96]. These highly divergent substrates for MT1-MMP make this enzyme a critical regulator of the pericellular environment.

1.3.3 Intracellular MMP targets

For a long time MMPs were viewed exclusively as ECM remodelers. More recently, there is evidence that MMPs cleave intracellular substrates, and that MMPs have been observed within cells in nuclear, mitochondrial, and various vesicular and cytoplasmic compartments, including the cytoskeletal intracellular matrix. Unbiased high-throughput degradomics approaches have demonstrated that many intracellular proteins are cleaved by MMPs, including apoptotic regulators, signal transducers, molecular chaperones, cytoskeletal proteins, systemic autoantigens, enzymes in carbohydrate metabolism and protein biosynthesis, transcriptional and translational regulators, and proteins in charge of protein clearance such as lysosomal and ubiquitination enzymes. Intracellular substrate proteolysis by MMPs is involved in innate immune defense and apoptosis, and affects oncogenesis and pathology of cardiac, neurological, protein conformational, and autoimmune diseases, including ischemia-reperfusion injury, cardiomyopathy, Parkinson's disease, cataract, multiple sclerosis, and systemic lupus erythematosus. Intracellular activation of MMPs strongly suggests that MMPs are responsible for proteolytic actions on intracellular substrates.

MMP-2 cleaves the cytoskeletal proteins desmin and α-actinin and colocalizes with α-actinin in cardiomyocytes [97]. MMP-2 and MMP-9-containing vesicles are aligned with the cytoskeleton in neurons and reactive astrocytes, and both gelatinases are found in cytoskeletal fractions from these cells [98, 99]. MT1-MMP and MT3-MMP are detected in cytoskeletal fractions of smooth muscle cells, where they cleave the cytoskeletal protein focal adhesion kinase (FAK) [100]. Moreover, cytoskeletal proteins constitute an important fraction of the intracellular degradomes of MMP-2, MMP-9, and MT1-MMP. Both pro- and activated MMP-1 are associated with the mitochondrial membrane in glial Müller cells, Tenon's capsule fibroblasts, corneal fibroblasts, and retinal pigment epithelial cells [101]. The mitochondrial localization of MMP-1 is found in resting cells, suggesting a physiological role for MMP-1 in cellular homeostasis. Both MMP-2 [102, 103] and MMP-9 [104] are detected in cardiac mitochondria during cardiac injury and increased levels of mitochondrial MMP-9 are associated with exacerbated mechanical dysfunction. Studies report nuclear localization of MMPs, including MMP-1, MMP-2, MMP-3, MMP-9, MMP-13, MMP-26, and MT1-MMP, and cleavage of nuclear matrix proteins. Nuclear translocation of MMP-3 was confirmed by Eguchi et al. [105], who showed that extracellular MMP-3 is taken up into chondrosarcoma cells and subsequently translocates to the nucleus where it induces transcription of the connective tissue growth factor (CTGF) gene. To avoid excessive proteolysis of nuclear proteins during cellular homeostasis, these nuclear MMPs may be under inhibition by TIMP-1 and TIMP-4, which are also present in the nucleus [106–109]. MMP-7 colocalizes with cryptdins (antimicrobial α-defensins, Crps) in mouse Paneth cells and mediates the processing and activation of various Crps *in vitro* [110]. MMP-7 cleaves pro-Crp-1, −6, and −15. MT1-MMP was shown to have an intracellular oncogenic function by cleaving the integral centrosomal protein, pericentrin [111, 112]. Pericentrin and pericentrin-2 (pericentrin-B or kendrin) are derived from splice variants of the same gene and are known to be essential for normal centrosome function by the anchorage of the γ-tubulin ring complex, which initiates microtubule nucleation, to the centrosome [113]. Besides its actions in the centrosomal compartment and at focal adhesions, activated MT1-MMP is also detected in the nuclei of hepatocellular carcinoma cells. Interestingly, liver cancer patients with nuclear MT1-MMP (and co-localized MMP-2) have a poor overall survival and large tumor size, whereas MT1-MMP is not found in nuclei of normal paralleled liver tissues and normal control livers [114]. Finally, MMP-1 was found to be strongly associated with mitochondrial membranes and nuclei and accumulated within cells during the mitotic phase of the cell cycle. The intracellular association of MMP-1 to mitochondria and nuclei conferred resistance to apoptosis, which may be a mechanism for tumor cells to escape from apoptosis [101].

MMP-2 was localized to sarcomeres in close association with the thin myofilaments in hearts subjected to ischemia/reperfusion (I/R) [102]. Interestingly, a different localization of the two gelatinases, MMP-2 and MMP-9, was observed in hearts of patients with dilated cardiomyopathy (DCM) compared to control hearts. In DCM hearts the gelatinases were localized exclusively within the cardiomyocytes in close association with the sarcomeric structure, whereas localization was mainly around the myocytes in control hearts. I/R injury is associated with the degradation of cytoskeletal proteins such as α-actinin, desmin, and spectrin [115]. This may constitute an additional intracellular function of MMP-2, as α-actinin and desmin (but not spectrin) were found to be *in vitro* substrates of MMP-2. Moreover, dopaminergic neuroblastoma cells under oxidative stress showed an upregulation of intracellular and secreted activated forms of MMP-3 and cleavage of α-synuclein, which was inhibited by an MMP inhibitor. Purified α-synuclein is cleaved by MMP-3 most efficiently, but also by MT1-MMP, MMP-2, MMP-1, and MMP-9 (ordered by decreasing efficiency) [116].

1.4 Selective mechanisms of action

1.4.1 Collagenolysis

Collagens are composed of three α chains of primarily repeating Gly-Xxx-Yyy triplets, which induce each α chain to adopt a left-handed polyPro II helix. Three chains then intertwine, staggered by one residue and coiled, to form a right-handed superhelix [117, 118]. Triple-helical structure provides collagens with exceptional mechanical strength and broad resistance to proteolytic enzymes. Interstitial collagens have long been recognized as being hydrolyzed by collagenolytic MMPs (MMP-1, MMP-8, MMP-13, and MT1-MMP) into one-fourth and three-fourth length fragments.

The 15 Å collagen triple-helix does not fit into the 5 Å MMP CAT domain active site cavity [119]. Models have generally accounted for the steric clash of the triple-helix with enzyme-active sites by (i) requiring active unwinding of the triple-helix by an MMP [119–121] and/or (ii) considering that the site of hydrolysis within collagen has a distinct conformation, or conformational flexibility, rendering it more susceptible to proteolysis than other regions in collagen [122].

A detailed mechanism of collagenolysis was developed from examination of structures and docking experiments of MMP-1 and MMP-1•triple-helical peptide (THP) complexes [16]. MMP-1 in solution is in equilibrium between open/extended and closed structures (Fig. 1.8(a)) [12]. The maximum occurrence (MO) of MMP-1 conformations in solution has recently been calculated, through paramagnetic NMR and small angle X-ray scattering [123]. Many of the MMP-1 conformations with the highest MO value (>35%) were found to have interdomain orientations and positions that could be grouped into a cluster [123]. Within this cluster, the collagen- binding residues of the HPX domain were solvent-exposed and the CAT domain correctly positioned for its subsequent interaction with the collagen. A approximately 50° rotation around a single axis of the CAT domain with respect to the HPX domain positioned the CAT domain right in front of the preferred cleavage site in interstitial collagen. The conformations belonging to this cluster can thus be seen as the antecedent step of collagenolysis.

The HPX domain then binds the leading chain (designated 1T) and the middle chain (designated 2T) of the THP and, due to the flexibility of the linker, the CAT domain is guided toward the Gly~Ile bond of chain 1T (Fig. 1.8(b)). Back-rotation of the CAT and HPX domains, to achieve the X-ray crystallographic closed MMP-1 conformation, resulted in visible perturbation of the THP (Fig. 1.8(c)). The domain movement drove chain 1T into the active site, allowing the polypeptide to establish a number of H-bonding interactions and the carbonyl oxygen of the cleavage site amide bond to coordinate the metal ion. This result is consistent with the experimentally observed weakening in NOEs for the interaction of chain 1T with chains 2T and 3T (the trailing chain) at the THP cleavage site. MMP-1 does not actively unwind the triple-helix [124]. Rather, MMP-1 shifts the equilibrium between native helical and locally unwound states, destabilizing the helical state and/or stabilizing the unwound state [124]. The position that the two peptide fragments would assume after cleavage in the present model (Fig. 1.8(d,e)) was almost superimposable to the

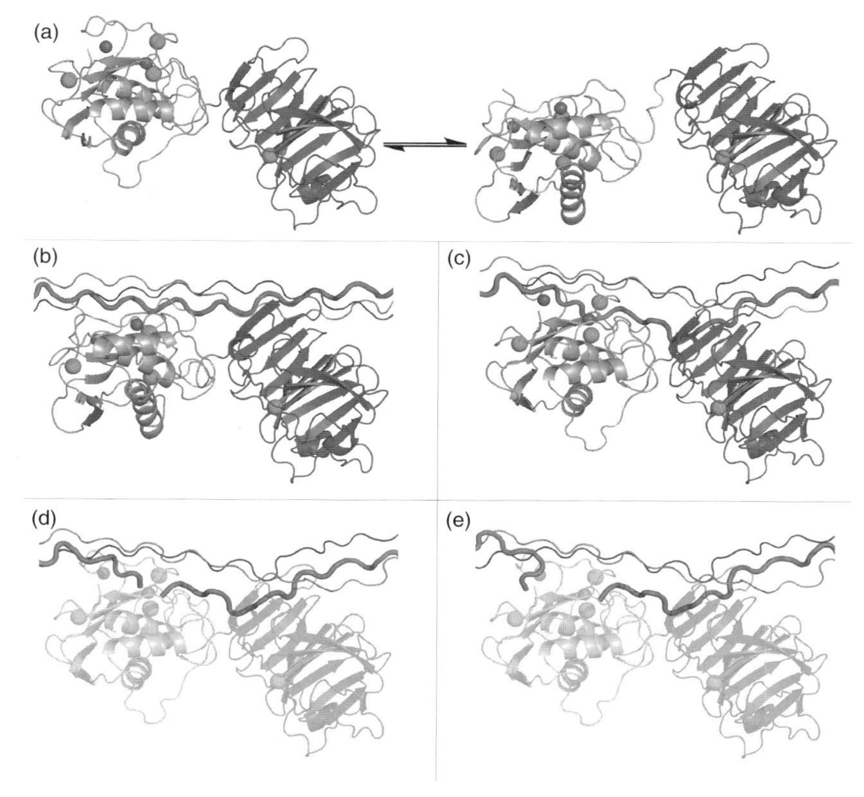

Figure 1.8 Mechanism of the initial steps of collagenolysis. (a) Closed (left) and open/extended (right) forms of MMP-1 in equilibrium. (b) The extended protein binds THP chains 1T-2T at Val23-Leu26 with the HPX domain and the residues around the cleavage site with the CAT domain. The THP is still in a compact conformation. (c) Closed FL-MMP-1 interacting with the released 1T chain (in magenta). (d) After hydrolysis, both peptide fragments (C- and N-terminal) are initially bound to the active site. (e) The C-terminal region of the N-terminal peptide fragment is released. (Reprinted with permission from [16]. Copyright (2012) American Chemical Society). (*See insert for color representation of this figure.*)

X-ray crystallographic structure of the complex between the MMP-12 CAT domain and the two fragments obtained by enzymatic cleavage of the α1(I) collagen model Pro-Gln-Gly-Ile-Ala-Gly hexapeptide at the Gly-Ile bond [125].

Binding sites for the triple-helix within the HPX domain have been identified [16, 126–128]. Initially, in MMP-1, Ile_{290} and Arg_{291} in the A-B loop of blade I were identified as key residues in collagenolysis [126]. Subsequently, Phe_{301}, Val_{319}, and Asp_{338} were implicated in collagen binding [127]. Phe_{320} was found to be an important contributor, along with Ile_{290} and Arg_{291}, to the S_{10}' binding pocket [128]. The S_{10}' binding pocket binds the P_{10}' subsite of collagen, which possesses a conserved Leu residue important for interaction of triple-helices with MMP-1 [127–129]. Other residues within the HPX domain may also participate in collagen binding [126–128].

Hydrolysis of collagen proceeds at the outer edge of the fibril [130, 131]. MMP-1 is a diffusion-based "Burnt Bridge" Brownian Ratchet capable of biased diffusion on the surface of collagen fibrils, where the bias is driven by proteolysis [11]. While on collagen fibrils, MMP-1 spends approximately 90% of its time in one of two distinct pause classes [132]. Class I occurs randomly along the fibril, while class II occurs periodically at 1.3 and 1.5 μm along the fibril and exhibits multistep escape kinetics [132]. Five percent of the class II pauses result in initiation of processive collagen degradation for approximately 15 consecutive cleavage events [132]. The temperature dependence of the pauses suggests local unfolding, but the low probability of hydrolysis (~5%) indicates that local unfolding is not sufficient for hydrolysis [132]. It has been proposed that local unfolding exposes the α2(I) chain, which is then reoriented by the MMP for hydrolysis to occur efficiently [133].

It has been noted that, while MMP-1 and MMP-8 have similar collagenolytic mechanisms, MMP-2, MMP-9, and MT1-MMP have mechanisms distinct from MMP-1 and MMP-8 [120, 138–140]. In the case of MMP-2 (and MMP-9 as well), interaction with collagen is primarily via the FN2 modules within the CAT domain, not the HPX domain [137, 138]. All three FN2 modules contribute to collagen binding, with the greatest effects observed for modules 2 and 3 [120, 139, 140]. Individual residues involved in collagen binding are primarily Arg (252, 296, and 368) and aromatics (Phe_{297}, Tyr_{302}, Tyr_{323}, Tyr_{329}, Trp_{374}, and Tyr_{381}) [139, 140]. It has been proposed that MMP-2 grossly distorts the triple-helix, followed by initial hydrolysis of the α2(I) chain [135].

1.4.2 Gelatinolysis

MMP-2 and MMP-9 gelatinolysis has been proposed to involve the three CBD modules. The contributions of the three CBD modules to MMP-2 activities on denatured α1(I) and α2(I) collagen chains were recently investigated [140]. The CBD was required for cleavage of both chains by MMP-2 as shown by the absence of cleavage in a triple point mutant of MMP-2. Eliminating binding to a single CBD modular binding site had a substantial effect on α1 chain cleavage, but had little effect on α2 chain cleavage, suggesting that positioning of the α2 chain is less specific to the modular binding sites and that this chain can utilize the binding site redundancies on the CBDs. Consistent with this is the observation that the α2 chain was preferentially cleaved by MMP-2 and that the K_M was 60-fold lower for the α1 chain compared with the α2 chain [135]. Both chains were most impacted by the elimination of the collagen binding site in module 3 [140]. On this basis, it was proposed that there is modular selectivity in α1 and α2 chain binding, which impacts their hydrolysis by MMP-2.

The strongest effects of substitution of single residues on MMP-2 activity were measured following the R368A modification in module 3 and the simultaneous substitutions of F297A and R368A in modules 2 and 3, respectively [140]. This is consistent with the results from prior NMR structural analyses that showed strong chemical shifts of R^{368} when recombinant proteins containing CBD modules 2 and 3 [141] or all three modules [143] were studied in complex with collagen-derived peptides. The CBD has been linked to the enzymatic activity of MMP-2 by the proposal that the CBD binds and positions substrate molecules relative to the active site for cleavage site by MMP-2 [142 –144].

Acknowledgments

We gratefully acknowledge the National Institutes of Health (CA98799, MH078948, AR063795, and NHLBI contract 268201000036C), the Multiple Sclerosis National Research Institute, and the Robert A. Welch Foundation for support of our research on MMPs.

References

1. Rawlings, N.D., Barrett, A.J. & Bateman, A. (2012) MEROPS: the database of proteolytic enzymes, their substrates and inhibitors. *Nucleic Acids Research*, 40 (Database issue), D343–50.
2. Itoh, Y., Kajita, M., Kinoh, H. *et al.* (1999) Membrane type 4 matrix metalloproteinase (MT4-MMP, MMP-17) is a glycosylphosphatidylinositol-anchored proteinase. *The Journal of Biological Chemistry*, 274 (48), 34260–6.
3. Butler, G.S., Tam, E.M. & Overall, C.M. (2004) The canonical methionine 392 of matrix metalloproteinase 2 (gelatinase A) is not required for catalytic efficiency or structural integrity: probing the role of the methionine-turn in the metzincin metalloprotease superfamily. *The Journal of Biological Chemistry*, 279 (15), 15615–20.
4. Bertini, I., Calderone, V., Fragai, M. *et al.* (2006) Snapshots of the reaction mechanism of matrix metalloproteinases. *Angewandte Chemie International Edition in English*, 45 (47), 7952–5.
5. Lovejoy, B., Hassell, A.M., Luther, M.A. *et al.* (1994) Crystal structures of recombinant 19-kDa human fibroblast collagenase complexed to itself. *Biochemistry*, 33 (27), 8207–17.
6. Stocker, W. & Bode, W. (1995) Structural features of a superfamily of zinc-endopeptidases: the metzincins. *Current Opinion in Structural Biology*, 5 (3), 383–90.
7. Collier, I.E., Wilhelm, S.M., Eisen, A.Z., Marmer, B.L. *et al.* (1988) H-ras oncogene-transformed human bronchial epithelial cells (TBE-1) secrete a single metalloprotease capable of degrading basement membrane collagen. *The Journal of Biological Chemistry*, 263 (14), 6579–87.
8. Morgunova, E., Tuuttila, A., Bergmann, U., Isupov, M. *et al.* (1999) Structure of human pro-matrix metalloproteinase-2: activation mechanism revealed. *Science*, 284 (5420), 1667–70.
9. Rosenblum, G., Cohen, S.R., Grossmann, J.G., Frenkel, J. *et al.* (2007) Insights into the structure and domain flexibility of full-length pro-matrix metalloproteinase-9/gelatinase B. *Structure*, 15 (10), 1227–36.
10. Overall, C.M. & Butler, G.S. (2007) Protease yoga: extreme flexibility of a matrix metalloproteinase. *Structure*, 15 (10), 1159–61.
11. Saffarian, S., Collier, I.E., Marmer, B.L. *et al.* (2004) Interstitial collagenase is a Brownian ratchet driven by proteolysis of collagen. *Science*, 306 (5693), 108–11.
12. Bertini, I., Fragai, M., Luchinat, C., Melikian, M. *et al.* (2009) Interdomain flexibility in full-length matrix metalloproteinase-1 (MMP-1). *The Journal of Biological Chemistry*, 284 (19), 12821–8.
13. Rosenblum, G., Meroueh, S., Toth, M., Fisher, J.F. *et al.* (2007) Molecular structures and dynamics of the stepwise activation mechanism of a matrix metalloproteinase zymogen: challenging the cysteine switch dogma. *Journal of the American Chemical Society*, 129 (44), 13566–74.
14. Bannikov, G.A., Karelina, T.V., Collier, I.E., Marmer, B.L. *et al.* (2002) Substrate binding of gelatinase B induces its enzymatic activity in the presence of intact propeptide. *The Journal of Biological Chemistry*, 277 (18), 16022–7.
15. Geurts, N., Martens, E., Proost, P., Opdenakker, G. *et al.* (2008) Beta-hematin interaction with the hemopexin domain of gelatinase B/MMP-9 provokes autocatalytic processing of the propeptide, thereby priming activation by MMP-3. *Biochemistry*, 47 (8), 2689–99.
16. Bertini, I., Fragai, M., Luchniat, C., Melikian, M. *et al.* (2012) Structural basis for matrix metalloproteinase 1 catalyzed collagenolysis. *Journal of the American Chemical Society*, 134, 2100–10.
17. Van den Steen, P.E., Van Aelst, I., Hvidberg, V., Piccard, H., Van den Steen, P.E. *et al.* (2006) The hemopexin and O-glycosylated domains tune gelatinase B/MMP-9 bioavailability via inhibition and binding to cargo receptors. *The Journal of Biological Chemistry*, 281 (27), 18626–37.
18. Lehti, K., Lohi, J., Lohi, M.M., Pei, D. *et al.* (2002) Oligomerization through hemopexin and cytoplasmic domains regulates the activity and turnover of membrane-type 1 matrix metalloproteinase. *The Journal of Biological Chemistry*, 277 (10), 8440–8.
19. Apweiler, R., Bairoch, A. & Wu, C.H. (2004) Protein sequence databases. *Current Opinion in Chemical Biology*, 8 (1), 76–80.

20. Gouy, M., Guindon, S. & Gascuel, O. (2010) SeaView version 4: A multiplatform graphical user interface for sequence alignment and phylogenetic tree building. *Molecular Biology and Evolution*, 27 (2), 221–4.
21. Troshin, P.V., Procter, J.B. & Barton, G.J. (2011) Java bioinformatics analysis web services for multiple sequence alignment--JABAWS:MSA. *Bioinformatics*, 27 (14), 2001–2.
22. Wu, Y.I., Munshi, H.G., Sen, R. *et al.* (2004) Glycosylation broadens the substrate profile of membrane type 1-matrix metalloproteinase. *The Journal of Biological Chemistry*, 279, 8278–8289.
23. Gohlke, U., Gomis-Rüth, F.X., Crabbe, T. *et al.* (1996) The C-terminal (haemopexin-like) domain structure of human gelatinase A (MMP2): structural implications for its function. *FEBS Letters*, 378 (2), 126–30.
24. Clark, I.M. & Cawston, T.E. (1989) Fragments of human fibroblast collagenase. Purification and characterization. *The Biochemical Journal*, 263 (1), 201–6.
25. Windsor, L.J., Birkedal-Hansen, H., Birkedal-Hansen, B. *et al.* (1991) An internal cysteine plays a role in the maintenance of the latency of human fibroblast collagenase. *Biochemistry*, 30 (3), 641–7.
26. Murphy, G., Allan, J.A., Willenbrock, F. *et al.* (1992) The role of the C-terminal domain in collagenase and stromelysin specificity. *The Journal of Biological Chemistry*, 267 (14), 9612–8.
27. Hirose, T., Patterson, C., Pourmotabbed T. *et al.* (1993) Structure-function relationship of human neutrophil collagenase: identification of regions responsible for substrate specificity and general proteinase activity. *Proceedings of the National Academy of Sciences of the United States of America*, 90 (7), 2569–73.
28. Knauper, V., Cowell, S., Smith, B. *et al.* (1997) The role of the C-terminal domain of human collagenase-3 (MMP-13) in the activation of procollagenase-3, substrate specificity, and tissue inhibitor of metalloproteinase interaction. *The Journal of Biological Chemistry*, 272 (12), 7608–16.
29. Wallon, U.M. & Overall, C.M. (1997) The hemopexin-like domain (C domain) of human gelatinase A (matrix metalloproteinase-2) requires Ca2+ for fibronectin and heparin binding. Binding properties of recombinant gelatinase A C domain to extracellular matrix and basement membrane components. *The Journal of Biological Chemistry*, 272 (11), 7473–81.
30. Itoh, Y., Takamura, A., Ito, N. *et al.* (2001) Homophilic complex formation of MT1-MMP facilitates proMMP-2 activation on the cell surface and promotes tumor cell invasion. *The EMBO Journal*, 20 (17), 4782–93.
31. Brooks, P.C., Silletti, S., von Schalscha, T.L. *et al.* (1998) Disruption of angiogenesis by PEX, a noncatalytic metalloproteinase fragment with integrin binding activity. *Cell*, 92 (3), 391–400.
32. Nisato, R.E., Hosseini, G. & Sirrenberg, C. (2005) Dissecting the role of matrix metalloproteinases (MMP) and integrin alpha(v)beta3 in angiogenesis in vitro: absence of hemopexin C domain bioactivity, but membrane-Type 1-MMP and alpha(v)beta3 are critical. *Cancer Research*, 65 (20), 9377–87.
33. Gálvez, B.G., Matías-Román, S., Yáñez-Mó, M. *et al.* (2002) ECM regulates MT1-MMP localization with beta1 or alphavbeta3 integrins at distinct cell compartments modulating its internalization and activity on human endothelial cells. *The Journal of Cell Biology*, 159, 509–521.
34. Mori, H. Tomari, T., Koshifumi, I. *et al.* (2002) CD44 directs membrane-type I matrix metalloproteinase to lamellipodia by associating with its hemopexin-like domain. *The EMBO Journal*, 21, 3949–3959.
35. Yañez-Mó, M. Barreiro, O., Gonzalo, P. *et al.* (2008) MT1-MMP collagenolytic activity is regulated through association with tetraspanin CD151 in primary endothelial cells. *Blood*, 112, 3217–3226.
36. Lafleur, M.A., Xu, D. & Hemler, M.E. (2009) Tetraspanin proteins regulate membrane type-1 matrix metalloproteinase-dependent pericellular proteolysis. *Molecular Biology of the Cell*, 20, 2030–2040.
37. Tomari, T., Koshikawa, N., Uematsu, T. *et al.* (2009) High throughput analysis of proteins associating with a proinvasive MT1-MMP in human malignant melanoma A375 cells. *Cancer Science*, 100, 1284–1290.
38. Fisher, K.E., Sacharidou, A., Stratman, A.N. *et al.* (2009) MT1-MMP- and Cdc42-dependent signaling co-regulate cell invasion and tunnel formation in 3D collagen matrices. *Journal of Cell Science*, 122, 4558–4569.
39. Sacharidou, A., Koh, W., Stratman, A.N. *et al.* (2010) Endothelial lumen signaling complexes control 3D matrix-specific tubulogenesis through interdependent Cdc42- and MT1-MMP-mediated events. *Blood*, 115, 5259–5269.
40. Murphy, G. & Nagase, H. (2011) Localizing matrix metalloproteinase activities in the pericellular environment. *The FEBS Journal*, 278, 2–15.
41. Zarrabi, K., Dufour, A. Li, J. *et al.* (2011) Inhibition of matrix metalloproteinase-14 (MMP-14)-mediated cancer cell migration. *The Journal of Biological Chemistry*, 286, 33167–33177.
42. Itoh, Y. & Seiki, M. (2006) MT1-MMP: a potent modifier of pericellular microenvironment. *Journal of Cellular Physiology*, 206, 1–8.

43. Tochowicz, A., Goettig, P., Evans, R. *et al.* (2011) The dimer interface of the membrane type 1 matrix metalloproteinase hemopexin domain: crystal structure and biological functions. *The Journal of Biological Chemistry*, 286, 7587–7600.
44. Jiang, A., Lehti, K., Wang, X. *et al.* (2001) Regulation of membrane-type matrix metalloproteinase 1 activity by dynamin-mediated endocytosis. *Proceedings of the National Academy of Sciences of the United States of America*, 98 (24), 13693–8.
45. Lehti, K., Valtanen, H., Wickstrom, S.A. *et al.* (2000) Regulation of membrane-type-1 matrix metalloproteinase activity by its cytoplasmic domain. *The Journal of Biological Chemistry*, 275 (20), 15006–13.
46. Uekita, T., Itoh, Y., Yana, I. *et al.* (2001) Cytoplasmic tail-dependent internalization of membrane-type 1 matrix metalloproteinase is important for its invasion-promoting activity. *The Journal of Cell Biology*, 155 (7), 1345–56.
47. Kojima, S., Itoh, Y., Matsumoto, S. *et al.* (2000) Membrane-type 6 matrix metalloproteinase (MT6-MMP, MMP-25) is the second glycosyl-phosphatidyl inositol (GPI)-anchored MMP. *FEBS Letters*, 480 (2–3), 142–6.
48. Hernandez-Barrantes, S., Bernardo, M., Toth, M. *et al.* (2002) Regulation of membrane type-matrix metalloproteinases. *Seminars in Cancer Biology*, 12 (2), 131–8.
49. Seiki, M., Mori, H., Kajita, M. *et al.* (2003) Membrane-type 1 matrix metalloproteinase and cell migration. *Biochemical Society Symposium*, 70, 253–62.
50. Ndinguri, M.W., Bhowmick, M., Tokmina-Roszyk, D. *et al.* (2012) Peptide-based selective inhibitors of matrix metalloproteinase-mediated activities. *Molecules*, 17 (12), 14230–48.
51. Roy, A., Kucukural, A. & Zhang, Y. (2010) I-TASSER: a unified platform for automated protein structure and function prediction. *Nature Protocols*, 5 (4), 725–38.
52. Gingras, D., Bousquet-Gagnon, N., Langlois, S. *et al.* (2001) Activation of the extracellular signal-regulated protein kinase (ERK) cascade by membrane-type-1 matrix metalloproteinase (MT1-MMP). *FEBS Letters*, 507 (2), 231–6.
53. Langlois, S., Gingras, D. & Beliveau, R. (2004) Membrane type 1-matrix metalloproteinase (MT1-MMP) cooperates with sphingosine 1-phosphate to induce endothelial cell migration and morphogenic differentiation. *Blood*, 103 (8), 3020–8.
54. Sounni, N.E., Roghi, C., Chabottaux, V. *et al.* (2004) Up-regulation of vascular endothelial growth factor-A by active membrane-type 1 matrix metalloproteinase through activation of Src-tyrosine kinases. *The Journal of Biological Chemistry*, 279 (14), 13564–74.
55. Rozanov, D.V., Ghebrehiwet, B., Ratnikov, B. *et al.* (2002) The cytoplasmic tail peptide sequence of membrane type-1 matrix metalloproteinase (MT1-MMP) directly binds to gC1qR, a compartment-specific chaperone-like regulatory protein. *FEBS Letters*, 527 (1–3), 51–7.
56. Uekita, T., Gotoh, I., Kinoshita, T. *et al.* (2004) Membrane-type 1 matrix metalloproteinase cytoplasmic tail-binding protein-1 is a new member of the Cupin superfamily. A possible multifunctional protein acting as an invasion suppressor down-regulated in tumors. *The Journal of Biological Chemistry*, 279 (13), 12734–43.
57. Sakamoto, T., Niiya, D. & Seiki, M. (2011) Targeting the Warburg effect that arises in tumor cells expressing membrane type-1 matrix metalloproteinase. *The Journal of Biological Chemistry*, 286, 14691–14704.
58. Nyalendo, C. *et al.* (2008) Impaired tyrosine phosphorylation of membrane type 1-matrix metalloproteinase reduces tumor cell proliferation in three-dimensional matrices and abrogates tumor growth in mice. *Carcinogenesis*, 29, 1655–1664.
59. Nyalendo, C., Sartelet, H., Barrette, S. *et al.* (2009) Identification of membrane-type 1 matrix metalloproteinase tyrosine phosphorylation in association with neuroblastoma progression. *BMC Cancer*, 9, 422.
60. Moss, N.M., Wu, Y.I., Liu, Y. *et al.* (2009) Modulation of the membrane type 1 matrix metalloproteinase cytoplasmic tail enhances tumor cell invasion and proliferation in three-dimensional collagen matrices. *The Journal of Biological Chemistry*, 284, 19791–19799.
61. Osenkowski, P., Toth, M. & Fridman, R. (2004) Processing, shedding, and endocytosis of membrane type 1-matrix metalloproteinase (MT1-MMP). *Journal of Cellular Physiology*, 200, 2–10.
62. Itoh, Y. & Seiki, M. (2004) MT1-MMP: an enzyme with multidimensional regulation. *Trends in Biochemical Sciences*, 29 (6), 285–9.
63. Lopez-Otin, C. & Overall, C.M. (2002) Protease degradomics: a new challenge for proteomics. *Nature Reviews. Molecular Cell Biology*, 3 (7), 509–519.
64. Jarvelainen, H., Sainio, A. Koulu, M. *et al.* (2009) Extracellular matrix molecules: potential targets in pharmacotherapy. *Pharmacological Reviews*, 61 (2), 198–223.
65. Schaefer, L. & Schaefer, R.M. (2010) Proteoglycans: from structural compounds to signaling molecules. *Cell and Tissue Research*, 339 (1), 237–46.

66. Alberts, B., Johnson, A., Lewis, J. *et al.* (2008) *Molecular Biology of the Cell*, 5th edn. Garland Science, New York, NY, USA, pp. 1616.
67. Rozario, T. & DeSimone, D.W. (2010) The extracellular matrix in development and morphogenesis: a dynamic view. *Developmental Biology*, 341 (1), 126–40.
68. Seiki, M. (2003) Membrane-type 1 matrix metalloproteinase: a key enzyme for tumor invasion. *Cancer Letters*, 194 (1), 1–11.
69. Overall, C.M. & Lopez-Otin, C. (2002) Strategies for MMP inhibition in cancer: Innovations for the post-trial era. *Nature Reviews. Cancer*, 2, 657–672.
70. Gialeli, C., Theocharis, A.D. & Karamanos, N.K. (2011) Roles of matrix metalloproteinases in cancer progression and their pharmacological targeting. *The FEBS Journal*, 278, 16–27.
71. Folgueras, A.R., Pendás, A.M., Sánchez, L.M. *et al.* (2004) Matrix metalloproteinases in cancer: from new functions to improved inhibition strategies. *The International Journal of Developmental Biology*, 48, 411–424.
72. Morrison, C.J., Butler, G.S., Rodríguez, D. *et al.* (2009) Matrix metalloproteinase proteomics: substrates, targets, and therapy. *Current Opinion in Cell Biology*, 21, 645–653.
73. Decock, J., Thirkettle, S., Wagstaff, L. *et al.* (2011) Matrix metalloproteinases: Protective roles in cancer. *Journal of Cellular and Molecular Medicine*, 15, 1254–1265.
74. Dufour, A. & Overall, C.M. (2013) Missing the target: matrix metalloproteinase antitargets in inflammation and cancer. *Trends in Pharmacological Sciences*, 34, 233–242.
75. Visse, R. & Nagase, H. (2003) Matrix metalloproteinases and tissue inhibitors of metalloproteinases: structure, function, and biochemistry. *Circulation Research*, 92 (8), 827–39.
76. Li, Q., Park, P.W., Wilson, C.L. *et al.* (2002) Matrilysin shedding of syndecan-1 regulates chemokine mobilization and transepithelial efflux of neutrophils in acute lung injury. *Cell*, 111 (5), 635–46.
77. Park, H.I., Ni, J., Gerkema, F.E. *et al.* (2000) Identification and characterization of human endometase (Matrix metalloproteinase-26) from endometrial tumor. *The Journal of Biological Chemistry*, 275 (27), 20540–4.
78. Marchenko, G.N., Ratnikov, B.I., Rozanov, D.V. *et al.* (2001) Characterization of matrix metalloproteinase-26, a novel metalloproteinase widely expressed in cancer cells of epithelial origin. *The Biochemical Journal*, 356 (Pt 3), 705–18.
79. Cauwe, B., Van den Steen, P.E. & Opdenakker, G. (2007) The biochemical, biological, and pathological kaleidoscope of cell surface substrates processed by matrix metalloproteinases. *Critical Reviews in Biochemistry and Molecular Biology*, 42 (3), 113–85.
80. Cauwe, B. & Opdenakker, G. (2010) Intracellular substrate cleavage: a novel dimension in the biochemistry, biology and pathology of matrix metalloproteinases. *Critical Reviews in Biochemistry and Molecular Biology*, 45 (5), 351–423.
81. Sbardella, D., Fasciglione, G.F., Gioia, M. *et al.* (2012) Human matrix metalloproteinases: An ubiquitarian class of enzymes involved in several pathological processes. *Molecular Aspects of Medicine*, 33 (2), 119–208.
82. Egeblad, M. & Werb, Z. (2002) New functions for the matrix metalloproteinases in cancer progression. *Nature Reviews. Cancer*, 2 (3), 161–174.
83. Sternlicht, M.D. & Werb, Z. (2001) How matrix metalloproteinases regulate cell behavior. *Annual Review of Cell and Developmental Biology*, 17, 463–516.
84. Mu, D., Cambier, S., Fjellbirkeland, L. *et al.* (2002) The integrin alpha(v)beta8 mediates epithelial homeostasis through MT1-MMP-dependent activation of TGF-beta1. *The Journal of Cell Biology*, 157 (3), 493–507.
85. Velasco-Loyden, G., Arribas, J. & Lopez-Casillas, F. (2004) The shedding of betaglycan is regulated by pervanadate and mediated by membrane type matrix metalloprotease-1. *The Journal of Biological Chemistry*, 279 (9), 7721–33.
86. Kajita, M., Itoh, Y., Chiba, T. *et al.* (2001) Membrane-type 1 matrix metalloproteinase cleaves CD44 and promotes cell migration. *The Journal of Cell Biology*, 153 (5), 893–904.
87. Endo, K., Takino, T., Miyamori, H. *et al.* (2003) Cleavage of syndecan-1 by membrane type matrix metalloproteinase-1 stimulates cell migration. *The Journal of Biological Chemistry*, 278 (42), 40764–70.
88. Deryugina, E.I., Ratnikov, B.I., Postnova, T.I. *et al.* (2002) Processing of integrin alpha(v) subunit by membrane type 1 matrix metalloproteinase stimulates migration of breast carcinoma cells on vitronectin and enhances tyrosine phosphorylation of focal adhesion kinase. *The Journal of Biological Chemistry*, 277 (12), 9749–56.
89. Rozanov, D.V., Ghebrehiwet, B., Postnova, T.I. *et al.* (2002) The hemopexin-like C-terminal domain of membrane type 1 matrix metalloproteinase regulates proteolysis of a multifunctional protein, gC1qR. *The Journal of Biological Chemistry*, 277 (11), 9318–25.

90. Strickland, D.K., Gonias, S.L. & Argraves, W.S. (2002) Diverse roles for the LDL receptor family. *Trends in Endocrinology and Metabolism*, 13 (2), 66–74.
91. Rozanov, D.V., Hahn-Dantona, E., Strickland, D.K. *et al.* (2004) The low density lipoprotein receptor-related protein LRP is regulated by membrane type-1 matrix metalloproteinase (MT1-MMP) proteolysis in malignant cells. *The Journal of Biological Chemistry*, 279 (6), 4260–8.
92. Hwang, I.K., Park, S.M., Kim, S.Y. *et al.* (2004) A proteomic approach to identify substrates of matrix metalloproteinase-14 in human plasma. *Biochimica et Biophysica Acta*, 1702, 79–87.
93. Tam, E.M., Morrison, C.J., Wu, Y.I *et al.* (2004) Membrane protease proteomics: Isotope-coded affinity tag MS identification of undescribed MT1-matrix metalloproteinase substrates. *Proceedings of the National Academy of Sciences of the United States of America*, 101, 6917–6922.
94. Egawa, N., Koshikawa, N., Tomari, T. *et al.* (2006) Membrane type 1 matrix metalloproteinase (MT1-MMP/MMP-14) cleaves and releases a 22-kDa extracellular matrix metalloproteinase inducer (EMMPRIN) fragment from tumor cells. *The Journal of Biological Chemistry*, 281, 37576–37585.
95. Butler, G.S., Dean, R.A., Tam, E.M. *et al.* (2008) Pharmacoproteomics of a metalloproteinase hydroxamate inhibitor in breast cancer cells: dynamics of membrane type 1 matrix metalloproteinase-mediated membrane protein shedding. *Molecular and Cellular Biology*, 28, 4896–4914.
96. Liu, G., Atteridge, C.L., Wang, X. *et al.* (2010) The membrane type matrix metalloproteinase MMP14 mediates constitutive shedding of MHC class I chain-related molecule A independent of A disintegrin and metalloproteinases. *The Journal of Immunology*, 184, 3346–3350.
97. Sung, M.M., Schulz, C.G., Wang, W. *et al.* (2007) Matrix metalloproteinase-2 degrades the cytoskeletal protein alpha-actinin in peroxynitrite mediated myocardial injury. *Journal of Molecular and Cellular Cardiology*, 43 (4), 429–36.
98. Sbai, O., Ferhat, L., Bernard, A. *et al.* (2008) Vesicular trafficking and secretion of matrix metalloproteinases-2, −9 and tissue inhibitor of metalloproteinases-1 in neuronal cells. *Molecular and Cellular Neurosciences*, 39 (4), 549–68.
99. Sbai, O., Ould-Yahoui, A., Ferhat, L. *et al.* (2010) Differential vesicular distribution and trafficking of MMP-2, MMP-9, and their inhibitors in astrocytes. *Glia*, 58 (3), 344–66.
100. Shofuda, T., Shofuda, K., Ferri, N. *et al.* (2004) Cleavage of focal adhesion kinase in vascular smooth muscle cells overexpressing membrane-type matrix metalloproteinases. *Arteriosclerosis, Thrombosis, and Vascular Biology*, 24 (5), 839–44.
101. Limb, G.A., Matter, K., Murphy, G. *et al.* (2005) Matrix metalloproteinase-1 associates with intracellular organelles and confers resistance to lamin A/C degradation during apoptosis. *The American Journal of Pathology*, 166 (5), 1555–63.
102. Wang, W., Schulze, C.J., Suarez-Pinzon, W.L. *et al.* (2002) Intracellular action of matrix metalloproteinase-2 accounts for acute myocardial ischemia and reperfusion injury. *Circulation*, 106 (12), 1543–9.
103. Kwan, J.A., Schulze, C.J., Wang, W. *et al.* (2004) Matrix metalloproteinase-2 (MMP-2) is present in the nucleus of cardiac myocytes and is capable of cleaving poly (ADP-ribose) polymerase (PARP) in vitro. *The FASEB Journal*, 18 (6), 690–2.
104. Moshal, K.S., Tipparaju, S.M., Vacek, T.P. *et al.* (2008) Mitochondrial matrix metalloproteinase activation decreases myocyte contractility in hyperhomocysteinemia. *American Journal of Physiology. Heart and Circulatory Physiology*, 295 (2), H890–7.
105. Eguchi, T., Kubota, S., Kawata, K. *et al.* (2008) Novel transcription-factor-like function of human matrix metalloproteinase 3 regulating the CTGF/CCN2 gene. *Molecular and Cellular Biology*, 28 (7), 2391–413.
106. Zhao, W.Q., Li, H., Yamashita, K. *et al.* (1998) Cell cycle-associated accumulation of tissue inhibitor of metalloproteinases-1 (TIMP-1) in the nuclei of human gingival fibroblasts. *Journal of Cell Science*, 111 (Pt 9), 1147–53.
107. Ritter, L.M., Garfield, S.H. & Thorgeirsson, U.P. (1999) Tissue inhibitor of metalloproteinases-1 (TIMP-1) binds to the cell surface and translocates to the nucleus of human MCF-7 breast carcinoma cells. *Biochemical and Biophysical Research Communications*, 257 (2), 494–9.
108. Zhang, J., Cao, Y.J., Zhao, Y.G. *et al.* (2002) Expression of matrix metalloproteinase-26 and tissue inhibitor of metalloproteinase-4 in human normal cytotrophoblast cells and a choriocarcinoma cell line, JEG-3. *Molecular Human Reproduction*, 8 (7), 659–66.
109. Si-Tayeb, K., Monvoisin, A., Mazzocco, C. *et al.* (2006) Matrix metalloproteinase 3 is present in the cell nucleus and is involved in apoptosis. *The American Journal of Pathology*, 169 (4), 1390–401.
110. Wilson, C.L., Ouellette, A.J., Satchell, D.P. *et al.* (1999) Regulation of intestinal alpha-defensin activation by the metalloproteinase matrilysin in innate host defense. *Science*, 286 (5437), 113–7.

111. Golubkov, V.S., Boyd, S., Savinov, A.Y. *et al.* (2005) Membrane type-1 matrix metalloproteinase (MT1-MMP) exhibits an important intracellular cleavage function and causes chromosome instability. *The Journal of Biological Chemistry*, 280 (26), 25079–86.
112. Golubkov, V.S., Chekanov, A.V., Doxsey, S.J. *et al.* (2005) Centrosomal pericentrin is a direct cleavage target of membrane type-1 matrix metalloproteinase in humans but not in mice: potential implications for tumorigenesis. *The Journal of Biological Chemistry*, 280 (51), 42237–41.
113. Takahashi, M., Yamagiwa, A., Nishimura, T. *et al.* (2002) Centrosomal proteins CG-NAP and kendrin provide microtubule nucleation sites by anchoring gamma-tubulin ring complex. *Molecular Biology of the Cell*, 13 (9), 3235–45.
114. Ip, Y.C., Cheung, S.T. & Fan, S.T. (2007) Atypical localization of membrane type 1-matrix metalloproteinase in the nucleus is associated with aggressive features of hepatocellular carcinoma. *Molecular Carcinogenesis*, 46 (3), 225–30.
115. Matsumura, Y., Saeki, E., Inoue, M. *et al.* (1996) Inhomogeneous disappearance of myofilament-related cytoskeletal proteins in stunned myocardium of guinea pig. *Circulation Research*, 79 (3), 447–54.
116. Sung, J.Y., Park, S.M., Lee, C.H. *et al.* (2005) Proteolytic cleavage of extracellular secretedα-synuclein via matrix metalloproteinases. *The Journal of Biological Chemistry*, 280 (26), 25216–24.
117. Bella, J., Eaton, M., Brodsky, B. *et al.* (1994) Crystal and molecular structure of a collagen-like peptide at 1.9 Å resolution. *Science*, 266, 75–81.
118. Holmgren, S.K., Taylor, K.M., Bretscher, L.E. *et al.* (1998) Code for collagen's stability deciphered. *Nature*, 392, 666–667.
119. Chung, L., Dinakarpandian, D., Yoshida, N. *et al.* (2004) Collagenase unwinds triple helical collagen prior to peptide bond hydrolysis. *The EMBO Journal*, 23, 3020–3030.
120. Tam, E.M., Moore, T.R., Butler, G.S. *et al.* (2004) Characterization of the distinct collagen binding, helicase and cleavage mechanisms of matrix metalloproteinase 2 and 14 (gelatinase A and MT1-MMP): The differential roles of the MMP hemopexin C domains and the MMP-2 fibronectin type II modules in collagen triple helicase activities. *The Journal of Biological Chemistry*, 279, 43336–43344.
121. Adhikari, A.S., Chai, J. & Dunn, A.R. (2011) Mechanical load induces a 100-fold increase in the rate of collagen proteolysis by MMP-1. *Journal of the American Chemical Society*, 133, 1686–1689.
122. Fields, G.B. (1991) A model for interstitial collagen catabolism by mammalian collagenases. *Journal of Theoretical Biology*, 153, 585–602.
123. Cerofolini, L., Fields, G.M., Fragai, M. *et al.* (2013) Examination of matrix metalloproteinase-1 (MMP-1) in solution: A preference for the pre-collagenolysis state. *The Journal of Biological Chemistry*, 288, 30659–30671.
124. Han, S., Makareeva, E., Kuznetsova, N.V. *et al.* (2010) Molecular mechanism of type I collagen homotrimer resistance to mammalian collagenases. *The Journal of Biological Chemistry*, 285, 22276–22281.
125. Bertini, I., Calderone, V., Fragai, M. *et al.* (2006) Snapshots of the reaction mechanism of matrix metalloproteinases. *Angewandte Chemie International Edition in English*, 45, 7952–7955.
126. Lauer-Fields, J.L., Chalmers, M.J., Busby, S.A. *et al.* (2009) Identification of specific hemopexin-like domain residues that facilitate matrix metalloproteinase collagenolytic activity. *The Journal of Biological Chemistry*, 284, 24017–24024.
127. Arnold, L.H., Butt, L., Prior, S.H. *et al.* (2011) The interface between catalytic and hemopexin domains in matrix metalloproteinase 1 conceals a collagen binding exosite. *The Journal of Biological Chemistry*, 286, 45073–45082.
128. Manka, S.W., Carafoli, F., Visse, R. *et al.* (2012) Structural insights into triple-helical collagen cleavage by matrix metalloproteinase 1. *Proceedings of the National Academy of Sciences of the United States of America*, 109, 12461–12466.
129. Robichaud, T.K., Steffensen, B. & Fields, G.B. (2011) Exosite interactions impact matrix metalloproteinase collagen specificities. *The Journal of Biological Chemistry*, 286, 37535–37542.
130. Welgus, H.G., Jeffrey, J.J. & Eisen, A.Z. (1981) Human skin fibroblast collagenase: Assessment of activation energy and deuterium isotope effect with collagenous substrates. *The Journal of Biological Chemistry*, 256, 9516–9521.
131. Perumal, S., Antipova, O. & Orgel, J.P.R.O. (2008) Collagen fibril architecture, domain organization, and triple-helical conformation govern its proteolysis. *Proceedings of the National Academy of Sciences of the United States of America*, 105, 2824–2829.
132. Sarkar, S.K., Marmer, B., Goldberg, G. *et al.* (2012) Single-molecule tracking of collagenase on native type I collagen fibrils reveals degradation mechanism. *Current Biology*, 22, 1047–1056.
133. Lu, K.G. & Stultz, C.M. (2013) Insight into the degradation of type-I collagen fibrils by MMP-8. *Journal of Molecular Biology*, 425 (10), 1815–25.

134. Minond, D., Lauer-Fields, J.L., Cudic, M. *et al.* (2006) The roles of substrate thermal stability and P2 and P1' subsite identity on matrix metalloproteinase triple-helical peptidase activity and collagen specificity. *Journal of Biological Chemistry*, 281, 38302–38313.
135. Giola, M., Monaco, S., Fasciglione, G.F. *et al.* (2007) Characterization of the mechanisms by which gelatinase A, neutrophil collagenase, and membrane-type metalloproteinase MMP-14 recognize collagen I and enzymatically process two α-chains. *Journal of Molecular Biology*, 368, 1101–1113.
136. Lauer-Fields, J.L., Whitehead, J.K., Li, S. *et al.* (2008) Selective modulation of matrix metalloproteinase 9 (MMP-9) functions via exosite inhibition. *Journal of Biological Chemistry*, 283, 20087–20095.
137. Allan, J.A., Docherty, A.J.P., Barker, P.J. *et al.* (1995) Binding of gelatinases A and B to type-I collagen and other matrix components. *The Biochemical Journal*, 309, 299–306.
138. Steffensen, B., Wallon, U.M. & Overall, C.M. (1995) Extracellular matrix binding properties of recombinant fibronectin type II-like modules of human 72-kDa gelatinase/type IV collagenase. High affinity binding to native type I collagen but not native type IV collagen. *Journal of Biological Chemistry*, 270, 11555–11566.
139. Xu, X., Mikhailova, M., Ilangovan, U. *et al.* (2009) Nuclear magnetic resonance mapping and functional confirmation of the collagen binding sites of matrix metalloproteinase-2. *Biochemistry*, 48, 5822–5831.
140. Mikhailova, M., Xu, X., Robichaud, T. *et al.* (2012) Identification of collagen binding domain residues that govern catalytic activities of matrix metalloproteinase-2 (MMP-2). *Matrix Biology*, 31, 380–388.
141. Briknarová, K., Gehrmann, M., Bányai, L. *et al.* (2001) Gelatin-binding region of human matrix metalloproteinase-2: Solution structure, dynamics, and function of the Col-23 two-domain construct. *Journal of Biological Chemistry*, 276, 27613–27621.
142. Overall, C.M. (2001) Matrix metalloproteinase substrate binding domains, modules and exosites: Overview and experimental strategies. In: Clark, I.M. (ed), Methods in Molecular Biology 151: Matrix Metalloproteinase Protocols. Humana Press, Totowa, NJ, pp. 79–120.
143. Overall, C.M. (2002) Molecular determinants of metalloproteinase substrate specificity. *Mol. Biotech.*, 22, 51–86.
144. Xu, X., Wang, Y., Lauer-Fields, J.L. *et al.* (2004) Contributions of the MMP-2 Collagen Binding Domain to Gelatin Cleavage: Substrate Binding via the Collagen Binding Domain is Required for MMP-2 Degradation of Gelatin But Not Short Peptides. *Matrix Biology*, 23, 171–181.

2 Dynamics and Mechanism of Substrate Recognition by Matrix Metalloproteases

Ivan E. Collier[1] and Gregory I. Goldberg[1,2]

[1] Departments of Medicine, Division of Dermatology, Saint Louis, MO, USA
[2] Biochemistry and Molecular Biophysics, Washington University School of Medicine, Saint Louis, USA

"... one century after Einstein's seminal work, Brownian Motion continues to be a subject of intricate and fascinating discussions." [1]

2.1 Introduction

All objects that are immersed in a fluid environment with a uniform temperature, T, are subject to collisions that impart fluctuations in kinetic energy of the order Boltzman's constant times the absolute temperature (kT). In consequence, the position of dissolved molecules or suspended microscopic particles exhibits apparently random fluctuations whose mean square displacement (MSD) is inversely proportional to their size. But that is not all. Complex macromolecules will also see their internal energy fluctuate by the same orders of magnitude. Living things and their components must come to terms with this fact of life. As a consequence, all the laws of biological dynamics, from chemical reaction rates to meiosis and muscle contraction, are affected by Brownian motion. The metabolism of collagen by matrix metalloproteinases is no exception.

For multi-cellular organisms, one strategy to control the displacement caused by Brownian motion is for cells and smaller macromolecules, like enzymes, to attach themselves to large insoluble or semi-solid complexes. The three dimensional scaffold of vertebrate extracellular matrix (ECM) is one such aggregation. Its composition, which includes collagen, proteoglycans, fibronectin, and laminin, provides tensile strength to the tissue [2, 3] and an anchor for constituent cells and secreted enzymes. The geometric properties of each particular ECM are determined largely by its component collagens. The resident cells of the ECM control the restructuring of the tissue during morphogenesis [4, 5], tissue repair [6, 7], angiogenesis [8, 9], uterine involution, and bone resorption [10] by secreting the specialized ECM metalloproteinases that degrade collagen.

Matrix Metalloproteinase Biology, First Edition. Edited by Irit Sagi and Jean P. Gaffney.

Here we focus on ECM metalloproteinases MMP-1, MMP-2, MMP-9, and MMP-14 (MT1 MMP) and their interactions with substrate. MMP-1, MMP-2, and MMP-9 are tethered in periplasmic space [11] while MT1-MMP is a trans-membrane proteinase. The enzymes were found tightly associated with ECM (MMP-1, MMP-2, and MMP-9 [11–13] and collagen in particular [14–18].

2.2 Conformational flexibility of MMPs is inexorably linked to collagen proteolysis

The MMP enzymes with few exceptions have a canonical structure consisting of two independently folded domains connected by a flexible linker region [19–21]. The roughly spherical catalytic domain is on the N-terminus of the molecules and contains the active center with an intrinsic Zn atom. In the zymogen form, the catalytic domain includes an N-terminal pro-peptide that obstructs access to the active site. The roughly cylindrical hemopexin-like C-terminal domain is folded into a four-bladed propeller [19, 22]. In addition, these enzymes have extramural binding sites (exosites) in both domains whose type and function are specific to the individual MMP [17, 23].

The flexible linker region of variable length (normally between 20 and 30 residues, but as large as 64 for MMP-9) connects the catalytic and C-terminal domains of all MMPs except for MMP-7 and MMP-26, which lack the C-terminal domain [23]. The absence of electron density for the linker region in crystal structures of full-length enzymes reflects the flexibility of this element that is rich in glycine and proline residues [19, 21, 23–25]. In the crystal structures of the full-length active MMPs, the two domains are in close proximity, often interacting with one another. However, in solution, the flexibility of this region permits relative mobility between the domains leading to multiple conformations centered around two global configurations referred to as open and closed. The open conformation is where the domains are well separated, and closed where the inter domain spacing is closer to that observed in the crystal structure.

Open and closed configurations have been demonstrated for gelatinase MMP-9 using atomic force microscopy [25]. A biphasic distribution of inter-domain spacing whose two maxima were separated by about 15 Å suggests the existence of two heterogeneous size populations centered around the open and closed configurations [25]. In the case of MMP-2, an interaction between the catalytic and C-terminal domains has been demonstrated by crystallography [19]. The first two blades of the MMP-2 C-terminal domain form hydrogen bonds with the first repeat of its fibronectin-like module, potentially limiting their relative mobility. The possibility of an equilibrium occurring in solution between this closed form and an open configuration has not been studied experimentally, although it is certainly possible, given the examples of MMP-1 and MMP-9.

The MMP-1 pro-enzyme is maintained in a closed configuration by hydrogen bonds and hydrophobic interactions between the pro-peptide of the catalytic domain and blade I of the C-terminal domain [20, 21]. In such a configuration, the MMP-1 collagen binding site that has been located in the C-terminal domain on blades I and II of the propeller [26–28] would be blocked [20, 29]. Activation of the enzyme is accomplished by removal of the pro-peptide. In solution, the active form exists in an equilibrium between an assortment of open and closed conformers [24, 27–29] in which the closed form is enthalpically driven by interaction between the catalytic and C-terminal domains and the open form entropically driven by conformational freedom conferred by the flexible linker [27]. The transition of MMP-1 from an open to closed conformation plays an important role in its enzymatic activity.

A variety of biophysical and biochemical studies address the cooperative motion between the two domains leading up to catalysis using model triple helical peptides (THP). Because the crystallographic closed form of the active enzyme would be sterically inhibited from binding to triple helical collagen (THC), it was initially suggested that the open form of the enzyme binds to collagen, setting in motion a series of conformational changes that result in the transition of the enzyme from open to closed form, accompanied by the local unwinding of the collagen helix [20]. A stereo-chemically specific series of stages in this transition were suggested to accomplish this task [27]. An open form of MMP-1 exists in solution where the catalytic and C-terminal domains are rotated relative to each other away from their positions in the crystal structure of the active form, breaking the interaction interface between them [24]. This form is available to bind specific chains of the THP via the C-terminal domain exosite, which is now exposed. Then, due to flexibility of the linker, the catalytic domain can specifically interact with the appropriate residues near the cleavage site. Though the isolated catalytic domain binds poorly to THPs, it binds strongly to its target peptide chain when cooperatively positioned by its C-terminus. Next, a reverse-rotation of the enzyme's two domains back to the closed position restores the favorable interactions between the two domains that were established by NMR and crystallography. This domain back-rotation is accompanied by local dissociation of one polypeptide from the THP and its insertion into the catalytic cleft of the enzyme. Molecular modeling [24] of these steps shows that they are energetically consistent with the relatively high activation energies for cleavage of THC and fibrils [30, 31].

Another proposed model of MMP-1 induced THP unfolding does not require the complete separation of the two domains during binding and consequent THP dissociation [28]. Although the sites of interaction with THP are the same in this proposal, the interface between the two MMP-1 domains is partially maintained during the much smaller MMP-1 conformational realignments that occur as the THP chain is dissociated during binding.

Both models require some degree of inter-domain flexibility. The essential disagreement between the two is the configuration of MMP-1 during its initial interaction with the THP [32]. Although previous work (see above) demonstrated open configurations on MMP-1 in solution, the likelihood of the ensemble of configurations containing any particular form remained unclear. Recent experimental and theoretical work [32] addressing this issue concludes that MMP-1 conformers in which catalytic and C-terminal domains of MMP-1 are not in tight contact dominate the ensemble of solution available forms. Thus it supports the notion that the first step in proteolysis is the binding of the open form to collagen as described in the first model of cleavage.

How then do such bound and tethered proteinases find their recognition sites in the ECM? Recent work demonstrated that they engage in Brownian motion constrained to the surface of the collagen substrate [15–18, 33].

2.3 Dynamics of MMP-2 and MMP-9 interaction with gelatin

Here we focus on interaction between MMP-2 and MMP-9 and one of their substrates, denatured type I collagen. These gelatinases are characterized by an insertion of three head-to-tail repeats in the catalytic domain. The repeats are homologous to the type 2 repeat found in fibronectin, conferring gelatin/collagen substrate binding properties to the pro and active forms of both enzymes [15, 34–38]. Despite being integrated in the catalytic domain, the repeats retain the characteristic fold of their homologues from fibronectin [19] and together are referred to as the fibronectin-like gelatin binding module. The hemopexin-like C-terminal domains of MMP-2 and MMP-9 each contain a specific binding site for the tissue inhibitors of metalloproteinases (TIMP), TIMP-2, and TIMP-1 correspondingly [39–41]. The C-terminal domain of MMP-2 has no collagen/gelatin binding activity [23] nor does it compete binding of MMP-2 to gelatin [15]. This is in contrast to the C-terminal domains of the collagenases, MMP-1, and MMP-14, which have exosites for collagen binding [23, 42–44].

2.4 Surface diffusion: a common mechanism for substrate interaction adapted by MMP-2 and MMP-9

The investigation of MMP-2 binding to gelatin-coated surfaces [17] produced a number of seemingly paradoxical observations. Saturation binding to gelatin coated surfaces of the isolated MMP-2 catalytic domain (trMMP-2), despite containing the gelatin binding module, is at most 2% that of the full-length enzyme. Nevertheless, trMMP-2 effectively competed the binding of full-length MMP-2. In contrast, the isolated C-terminal domain [22] does not significantly compete MMP-2 binding even at high molar ratios. More detailed investigation of binding kinetics revealed that nearly irreversible binding of MMP-2 (dissociation constant 7.5×10^{-4}/min) is accompanied by a relatively weak binding constant of 5.2×10^{5}/M. Conventional kinetics would require that a reaction with such a weak binding constant have a dissociation constant 10^{2} fold larger than that which was obtained experimentally. In addition, the shape of the MMP-2 absorption time courses fit poorly with pseudo-first order kinetics. These adsorption data, however, can be accounted for in detail by a fractal kinetic mechanism [45–47] in which rate-constants are replaced by time-dependent rate multipliers. These have a power law form, k t, where k and h are both positive constants with h values restricted to between zero and one.

The fractal kinetic mechanism infers that the binding of MMP-2 to gelatin is a self-limiting reaction where the rate limiting process is confined to the gelatin surface rather than adsorption from solution to gelatin. There are various mechanisms that can be proposed to account for this phenomenon with the simplest being surface diffusion of the ligand on the gelatin layer. Fluorescence photobleaching recovery (FPR) [48] experiments have shown that the bulk of the MMP-2 binding to gelatin layers indeed depends on the enzyme's ability to diffuse laterally on the substrate surface. The C-terminal domain of the enzyme greatly facilitates the two-dimensional lateral diffusion whereas the specificity of binding resides with the fibronectin-like gelatin-binding module [15].

Table 2.1 Motion parameters of MMPs on substrate surfaces.

	MMP mobility on Gelatin	MMP mobility on Collagen Fibrils		
Method:	**FPR**	**FCS**		**MSD**
Protease	***D*** **(cm^2/s) × 10^{-8}**	***D*** **(cm^{2}s^{-1}) × 10^{-8}**	***V*** **(μm/s)**	***D*** **(cm^2/s) × 10^{-8}**
MMP-1		0.8 ± 0.15 [16]	4.5 ± 0.4 [16]	0.7 ± 0.01 [18]
E_{219}Q MMP-1		0.67 ± 0.15 [16]	n.d*. [16]	
MT-1 MMP		0.6 ± 0.05 [17]	5.8 ± 0.2 [17]	
E_{240}A MT-1 MMP		1.1 ± 0.04 [17]	n.d. [17]	
MMP-2	0.23 ± 0.06 [17]	1.29 ± 0.05 [17]	n.d. [17]	
MMP-2/TIMP-2	0.15 ± 0.02 [17]	1.8 ± 0.1 [17]	n.d. [17]	
MMP-9	0.25 ± 0.06 [17]	0.6 ± 0.02 [17]	n.d. [17]	
MMP-9/TIMP-1	0.23 [17]	1.2 ± 0.1 [17]	n.d. [17]	
MMP-9 Homodimer	Immobile on both substrates			

*n.d. = none detected; Numbers in parenthesis refer to a source as in list of references.

While the reduced capacity of the C-terminal truncated mutant to diffuse explains its binding behavior, it raises the question: How does the C-terminal domain enable the diffusion process and promote high-level binding? Though the isolated C-terminal domain of MMP-2 may bind gelatin or collagen poorly or not at all, once it is cooperatively positioned on collagen or gelatin by the catalytic domain, favorable interactions between the C-terminal domain and substrates may be possible, as suggested by the crystal structure of MMP-2 [19] and molecular modeling studies [49]. In the MMP-2/TIMP-2 complex, the inhibitor binds to sites on blades III and IV of the propeller of the C-terminal domain. The inhibitor does not interact with the catalytic site and can serve as a linker to other metalloproteinases, MMP-14 in particular [41, 50]. In keeping with this arrangement, complexing MMP-2 with TIMP-2 had no significant effect on its mobility [17]. Similar average diffusion coefficients of $2.3 \pm 0.06 \times 10^{-9}$ and $1.5 \pm 0.02 \times 10^{-9}$ cm^2/s were obtained for MMP-2 and MMP-2/TIMP-2 complex respectively (Table 2.1).

MMP-2 and MMP-9 are closely related enzymes and it might be expected that MMP-9 would bind to gelatin layers by a mechanism similar to that of MMP-2. However, there are significant structural differences between MMP-2 and MMP-9 that could affect substrate binding and mobility. First, the flexible linker region of MMP-9 (64 amino acids) is 41 amino acids longer than that of MMP-2 (23 amino acids). Second, MMP-9 has an additional two non-conserved cysteine residues, Cys^{468} in its flexible linker region and Cys^{674} in its C-terminal domain, that permit it to form a unique covalent MMP-9 homodimer with an inter-molecular disulfide bridge which is formed between the Cys^{468} residues of each constituent monomer [17]. The MMP-9 homodimer can be expected to have a higher binding constant for gelatin surfaces than that of MMP-9 if both of the fibronectin-like gelatin binding modules can bind gelatin simultaneously. Analysis of MMP-9 binding yielded simple non-cooperative binding isotherms for the monomer and dimer with binding constants of 3.5×10^5/M and 5×10^6/M (IEC and GIG unpublished) respectively, suggesting that both of its gelatin binding modules could simultaneously interact with the gelatin substrate. However, the saturation of gelatin binding of the MMP-9 homodimer was only 5–10% t that of MMP-9 monomer. The MMP-9 homodimer effectively competed with the binding of the monomer suggesting that, like trMMP-2, the dimer binding is restricted to a small number of binding sites accessible from solution while the bulk of the MMP-9 binding depends on the lateral diffusion. FPR measurements confirmed this expectation and revealed a diffusion-dependent recovery of fluorescently labeled MMP-9 [17] with an average diffusion coefficient of 2.5×10^{-9} cm^2/s (Table 2.1) that does not differ significantly from that of MMP-2. The gelatin bound MMP-9 homodimer remained immobile [17]. The recovery of the MMP-9/TIMP-1 complex has a diffusion coefficient of 2.3×10^{-9} cm^2/s (Table 2.1) showing that the diffusion of MMP-9 is also not affected by complex formation with TIMP-1. This is consistent with the fact that binding of TIMP-1 to MMP-9 through sites in blades III and IV of the propeller [39] is analogous to that of TIMP-2 to MMP-2.

There are at least two explanations as to why the MMP-9 homodimer is immobile on gelatin surface. First, the inter-molecular disulfide bond in the linker region adjacent to blades I and II of the C-terminal domain is likely to alter the cooperative interaction between the catalytic and C-terminal domains necessary for diffusion. Second, the tighter binding (ten-fold higher binding constant) exhibited by MMP-9 homodimer might alone be sufficient to preclude surface mobility.

These results demonstrate that both gelatinases adopt a surface diffusion as a mechanism for substrate interaction.

2.5 Dynamics of MMP interaction with collagen fibrils

The results discussed above that describe the interaction of MMP-2 and MMP-9 with disordered gelatin layers leave two questions unanswered. Does surface diffusion with reduced dimensionality play a role in enzyme interaction with a physiologically relevant and highly ordered linear substrate and collagen fibrils, and how widely spread is this mechanism among the enzymes of MMP family? Triple helical collagen monomers (THMs) are tightly packed into fibrils that are highly resistant to proteolytic degradation. The interstitial collagenase (MMP-1) [30, 51–54] and the membrane-type metalloproteinase (MT1-MMP) [55–59] are among the few enzymes capable of collagen fibril digestion. Both of these enzymes cleave accessible THC monomers of the fibril at the classical cleavage site [60–62] approximately three-fourths of the way from the monomer's N-terminus, thus destabilizing the monomer helix and eventually leading to fibril dissolution. Both of these enzymes have the canonical catalytic domain/C-terminal domain architecture, but MT1-MMP also contains the trans-membrane domain and an intracellular C-terminus. The extracellular portion of the enzyme consists of the catalytic and hemopexin-like domains connected by a flexible linker.

2.6 Mechanism of interaction of MMP-1, MMP-2, MMP-9, and MMP-14 with collagen substrate involves surface diffusion

Fluorescence correlation spectroscopy (FCS) [48] was used to investigate the behavior of single enzyme molecules bound to an individual collagen fibril in reconstituted collagen gels [16, 17]. The record of the large spikes of fluorescence derived from single enzyme molecules passing through the laser beam was subjected to correlation function analysis. The experimental correlation functions obtained for MMP-2 and MMP-9, and inactive mutants of MMP-1 and MMP-14 fitted well into a one-dimensional diffusion model. The correlation functions obtained from activated wild type MMP-1 and the extracellular portion of MMP-14 fit well into a one-dimensional diffusion plus flow model [63]. In this model, particles exhibit diffusive behavior at fast time scales but have a low probability of returning to the same spot at longer times due to the directional flow. The local diffusion coefficient $D = 8 \pm 1.5 \times 10^{-9}$ cm^2/s and the transport velocity $V = 4.5 \pm 0.36$ µm/sec were determined from the fit of the correlation function obtained for the wild type activated MMP-1 [16]. The transport characteristics for the MT1-MMP enzyme were similar [17]. These results show that in the absence of collagenolytic activity, all investigated MMP enzymes move processively on the surface of collagen fibrils in one-dimensional Brownian diffusion. Active collagenolysis in the case of MMP-1 and MMP-14 adds an additional bias component (Table 2.1) to the motion, a highly unusual feature for an isothermal system and a first example of such in the extracellular space.

The two-photon excitation FCS experiments report the properties of enzyme motion in a microscopically small observation volume. To determine whether the biased component dominates the transport process on a macroscopic scale, the flux of single MMP-1 molecules was measured around a "no transport" block created on a collagen fibril by local exposure to high intensity laser radiation that damages the fibril and thus blocks transport across the area. For unbiased diffusion, the average number of single enzyme molecules at the left (CL) and the right (CR) flanks of the "no transport" block are equal while a difference CL and CR would indicate directional transport of the enzyme. The results demonstrate that the flux is highly asymmetric for the wild type active enzyme, while the inactive mutants showed complete symmetry. These results support the conclusion that active collagenases bound to a collagen fibril undergo proteolysis-dependant directional transport. In addition, similar experiments in the presence of inhibitors show that the degree of asymmetry in the flux depends on the efficiency of proteolysis [16, 17].

Intact collagen fibrils are not cleaved by MMP-2 and MMP-9 and accordingly these enzymes exhibit pure Brownian diffusion. The diffusive motion of MMP-9 on THC monomers was also demonstrated using atomic force microscopy [33]. Specific complex formation with inhibitors TIMP-2 and TIMP-1 does not interfere with the mobility of the enzymes on the surface of the fibril, while dimerization of MMP-9 renders the enzyme immobile, as in the case with gelatin layer diffusion [17]. These observations are of particular significance as they demonstrate that all components of the cell membrane tri-molecular activation complex, MMP-14/TIMP-2/MMP-2 [50], are capable of diffusion on the surface of the underlying collagen fibril [17]. This provides an efficient mechanism of substrate recognition for relatively immobile structures of ECM by the cell membrane enzymatic complex.

Particles that diffuse in an anisotropic environment cannot produce work as long as the system is isothermal, but if a thermal gradient is applied across the system it can exhibit biased diffusion [64]. Thus, biased diffusion is a characteristic of a molecular motor and requires energy dissipation. This illustration was of considerable importance in exploring the plausibility of microscopic machines. Harnessing work from Brownian motion has been both an exciting and a controversial topic [65]. Theoretical physicists have produced several models of biased diffusion without a need for a system-wide gradient or a field. For instance, coupling to external fluctuations can create machines known as "thermal ratchets" [66, 67] that harvest the energy from colored noise [68]. A "Brownian Ratchet" can be powered by coupling to a non-equilibrium chemical reaction driving the particle between two states [69–71]. In a "Burnt Bridge" theoretical model of a Brownian ratchet the diffusion bias is created because a moving particle may destroy weak places on its track in a way that inhibits its ability to diffuse back.

A modified version of this model explains the biased diffusion of active collagenases on a fibril surface. When the proteolytic activity of MMP-1 is inhibited, the enzyme molecules perform a random walk in one dimension along the fibril. When an active enzyme walker encounters a cleavage recognition site, a successful cleavage may occur with a probability (Pc), and the enzyme molecule responsible for the cleavage always ends up on the same side of the cleaved peptide bond. Although molecules are not allowed to cross the cleaved collagen helix, they are allowed to jump to a neighboring triple-helix track with a small probability (Pj). This mechanism of a constrained random walk results in net transport [16] with velocity (V) that depends on the diffusion coefficient, the probabilities defined above, and the spatial distribution of the cleavage sites along the track. The Monte Carlo simulations of the correlation functions and the asymmetry ratios based on these rules accurately predict the experimental results for both active and inhibited enzymes, depending on the value of a single parameter, Pc. Simulations also demonstrate that the asymmetry ratio depends exponentially on the probability of cleavage, Pc, so that two orders of magnitude decrease in the cleavage probability results in a 50% decline in the asymmetry ratio.

The MMP-1/collagen system is the first example of an ATP-independent biased transport operating extracellularly. The mechanism of this transport is akin to a Brownian ratchet that is able to rectify Brownian forces into a propulsion mechanism by coupling to an energy source, in this case, collagen proteolysis.

2.7 Mechanism of MMP-1 diffusion on native collagen fibrils

Recent single molecule tracking experiments [18] of MMP-1 bound to native rat tail collagen fibrils revealed a pattern of enzyme motion that links its trajectory on the fibril to THC cleavage. MMP-1 moved along the axis of the fibrils showing negligible motion across the fibril confirming the one-dimensional diffusion observed with FCS. Average diffusion coefficients of 0.7×10^{-8} cm^2/s for MMP-1 as well as 0.7×10^{-8} and 1.0×10^{-8} cm^2/s for the inactive mutant of MMP-1 and MMP-9, respectively, were obtained from the slopes of MSD of the tracked particles. These diffusion coefficients are comparable with those obtained in FCS experiments (Table 2.1). Furthermore, the MSDs of MMP-1 observed at 20 °C are linear, in contrast to those taken at 37 °C that had an increasing slope at longer times. These results substantiate the cleavage-dependent, biased motion observed with FCS. In addition, the diffusion of wild type MMP-1 at 37 °C was substantially hindered due to the barriers erected by prior cleavage of the THC monomers.

Further analysis of individual single molecule trajectories provided a great amount of information regarding the pattern of motion obscured in FCS experiments. First, it became clear that MMP-1 moving along a collagen fibril spends 90% of the time in pauses. As an immediate consequence of that, the intrinsic diffusion coefficient (7.5×10^{-8} cm^2/s), which was determined from simulations of the enzyme motion, is higher than the average observed by FCS or calculated from the MSD curves. Second, pauses observed for MMP-1 and its inactive mutant are of two classes, distinguishable by their dwell time distributions. The dwell times of Class I pauses are characterized by a single exponential decay with a pause-escape time of 2.5/s. The dwell time distribution of Class II pauses is symmetric with the peak around one second pause-escape time. The Class II pauses cannot be described by a single or a sum of several exponential functions but are closely approximated by a gamma function distribution [18, 72], implying that escape from these pauses involves a sequence of about 13 consequent kinetic steps of a similar or identical rate of 15/s. For wild type MMP-1, the relative amplitude of the Class II pauses increased three-fold with the temperature increase from 20 to 37 °C suggesting a significant energy barrier to entry. For the inactive MMP-1 mutant, the amplitude of the Class II pauses was also temperature-dependent, with amplitude about 30–50% that of wild type enzyme across the temperature range. The MMP-9 exhibits only Class I pauses with a pause-escape rate of 3.6/s. Finally, the Class II pauses are spaced periodically 1.3 and 1.5 μm apart [18], a distance that does not correspond to any known periodic structural feature of fibrillar collagen [18, 73–75].

The motion of MMP-1 after escape from Class II pauses differed from that of the inactive mutant suggesting a connection between these types of pauses and fibril proteolysis. The mutant enzyme escaped in either direction to continue Brownian diffusion until the next pause was reached. The active enzyme escaped in a similar fashion with about 5% of trajectories showing an exception. In those trajectories the enzyme moved faster and further than MMP-1 mutant and the post Class II pauses motion was polarized. Simulations of wild type MMP-1 motion on fibrils at 37 °C also indicated that these polar runs are the source of the positive inflection of the MSD curves and thus of biased motion of MMP-1. The faster and longer polarized runs can be explained if the cleavage initiated during a productive Class II pause is followed by a fast burst of approximately 15 subsequent cleavages 67 nm apart during which the rules of "Burnt Bridge" Brownian ratchet are enforced so that diffusion is biased for a distance of about one micron.

2.8 Triple helical collagen cleavage–diffusion coupling

The uniqueness of the mammalian collagenase cleavage site in native THC types I, II, and III lies not just in the primary amino acid sequence around the cleavage, but also in unique physico–chemical properties of the surrounding area. Cleavage occurs after the Gly in the partial sequence Gly-[Ile or Leu]-[Ala or Leu] about three-fourths of the way from the amino terminus. This partial sequence is distinguished from 31 others by: *"(a) a low side-chain molal volume-, high imino acid (>33%)-containing region that is tightly triple-helical, consisting of four Gly-X-Y triplets preceding the cleavage site, (b) a low imino acid-containing (<17%), loosely triple-helical region consisting of four Gly-X- Y triplets following the cleavage site, and (c) a maximum of one charged residue for the entire 25 residue cleavage site region, which is always an Arg that follows the cleavage site in subsite P′5 or P′8. In addition, the high imino acid-containing region cannot have an imino acid adjacent to the cleaved Gly-[Ile or Leu] bond (i.e., in subsite P2)."* [61] Furthermore, collagenase must cleave three covalent bonds in a single THC cleavage event. To accomplish the triple cleavage, the enzyme must access the scissile peptide bond in each of the three chains sequentially. Some local unwinding must occur, since the 5 Å wide MMP-1 active site cleft can accommodate only one polypeptide chain of the 15 Å diameter THC molecule [76, 77].

The low imino acid content on the C-terminal side of the cleavage site might permit THC to "micro-unfold" producing a 20–30-residue loop that MMP-1 could cleave at physiological temperatures [61], thus resolving the issue. Recent structural calculations and molecular modeling confirm this possibility [74, 78, 79]. Micro-unfolding of THC [80–83] is a localized and temporary unwinding revealed by its vulnerability to trypsin/chymotrypsin digestion [81]. At temperatures as low as 35 °C type I THC monomers were vulnerable to the combined action of trypsin and chymotrypsin at high concentrations. Recent work [84] using ultra slow micro-calorimetry and isothermal circular dichroism demonstrated that THC unfolding is an equilibrium process with a much lower melting point (between 28 °C and 35 °C for rat tail collagen) than previously thought. Due to its cooperative nature the melting (and subsequent re-annealing) rate of THC at temperatures less than 37 °C is extremely slow and starts with "micro-unfolding" at more labile local regions, like the collagenase cleavage site. Furthermore, micro unfolding can also occur when THC monomers are stabilized by packing into fibrils [82, 84].

Recent observations demonstrated an MMP-1 dependent THC unfolding at 25 °C, a temperature where the collagen molecule is significantly more stable [76]. An inactive MMP-1 mutant bound to type I THC rendered it susceptible to cleavage by the isolated catalytic domain of MMP-1 at 25 °C. Two explanations of the MMP-1 dependent unwinding of THC have been advanced. The "helicase activity" hypothesis postulates that a localized specific triple helical unwinding activity is associated with full-length collagenases like MMP-1, MMP-8, and MT1-MMP. This unwinding activity was proposed to occur by a mechanism that might induce a conformational change in THC structure by direct generation of force through "molecular tectonics" [23]. Alternatively, a "thermal fluctuation" hypothesis proposes that "*... collagen, like all other biological hetero-polymers, undergoes thermal fluctuations that cause it to sample distinct structures in the neighborhood of the native state, and collagenolysis occurs when collagenases recognize the appropriate unwound conformers.*" [85]. The role of the C-terminal domain is crucial for both hypotheses, because without it the MMP-1 catalytic domain alone binds very poorly to Type I THC and consequently digestion is observed only at high enzyme concentrations and after long incubations [86]. In the "helicase" hypothesis the unwinding activity arises from cooperativity between the catalytic and C-terminal domains [76]. While in the thermal fluctuation hypothesis, the C-terminal domain positions the enzyme to bind to the transiently unwound chain.

It is hard to understand how a direct force or torque [23] can be created by the multi-point binding and subsequent correlated domain movement required by the "molecular tectonics" model. Forces generated at the nanometer scale are rapidly damped due to the surrounding medium: after about 45 ns any inertial motion of a particle the size of MMP-1 is completely lost [87, 88]. This means that vectorial motion would last only as long as the applied force, which would have to be associated with a large scale conformational shift of MMP-1. Unwinding sufficient to allow cleavage would require a 6–8 Å displacement of the α2(1) chain [27, 79] from its equilibrium position. The dynamics of conformational transitions have not been studied for MMP-1, but in general the intramolecular conformational changes in other proteins have relaxation times that range from tens of microseconds to milliseconds [89]. The forces responsible for binding as well as protein ternary and quaternary structure arise from weak non-covalent interactions like hydrogen and van der Waals bonds. The forces giving rise to these bonds are short ranged and become weak at distances well below 1 nm [90]. Even electrostatic forces are short-ranged because ionic screening prevails at physiological conditions [88]. It is difficult to see how relatively short-range interactions could cause a force that would persist for tens of microseconds over distances of 6–8 Å.

There are more physically realistic mechanisms for MMP-1 dependent unwinding. The "*stochastic unwinding*" model proposes that MMP1 "*promotes reversible unwinding by shifting the equilibrium between the native helical* ... *and locally unwound states of collagen*" by either "*destabilizing the native helical state* ... *and/or stabilizing the unwound state.*" [91]. This establishes a range of possibilities for involvement of thermal fluctuations in proteolysis of THC. A recently proposed mechanism suggests that MMP-1 can promote THC unwinding by weakening the non-covalent, interchain bonds at the points of MMP-1 exosite binding, based on the structure of complexes of MMP-1 with THPs [27, 28]. The weakened bonds could lead to completely (but locally) dissociated chains or make them more susceptible to thermally induced unwinding. A model at the other end of the range proposes that MMP-1 facilitates the adaptation of a thermally dissociated α2 chain to the active site cleft [79]. The transition between open and closed forms of the collagenase during unwinding (as discussed above) is almost certainly required by either model. The closed configurations, in which the catalytic and C-terminal domain are within crystallographic distances, do not bind to THC because of steric hindrance by the catalytic domain [29, 32]. Structural studies of the pre-catalytic complex with THPs show collagenase to be in the closed form [27, 28]. The central difference between these two models is the extent to which local unwinding of THC is driven by MMP-1 binding or thermal energy. The ability of the inactive MMP-1 mutant to promote THC cleavage by non-specific proteases at 25 °C [76], a temperature well below those where micro-unfolding was observed [81], argues qualitatively that MMP-1 binding is to some degree involved in destabilizing the helix.

The energy of activation (Ea) for MMP-1 collagenolysis over the temperature range, 14–34 °C, provides some insight into the possible relative contributions of MMP-1 induced and thermally induced instability to the unwinding. The melting of

Type I collagen at temperatures less than 37 °C is extremely slow [84] and mimics an irreversible reaction with a high Ea [92]. Classical catalysis theory asserts that catalysts enhance the reaction rate by lowering activation energy of the reaction. Therefore, the promotion of unwinding might be detected as relatively low activation energy. The Ea for MMP-1 digestion of gelatin (heat denatured type I collagen) is 13,000 kcal/mol, a value typical for enzymatic catalysis [30, 31]. All Ea for MMP-1 are from unless otherwise indicated. Thus, if the cleavage site were completely unwound as a consequence of MMP-1 binding, an Ea of that order would be expected (the relatively reduced configurational entropy available to a locally unwound chain could make it higher). Thus the exceptionally large Ea of 47,000 kcal/mol for human type I THC digestion by MMP-1 demonstrates a substantial role for thermal energy in THC unwinding. Furthermore, the large energies of activation strongly correlate with the stability of collagen structure. For human type II THC collagen, which is significantly more stable than type I [91], the Ea was 61,000 kcal/mol. The highest Ea obtained, 101,000 kcal/mol, was for highly thermostable fibrillar collagen (pig type I). In addition, a recent study [91] comparing homotrimeric type I THC (3 three $\alpha1(1)$ chains) with an apparent melting temperature 2.5 °C higher [93] than normal type I collagen found that the Ea of MMP-1 homotrimeric cleavage was 50% greater than that of type I collagen.

The unusually large Ea for THC cleavage by MMP-1, as well the correlation between Ea and the thermal stability of the THC substrate, demonstrates a substantial role for thermal energy (structural fluctuations) in THC unwinding necessary for MMP-catalysis. Thus the question is raised: Is there similar evidence for MMP-1 dependent unwinding? An Ea of 42,500 kcal/mole was obtained for MMP-1 digestion of type III THC while the Ea for trypsin digestion of this same substrate was 70,000 kcal/mole [31]. Thus the substantially lower Ea for MMP-1 compared to trypsin may *in part* reflect the contribution of MMP-1 dependent destabilization of triple helix.

It appears that data on MMP-1, MMP-8, and MT1-MMP are best described by the proposed "stochastic unwinding" [91] mechanism in which some combination of localized MMP-1 dependent THC destabilization makes the THC cleavage site more labile to a thermally driven micro-unfolded state [78] which is then stabilized in presence of MMP-1. The recently proposed mechanism suggesting that inter-chain bonds, weakened by MMP-1 binding, make the THC structure more labile to thermal unwinding is physically reasonable and supported by structural evidence [27, 28].

2.9 Conclusions

The digestion of collagen fibrils is significantly different from that of isolated THC monomers. Fibrillar collagen is a large supra-molecular structure where a quasi crystalline packing of THC monomers further enhances its resistance to proteolytic degradation [74]. Specific cleavage sites staggered with 67 nm periodicity are protected by the telopeptide of its N-terminal neighbor [74]. The telopeptide must be displaced by proteolysis, mechanical damage, thermal fluctuation or other cause to expose the cleavage site [74]. Nevertheless, even packed THC monomers in the fibril are subject to "breathing" or micro-unfolding [84] presenting a rare and fluctuating target vulnerable to proteolytic digestion by collagenases. Thus it appears that substrate surface diffusion of MMPs presents a most efficient mechanism of substrate recognition. In the conventional, mass action formulation of cleavage site recognition, described by pseudo-first order kinetics, an enzyme diffuses in three dimensions in solution until it encounters the substrate. How efficient is this? Not very: a ligand on an infinite random walk in three-dimension has a vanishing probability of visiting any one given site out of all the possible ones. The number of locations visited during a random walk in one dimension is dramatically reduced [94] as is the case for Brownian motion of MMP constrained to the surface of the fibril.

Brownian motion of MMP-1 is interrupted by the stochastic pauses of type I with exponential time of escape distribution and type II pauses at "hot spot" binding sites that occur at 1.3 and/or 1.5 μm intervals along the fibril [18] implying that a periodic structural feature of unknown nature may exist on the fibril. The latter longer pauses have a symmetrical distribution of dwell times centered at 1 s and are characterized by a gamma-function, which describes a process that includes multiple sequential kinetic steps of comparable rates. The longer pauses appear to be the precursor of proteolysis and are characteristic of MMP-1 but are not detected in the case of MMP-9 diffusion. The probability of the enzyme entering the type II pause increases with temperature, suggesting a role for thermally driven rearrangement of the "hot spot" sites encouraging interaction with the enzyme. Once an initial cleavage is made, the cleavage site of the succeeding THC monomer is exposed and a fast cascade of cleavages follows. Simulations indicated that 15 subsequent cleavages would account for the average behavior of MMP-1 following escape from a Class II pause. It is noteworthy that 15 ± 4 consecutive cleavages correspond to a distance of 1 ± 0.3 μm which overlaps the periodic spacing of "hot spot" binding site at 1.3 ± 0.2 μm.

Any hypothesis about the molecular basis of MMP substrate diffusion must explain the states of motion observed for the MMPs [18] on the fibril: type I pauses, type II pauses, and diffusive runs with a diffusion coefficient 7.5×10^{-8} cm^2/s. It is tempting to speculate that there is a connection between the multiple sequential kinetic steps preceding the escape from type II pauses and the coordinated series of cooperative pre-catalytic steps inferred from NMR and crystallography of THP-MMP-1 complexes [24, 27, 28]. In addition, it is not clear what role if any the fluctuations in the distance between the domains enabled by the flexibility of the linker region play in the process of diffusion.

The diffusion coefficient in water for MMP-1 is between 6.7 and 8.5×10^{-7} cm^2/s based on its radius of gyration. Thus the friction associated with MMP-1's intrinsic diffusion coefficient on fibrils represents only a 10-fold increase over that associated with diffusion in water, suggesting a very loose association of MMP-1 with the fibril during the diffusive runs. Hydrogen bonds [95] or ionic interactions [96] between surface diffusing macromolecules and their substrate can lower their mobility by a factor ranging from one hundred to several thousand. The enzyme cellulase, for example, has a two dimensional diffusion coefficient on crystalline cellulose of 6.0×10^{-11} cm^2/s, more than four orders of magnitude less than the free solution diffusion constant [97]. This enzyme has an analogous flexible linker remarkably homologous to that of MMP-9 [25] that plays an essential role in cellulase diffusion on its substrate [98, 99].

A theoretical study of the friction encountered by peptides, 15 amino acids in length, adsorbed to an ideal hydrophobic surface finds only a three-fold increase in friction over that of water. It concluded that "... *polypeptide friction forces on hydrophobic and hydrophilic surfaces are vastly different, even though the adhesion strength on both surfaces is rather similar. On hydrophobic surfaces we find good lubrication with peptide mobilities close to bulk water. In contrast, for a hydrophilic surface hydrogen bonds transiently lock the peptide, leading to a stick-slip type of motion and to mobility coefficients orders of magnitude lower* ... " [95]. This theory would suggest that the diffusing MMP enzymes are associated predominantly with hydrophobic residues of collagen like glycine and proline, which are prominent in the sequence as well as leucine, isoleucine, and valine, which are also notably present. In addition, the flexibility between two domains may enhance overall mobility because it would effectively separate them allowing the combined friction to reflect the average of the two rather than the sum, as would be the case for two rigidly positioned domains [95]. In addition, the flexibility between the domains may allow the enzyme to maintain a presence on hydrophobic regions of the collagen surface in the same way that one may walk across a creek by stepping only on stones.

Answering the questions regarding the role of conformational dynamics of MMPs during diffusive runs and pauses is critical for understanding the mechanism of collagenolysis and can be addressed experimentally by creating the enzymes labeled by a specific fluorescent FRET pair and positioned within the protein to report the distance between the domains. The frequency of the distance fluctuation can then be measured using FCS. Furthermore, combining the high temporal resolution of FCS with high positional resolution of the single molecule tracking it should be possible to measure the frequency of these fluctuations during the stages of the enzyme motion, diffusive runs, and type II pauses.

References

1. Cecconi, F., Cencini, M., Falcioni, M. & Vulpiani, A. (2005) Brownian motion and diffusion: From stochastic processes to chaos and beyond. *Chaos*, 15 (2), 026102 026101–026102 026109.
2. Yurchenko, P.D., Birk, D.E. & Mecham, R.P. (1994). In: Mecham, R.P. (ed), *Extracellular Matrix Assembly and Structure*. Biology of extracellular matrix series. Academic Press, Inc., San Diego,California, pp. 468.
3. Kreis, T. & Vale, R. (1999) *Guidebook to the Extracellular Matrix, Anchor and Adhesion Proteins*, 2nd edn. A Sambrose and Tooze Publication at Oxford University Press, New York, pp. 568.
4. Werb, Z. & Chin, J.R. (1998) Extracellular matrix remodeling during morphogenesis. *The Annals of the New York Academy of Sciences*, 857, 110–118.
5. Damsky, C.H., Moursi, A., Zhou, Y., Fisher, S.J. & Globus, R.K. (1997) The solid state environment orchestrates embryonic development and tissue remodeling. *Kidney International*, 51 (5), 1427–1433.
6. Trojanowska, M., LeRoy, E.C., Eckes, B. & Krieg, T. (1998) Pathogenesis of fibrosis: type 1 collagen and the skin. *Journal of Molecular Medicine*, 76 (3–4), 266–274.
7. Chiquet, M. (1999) Regulation of extracellular matrix gene expression by mechanical stress. *Matrix Biology*, 18 (5), 417–426.
8. Norrby, K. (1997) Angiogenesis: new aspects relating to its initiation and control. *APMIS*, 105 (6), 417–437.
9. Friedl, P. & Brocker, E.B. (2000) The biology of cell locomotion within three-dimensional extracellular matrix. *Cellular and Molecular Life Sciences*, 57 (1), 41–64.
10. Karsenty, G. (1999) The genetic transformation of bone biology. *Genes and Development*, 13 (23), 3037–3051.
11. Murphy, G. & Nagase, H. (2011) Localizing matrix metalloproteinase activities in the pericellular environment. *FEBS Journal*, 278 (1), 2–15.

12. Ryan, J.N. & Woessner, J.F. (1971) Mammalian collagenase: direct demonstration in homogenates of involuting rat uterus. *Biochemical and Biophysical Research Communications*, 44 (1), 144–149.
13. Woessner, J.F. & Ryan, J.N. (1973) Collagenase activity in homogenates of the involuting rat uterus. *Biochimica et Biophysica Acta*, 309 (2), 397–405.
14. Welgus, H.G., Jeffrey, J.J., Stricklin, G.P., Roswit, W.T. & Eisen, A.Z. (1980) Characteristics of the action of human skin fibroblast collagenase on fibrillar collagen. *The Journal of Biological Chemistry*, 255 (14), 6806–6813.
15. Collier, I.E., Saffarian, S., Marmer, B.L., Elson, E.L. & Goldberg, G.I. (2001) Substrate recognition by gelatinase a: the c-terminal domain facilitates surface diffusion. *Biophysical Journal*, 81 (4), 2370–2377.
16. Saffarian, S., Collier, I.E., Marmer, B.L., Elson, E.L. & Goldberg, G. (2004) Interstitial Collagenase is a Brownian ratchet driven by proteolysis of collagen. *Science*, 306 (5693), 108–111.
17. Collier, I.E., Legant, W., Marmer, B., Lubman, O., Saffarian, S. *et al.* (2011) Diffusion of MMPs on the surface of collagen fibrils: the mobile cell surface-collagen substratum inter interface. *PLoS ONE*, 6 (9), e24029.
18. Sarkar, S.K., Marmer, B., Goldberg, G. & Neuman, K.C. (2012) Single-molecule tracking of collagenase on native type I collagen fibrils reveals degradation mechanism. *Current Biology*, 22 (12), 1047–1056.
19. Morgunova, E., Tuuttila, A., Bergmann, U., Isupov, M., Lindqvist, Y. *et al.* (1999) Structure of human pro-matrix metalloproteinase-2: activation mechanism revealed [see comments]. *Science*, 284 (5420), 1667–1670.
20. Jozic, D., Bourenkov, G., Lim, N.H., Visse, R., Nagase, H. *et al.* (2005) X-ray structure of human proMMP-1 – new insights into procollagenase activation and collagen binding. *The Journal of Biological Chemistry*, 280 (10), 9578–9585.
21. Li, J., Brick, P., O'Hare, M., Skarzynski, T., Lloyd, L. *et al.* (1995) Structure of full-length porcine synovial collagenase reveals a C-terminal domain containing a calcium-linked, four-bladed α-propeller. *Structure*, 3 (6), 541–549.
22. Libson, A.M., Gittis, A.G., Collier, I.E., Marmer, B.L., Goldberg, G.I. *et al.* (1995) Crystal structure of the haemopexin-like C-terminal domain of gelatinase A. *Natural Structural Biology*, 2 (11), 938–942.
23. Overall, C.M. (2002) Molecular determinants of metalloproteinase substrate specificity: matrix metalloproteinase substrate binding domains, modules, and exosites. *Molecular Biotechnology*, 22 (1), 51–86.
24. Bertini, I., Fragai, M., Luchinat, C., Melikian, M., Mylonas, E. *et al.* (2009) Interdomain flexibility in full-length matrix metalloproteinase-1 (MMP-1). *The Journal of Biological Chemistry*, 284 (19), 12821–12828.
25. Rosenblum, G., Van den Steen, P.E., Cohen, S.R., Grossmann, J.G., Frenkel, J. *et al.* (2007) Insights into the structure and domain flexibility of full-length pro-matrix metalloproteinase-9/gelatinase B. *Structure*, 15 (10), 1227–1236.
26. Lauer-Fields, J.L., Chalmers, M.J., Busby, S.A., Minond, D., Griffin, P.R. *et al.* (2009) Identification of specific hemopexin-like domain residues that facilitate matrix metalloproteinase collagenolytic activity. *The Journal of Biological Chemistry*, 284 (36), 24017–24024.
27. Bertini, I., Fragai, M., Luchinat, C., Melikian, M., Toccafondi, M. *et al.* (2011) Structural basis for matrix metalloproteinase 1-catalyzed collagenolysis. *Journal of the American Chemical Society*, 134 (4), 2100–2110.
28. Manka, S.W., Carafoli, F., Visse, R., Bihan, D., Raynal, N. *et al.* (2012) Structural insights into triple-helical collagen cleavage by matrix metalloproteinase 1. *Proceedings of the National Academy of Sciences of the United States of America*, 109 (31), 12461–12466.
29. Arnold, L.H., Butt, L.E., Prior, S.H., Read, C.M., Fields, G.B. *et al.* (2011) The interface between catalytic and hemopexin domains in matrix metalloproteinase-1 conceals a collagen binding exosite. *The Journal of Biological Chemistry*, 286 (52), 45073–45082.
30. Welgus, H.G., Jeffrey, J.J. & Eisen, A.Z. (1981) Human skin fibroblast collagenase. Assessment of activation energy and deuterium isotope effect with collagenous substrates. *The Journal of Biological Chemistry*, 256 (18), 9516–9521.
31. Jeffrey, J.J., Welgus, H.G., Burgeson, R.E. & Eisen, A.Z. (1983) Studies on the activation energy and deuterium isotope effect of human skin collagenase on homologous collagen substrates. *The Journal of Biological Chemistry*, 258 (18), 11123–11127.
32. Cerofolini, L., Fields, G.B., Fragai, M., Geraldes, C.F.G.C., Luchinat, C. *et al.* (2013) Examination of matrix metalloproteinase-1 in solution: a preference for the pre-collagenolysis state. *The Journal of Biological Chemistry*, 288 (42), 30659–30671.
33. Rosenblum, G., Van den Steen, P.E., Cohen, S.R., Bitler, A., Brand, D.D. *et al.* (2010) Direct visualization of protease action on collagen triple helical structure. *PLoS ONE*, 5 (6), e11043.
34. Allan, J.A., Docherty, A.J. & Murphy, G. (1994) The binding of gelatinases A and B to type I collagen yields both high and low affinity sites. *The Annals of the New York Academy of Sciences*, 732, 365–366.

35. Allan, J.A., Docherty, A.J., Barker, P.J., Huskisson, N.S., Reynolds, J.J. *et al.* (1995) Binding of gelatinases A and B to type-I collagen and other matrix components. *The Biochemical Journal*, 309 (Pt 1), 299–306.
36. Collier, I.E., Krasnov, P.A., Strongin, A.Y., Birkedal-Hansen, H. & Goldberg, G.I. (1992) Alanine scanning mutagenesis and functional analysis of the fibronectin-like collagen-binding domain from human 92-kDa type IV collagenase. *The Journal of Biological Chemistry*, 267 (10), 6776–6781.
37. Banyai, L., Tordai, H. & Patthy, L. (1994) The gelatin-binding site of human 72 kDa type IV collagenase (gelatinase A). *The Biochemical Journal*, 298 (Pt 2), 403–407.
38. Collier, I.E., Wilhelm, S.M., Eisen, A.Z., Marmer, B.L., Grant, G.A. *et al.* (1988) H-ras oncogene-transformed human bronchial epithelial cells (TBE-1) secrete a single metalloprotease capable of degrading basement membrane collagen. *The Journal of Biological Chemistry*, 263 (14), 6579–6587.
39. Cha, H., Kopetzki, E., Huber, R., Lanzendorfer, M. & Brandstetter, H. (2002) Structural basis of the adaptive molecular recognition by MMP9. *Journal of Molecular Biology*, 320 (5), 1065–1079.
40. Goldberg, G.I., Marmer, B.L., Grant, G.A., Eisen, A.Z., Wilhelm, S. *et al.* (1989) Human 72-kilodalton type IV collagenase forms a complex with a tissue inhibitor of metalloproteases designated TIMP-2. *Proceedings of the National Academy of Sciences of the United States of America*, 86 (21), 8207–8211.
41. Morgunova, E., Tuuttila, A., Bergmann, U. & Tryggvason, K. (2002) Structural insight into the complex formation of latent matrix metalloproteinase 2 with tissue inhibitor of metalloproteinase 2. *Proceedings of the National Academy of Sciences of the United States of America*, 99 (11), 7414–7419.
42. Bigg, H., Shi, Y., Liu, Y., Steffensen, B. & Overall, C. (1997) Specific, high affinity binding of tissue inhibitor of metalloproteinases-4 (TIMP-4) to the COOH-terminal hemopexin-like domain of human gelatinase A. TIMP-4 binds progelatinase A and the COOH-terminal domain in a similar manner to TIMP-2. *The Journal of Biological Chemistry*, 272 (24), 15496–15500.
43. Windsor, L.J., Birkedal-Hansen, H., Birkedal-Hansen, B. & Engler, J.A. (1991) An internal cysteine plays a role in the maintenance of the latency of human fibroblast collagenase. *Biochemistry*, 30, 641–647.
44. Clark, I.M. & Cawston, T.E. (1989) Fragments of human fibroblast collagenase. Purification and characterization. *The Biochemical Journal*, 263 (1), 201–206.
45. Kopelman, R. (1988) Fractal Reaction Kinetics. *Science*, 241, 1620–1626.
46. Kopelman, R. (1989) Diffusion-controlled reaction kinetics. In: Avnir, D. (ed), *The Fractal Approach to Heterogeneous Chemistry*. John Wiley and sons, New York.
47. Havlin, S. (1989) Molecular diffusion and reactions. In: Avnir, D. (ed), *The Fractal Approach to Heterogeneous Chemistry*. John Wiley and sons, New York.
48. Petersen, N.O. & Elson, E.L. (1986) Measurements of diffusion and chemical kinetics by fluorescence photobleaching recovery and fluorescence correlation spectroscopy. *Methods in Enzymology*, 130, 454–484.
49. Falconi, M., Altobelli, G., Iovino, M.C., Politi, V. & Desideri, A. (2003) Molecular dynamics simulation of matrix metalloproteinase 2: fluctuations and time evolution of recognition pockets. *Journal of Computer-Aided Molecular Design*, 17 (12), 837–848.
50. Strongin, A.Y., Collier, I., Bannikov, G., Marmer, B.L., Grant, G.A. *et al.* (1995) Mechanism of cell surface activation of 72-kDa type IV collagenase. Isolation of the activated form of the membrane metalloprotease. *The Journal of Biological Chemistry*, 270 (10), 5331–5338.
51. Goldberg, G.I., Wilhelm, S.M., Kronberger, A., Bauer, E.A., Grant, G.A. *et al.* (1986) Human fibroblast collagenase. Complete primary structure and homology to an oncogene transformation-induced rat protein. *The Journal of Biological Chemistry*, 261 (14), 6600–6605.
52. Eisen, A.Z. (1969) Human skin collagenase: localization and distribution in normal human skin. *The Journal of Investigative Dermatology*, 52, 442–448.
53. Bauer, E.A., Gordon, J.M., Reddick, M.E. & Eisen, A.Z. (1977) Quantitation and immunocytochemical localization of human skin collagenase in basal cell carcinoma. *The Journal of Investigative Dermatology*, 69, 363–367.
54. Welgus, H.G., Jeffrey, J.J. & Eisen, A.Z. (1981) The collagen substrate specificity of human skin fibroblast collagenase. *The Journal of Biological Chemistry*, 256 (18), 9511–9515.
55. Itoh, Y., Takamura, A., Ito, N., Maru, Y., Sato, H. *et al.* (2001) Homophilic complex formation of MT1-MMP facilitates proMMP-2 activation on the cell surface and promotes tumor cell invasion. *The EMBO Journal*, 20 (17), 4782–4793.
56. Kinoshita, T., Sato, H., Okada, A., Ohuchi, E., Imai, K. *et al.* (1998) TIMP-2 promotes activation of progelatinase A by membrane-type 1 matrix metalloproteinase immobilized on agarose beads. *The Journal of Biological Chemistry*, 273 (26), 16098–16103.
57. Sato, H., Takino, T., Okada, Y., Cao, J., Shinagawa, A. *et al.* (1994) A matrix metalloproteinase expressed on the surface of invasive tumour cells. *Nature*, 370 (6484), 61–55.
58. Lehti, K., Lohi, J., Juntunen, M.M., Pei, D. & Keski-Oja, J. (2002) Oligomerization through hemopexin and cytoplasmic domains regulates the activity and turnover of membrane-type 1 matrix metalloproteinase. *The Journal of Biological Chemistry*, 277 (10), 8440–8448.

59. Ohuchi, E., Imai, K., Fujii, Y., Sato, H., Seiki, M. *et al.* (1997) Membrane type 1 matrix metalloproteinase digests interstitial collagens and other extracellular matrix macromolecules. *The Journal of Biological Chemistry*, 272 (4), 2446–2451.
60. Gross, J. & Nagai, Y. (1965) Specific degradation of the collagen molecule by tadpole collagenolytic enzyme. *Proceedings of the National Academy of Sciences of the United States of America*, 54 (4), 1197–1204.
61. Fields, G.B. (1991) A Model for Interstitial Collagen Catabolism by Mammalian Collagenases. *Journal of Theoretical Biology*, 153 (4), 585–602.
62. Eisen, A.Z., Jeffrey, J.J. & Gross, J. (1968) Human skin collagenase, isolation and mechanism of attack on the collagen molecule. *Biochimica et Biophysica Acta*, 151 (3), 637–645.
63. Magde, D., Webb, W.W. & Elson, E.L. (1978) Fluorescence correlation spectroscopy.III. Uniform translation and laminar flow. *Biopolymers*, 17 (2), 361–376.
64. Schliwa, M. & Woehlke, G. (2003) Molecular motors. *Nature*, 422 (6933), 759–765.
65. Astumian, R.D., Chock, P.B., Tsong, T.Y., Chen, Y.D. & Westerhoff, H.V. (1987) Can free energy be transduced from electric noise? *Proceedings of the National Academy of Sciences of the United States of America*, 84 (2), 434–438.
66. Julicher, F., Ajdari, A. & Prost, J. (1997) Modeling molecular motors. *Reviews of Modern Physics*, 69 (4), 1269–1281.
67. Magnasco, M.O. (1993) Forced thermal ratchets. *Physical Review Letters*, 71 (10), 1477–1481.
68. Peskin, C.S., Odell, G.M. & Oster, G.F. (1993) Cellular motions and thermal fluctuations: the Brownian ratchet. *Biophysical Journal*, 65 (1), 316–324.
69. Fox, R.F. (1998) Rectified Brownian movement in molecular and cell biology. *Physical Review E*, 57 (2), 2177–2203.
70. Prost, J., Chauwin, J.F., Peliti, L. & Ajdari, A. (1994) Asymmetric pumping of particles. *Physical Review Letters*, 72 (16), 2652–2655.
71. Mai, J., Sokolov, I.M. & Blumen, A. (2001) Directed particle diffusion under "burnt bridges" conditions. *Physical Review E*, 64 (1–1), 011102.
72. Neuman, K.C., Saleh, O.A., Lionnet, T., Lia, G., Allemand, J.-F. *et al.* (2005) Statistical determination of the step size of molecular motors. *Journal of Physicsn Condensed Matter*, 17 (47), S3811–S3820.
73. Orgel, J.P., Miller, A., Irving, T.C., Fischetti, R.F., Hammersley, A.P. *et al.* (2001) The in situ supermolecular structure of type I collagen. *Structure*, 9 (11), 1061–1069.
74. Perumal, S., Antipova, O. & Orgel, J.P. (2008) Collagen fibril architecture, domain organization, and triple-helical conformation govern its proteolysis. *Proceedings of the National Academy of Sciences of the United States of America*, 105 (8), 2824–2829.
75. Orgel, J.P., Irving, T.C., Miller, A. & Wess, T.J. (2006) Microfibrillar structure of type I collagen in situ. *Proceedings of the National Academy of Sciences of the United States of America*, 103 (24), 9001–9005.
76. Chung, L., Dinakarpandian, D., Yoshida, N., Lauer-Fields, J.L., Fields, G.B. *et al.* (2004) Collagenase unwinds triple-helical collagen prior to peptide bond hydrolysis. *The EMBO Journal*, 23 (15), 3020–3030. Epub 2004 Jul 3015.
77. Bode, W., Reinemer, P., Huber, R., Kleine, T., Schnierer, S. *et al.* (1994) The X-ray crystal structure of the catalytic domain of human neutrophil collagenase inhibited by a substrate analogue reveals the essentials for catalysis and specificity. *The EMBO Journal*, 13 (6), 1263–1269.
78. Stultz, C.M. (2002) Localized unfolding of collagen explains collagenase cleavage near imino-poor sites. *Journal of Molecular Biology*, 319 (5), 997–1003.
79. Lu, K.G. & Stultz, C.M. (2013) Insight into the degradation of type-I collagen fibrils by MMP-8. *Journal of Molecular Biology*, 425 (10), 1815–1825.
80. Privalov, P.L. (1979) Stability of proteins: small globular proteins. *Advances in Protein Chemistry*, 33, 167–241.
81. Ryhänen, L., Zaragoza, E.J. & Uitto, J. (1983) Conformational stability of type I collagen triple helix: Evidence for temporary and local relaxation of the protein conformation using a proteolytic probe. *Archives of Biochemistry and Biophysics*, 223 (2), 562–571.
82. Privalov, P.L. (1982) Stability of proteins. Proteins which do not present a single cooperative system. *Advances in Protein Chemistry*, 35, 1–104.
83. Bachinger, H.P., Morris, N. & Davis, J. (1993) Thermal stability and folding of the collagen triple helix and the effects of mutations in osteogenesis imperfecta on the triple helix of type I collagen. *American Journal of Medical Genetics*, 45 (2), 152–162.
84. Leikina, E., Mertts, M.V., Kuznetsova, N. & Leikin, S. (2002) Type I collagen is thermally unstable at body temperature. *Proceedings of the National Academy of Sciences of the United States of America*, 99 (3), 1314–1318.

85. Nerenberg, P.S. & Stultz, C.M. (2008) Differential unfolding of alpha 1 and alpha 2 chains in type I collagen and collagenolysis. *Journal of Molecular Biology*, 382 (1), 246–256.
86. Salsas-Escat, R., Nerenberg, P.S. & Stultz, C.M. (2010) Cleavage site specificity and conformational selection in type I collagen degradation. *Biochemistry*, 49 (19), 4147–4158.
87. Lukic, B., Jeney, S., Tischer, C., Kulik, A.J., Forro, L. *et al.* (2005) Direct observation of nondiffusive motion of a Brownian particle. *Physical Review Letters*, 95 (16), 160601–160604.
88. Vologodskii, A. (2006) Energy transformation in biological molecular motors. *Physics of Life Reviews*, 3 (2), 119–132.
89. Chattopadhyay, K., Saffarian, S., Elson, E.L. & Frieden, C. (2002) Measurement of microsecond dynamic motion in the intestinal fatty acid binding protein by using fluorescence correlation spectroscopy. *Proceedings of the National Academy of Sciences of the United States of America*, 99 (22), 14171–14176.
90. Schulz, G.E. & Schirmer, R.H. (1979) *Principles of Protein Structure*. Springer, pp. 314.
91. Han, S., Makareeva, E., Kuznetsova, N.V., DeRidder, A.M., Sutter, M.B. *et al.* (2010) Molecular mechanism of type I collagen homotrimer resistance to mammalian collagenases. *The Journal of Biological Chemistry*, 285 (29), 22276–22281.
92. Miles, C.A., Burjanadze, T.V. & Bailey, A.J. (1995) The kinetics of the thermal-denaturation of collagen in unrestrained rat tail tendon determined by differential scanning calorimetry. *Journal of Molecular Biology*, 245 (4), 437–446.
93. Kuznetsova, N.V., McBride, D.J. Jr. & Leikin, S. (2003) Changes in thermal stability and microunfolding pattern of collagen helix resulting from the loss of alpha2(I) chain in osteogenesis imperfecta murine. *Journal of Molecular Biology*, 331 (1), 191–200.
94. Adam, G. & Delbrück, M. (1968) Reduction of dimensionality in biological diffusion processes. In: Rich, A. & Davidson, N. (eds), *Structural Chemistry and Molecular Biology*. W. H. Freeman and Co., San Francisco.
95. Serr, A., Horinek, D. & Netz, R.R. (2008) Polypeptide Friction and Adhesion on Hydrophobic and Hydrophilic Surfaces: A Molecular Dynamics Case Study. *Journal of the American Chemical Society*, 130 (37), 12408–12413.
96. Pastre, D., Pietrement, O., Zozime, A. & Le Cam, E. (2005) Study of the DNA/ethidium bromide interactions on mica surface by atomic force microscope: Influence of the surface friction. *Biopolymers*, 77 (1), 53–62.
97. Jervis, E.J., Haynes, C.A. & Kilburn, D.G. (1997) Surface diffusion of cellulases and their isolated binding domains on cellulose. *The Journal of Biological Chemistry*, 272 (38), 24016–24023.
98. Receveur, V., Czjzek, M., Schulein, M., Panine, P. & Henrissat, B. (2002) Dimension, shape, and conformational flexibility of a two domain fungal cellulase in solution probed by small angle X-ray scattering. *The Journal of Biological Chemistry*, 277 (43), 40887–40892.
99. von Ossowski, I., Eaton, J.T., Czjzek, M., Perkins, S.J., Frandsen, T.P. *et al.* (2005) Protein disorder: conformational distribution of the flexible linker in a chimeric double cellulase. *Biophysical Journal*, 88 (4), 2823–2832.

3 Matrix Metalloproteinases: From Structure to Function

Marco Fragai[1,2] and Claudio Luchinat[1,2]

[1] *Magnetic Resonance Center (CERM), University of Florence, Florence, Italy*
[2] *Department of Chemistry, University of Florence, Florence, Italy*

3.1 Introduction

The structural diversity of the proteins and glycoproteins forming the extracellular matrix (ECM) has prompted the evolution and diversification of the human matrix metalloproteinases (MMPs). They are a family of multidomain endopeptidases involved in several physiological and pathological processes. They have been extensively investigated to decrypt their real pathological role and to design inhibitors as potential drug candidates since the discovery of their involvement in cancer progression about 30 years ago. The early enthusiasm for the discovery of potent inhibitors was rapidly replaced by disappointment and resignation, particularly for pharmaceutical companies, when all molecules failed the clinical trials. Nevertheless, the interest in these biologically relevant metalloproteases has not decreased over the years and the extensive structural and functional analyses performed on several members of the family have now provided the basis for a better understanding of their mechanism of action and a more rational approach to design therapeutic strategies focused on MMPs [1]. Any serious analysis of the structural and functional aspects of MMPs cannot ignore the structural features of the substrates. For a long time ECM scaffolds had been considered the main biological targets of MMPs [2–6]. Now it is known that MMPs hydrolyze a variety of substrates, including membrane receptors, cytokines, growth factors, and other proteases in the extracellular space [7–12]. For some MMPs, intracellular activities have also been reported or suggested [13–17]. The structural heterogeneity of the proteins and glycoproteins forming the ECM reflects the variety of its biological functions, where the mechanical support provided to tissues is flanked by a complex regulation of the cellular activity [18–20]. Besides the tight binding to the cell cytoskeleton through integrins [21, 22], ECM interplays with cells by binding, storing, and delivering cytokines and growth factors, often released by the proteolytic activity of MMPs [23].

Collagen and elastin are the two main components of the ECM with important structural and mechanical roles [24–29]. Elastin allows the reversible deformation

Matrix Metalloproteinase Biology, First Edition. Edited by Irit Sagi and Jean P. Gaffney.

of tissues thanks to a network of interconnected elastic fibers [29–34]. The fibers are formed through a multistep process where the soluble precursor tropoelastin undergoes a thermodynamic driven coacervation followed by microfibrillar deposition and enzymatic multi-sites cross-linking [35]. The structural analysis of tropoelastin suggests that the extended coil region extending from domain 2 to domain 18 confers elasticity to the protein, while the so-called foot-like region binds cells through integrins [36]. Conversely, collagens provide mechanical strength to tissues. The extraordinary mechanical strength of collagens is related to a well-designed structural organization. In collagens, the single left-handed helical α polypeptide chains are rich in repeating Gly-Xaa-Yaa triplets and form right-handed triple helices, which in turn assemble to form fibrils and fibers [28].

3.2 Classification and structural features

For MMPs, different classifications have been proposed over the years. From the very beginning, they were classified in different groups according to their substrate preference. Collagenases, gelatinases, stromelysins, and matrilysins constitute the four most important groups. In particular, collagen types I–III are efficiently degraded by the soluble collagenases MMP-1, MMP-8, MMP-13, and by two non-collagenases MMPs, the membrane bound MMP-14 [37] and the gelatinase MMP-2 [38, 39]. At the same time, collagenases degrade other ECM components and extracellular proteins with a broad proteolytic specificity. The group of gelatinases that includes MMP-2 and MMP-9, degrades gelatin, a form of partially degraded collagen, and type IV collagen. MMP-7 and MMP-26, forming the group of matrilysins, hydrolyze some components of the extracellular matrix such as fibronectin, gelatin, and type IV collagen but not the triple helical collagen [6]. Stromelysins, MMP-3, -10, and -11, degrade several ECM proteins such as proteoglycans, laminin, fibronectin, gelatin, and the non-triple helix region of collagen IV [40, 41]. Finally, the cross-linked elastin that together with fibronectin-like molecules provides flexibility and elasticity to tissues is quickly hydrolyzed by MMP-12 [31, 42–44]. All MMPs not included in these subclasses show a less defined substrate preference.

The sequencing of the human genome and bioinformatics allowed a new classification of MMPs. The genes encoding for proteases include those codifying the 23 different MMPs, the additional five shorter isoforms of MMP-16, MMP-19, MMP-21, MMP-23, and MMP-28, and the catalytically inactive MMP-1-like [45, 46]. MMPs show a common structural scheme (Fig. 3.1) with a prodomain of 66–120 AA, a catalytic domain of about 160 AA and a hemopexin-like domain of about 210 AA. The catalytic domain and the hemopexin-like domain are connected by a linker of a variable length. The hydrolysis of the peptide chain occurs at the active site in the catalytic domain [47–52]. Conversely, the role of the C-terminal hemopexin-like domain present in all MMPs but three (in MMP-23, that binds the cell membrane at the N-terminus, the hemopexin-like domain is replaced by an immunoglobulin-like domain, while MMP-7 and MMP-26 do not have the C-terminal domain) is poorly understood and probably different in the various members of the family [53]. The pro-domain masks the active site and prevents the interaction with substrates through

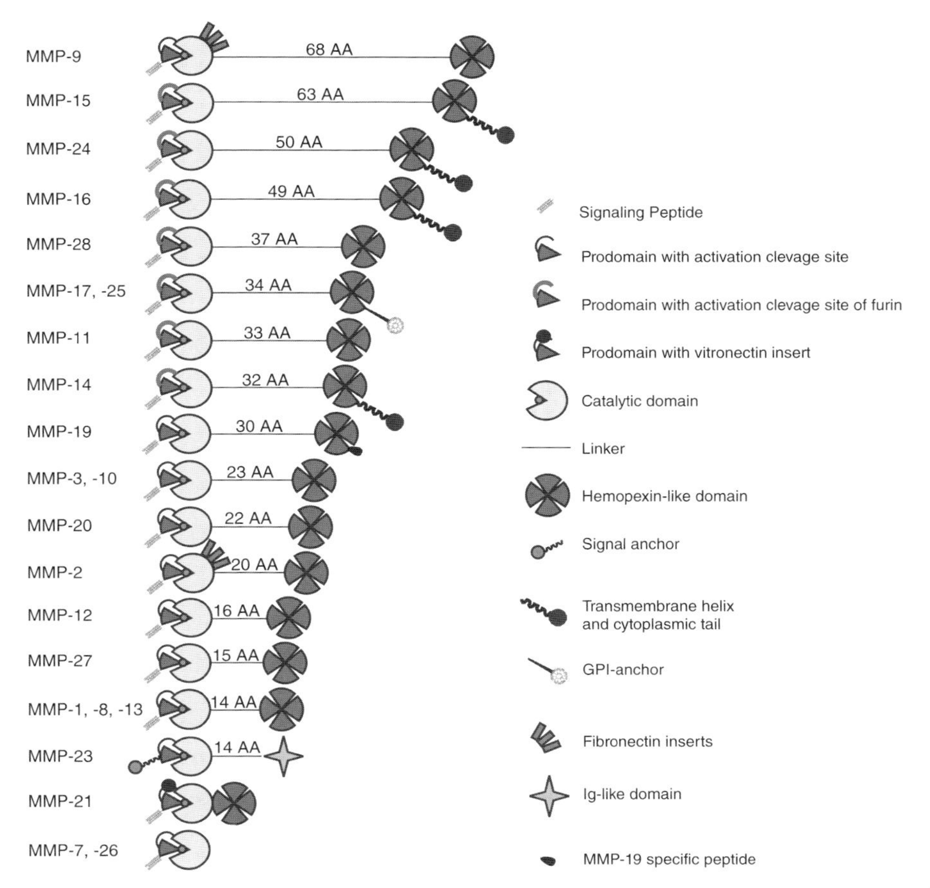

Figure 3.1 Structural organization of human MMPs with the corresponding linker length. (*See insert for color representation of this figure.*)

a cysteine switch PRCXXPD consensus sequence (PHCXXPD in MMP-26 [54]). The prodomain is removed upon the proteolytic activation by the very same catalytic domain or by other proteases [6, 55, 56]. All MMPs but one (MMP-23) are driven on the physiological cellular location by an N-terminal signal peptide of variable length (from 16 AA to 52 AA).

According to the analysis of the phylogenetic tree generated from the multiple alignment of the entire sequences, MMPs can be classified in five different groups [46]. The non-furine regulated MMPs are the largest group, with all proteins deriving from the same chromosome 11 [46]. This group comprises MMP-1, MMP-3, MMP-7, MMP-8, MMP-10, MMP-12, MMP-13, MMP-20, MMP-27 [22], and do not have in the prodomain a recognition sequence for furin-like serine proteases that makes the enzymes susceptible to activation in the trans-Golgi network [6, 57]. Soluble gelatinases, MMP-2 (Fig. 3.2) and MMP-9, constitute a second group. These proteins are characterized by the presence of three fibronectin-like inserts on the catalytic domain that have been reported to bind the triple helical regions of collagen [46, 58–60].

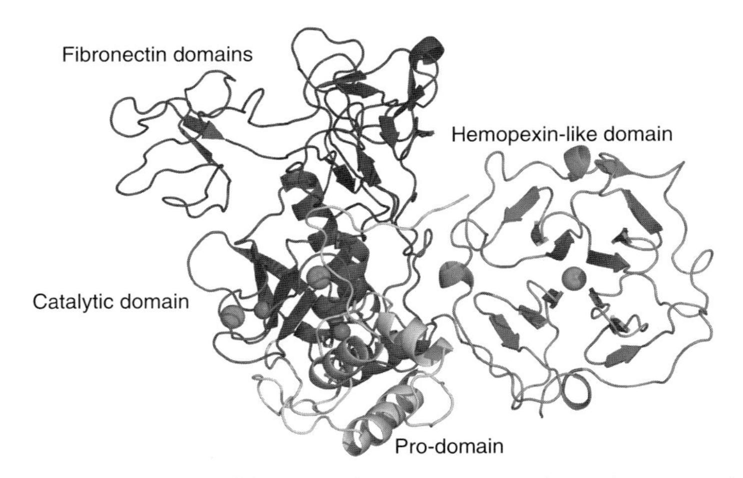

Figure 3.2 Ribbon representation of the inactive human proMMP-2. The prodomain, catalytic domain, fibronectin domains, and hemopexin domain are shown in yellow, red, blue, and orange, respectively. The catalytic and the structural zinc ions are represented as magenta spheres and calcium ions as green spheres. (*See insert for color representation of this figure.*)

The membrane-anchored MMPs, which remain bound to the outer cell membrane and are not secreted in the extracellular space, form two distinct groups, the transmembrane MMPs (MMP-14, MMP-15, MMP-16, and MMP-24) bearing a transmembrane helix with a small cytoplasmic domain at the C-terminal, and the glycosylphosphatidylinositol (GPI)-anchored MMPs with a C-terminal GPI moiety (MMP-17, MMP-25, and MMP-1-like) [46]. A separate group comprises all other MMPs (MMP-19, MMP-21, MMP-23, MMP-26, and MMP-28).

All the catalytic domains of MMPs show similar structural features with three α-helices, a twisted five-stranded β-sheet composed of four parallel (β2-β1-β3-β5) and one antiparallel (β4) strand, and eight intervening loops [61]. The long loop L8 between helices α2 and α3 is a region of relatively large variability among MMPs and constitutes the outside wall of the S_1' pocket [62]. This hydrophobic cavity is critical for the catalytic activity of MMPs and constitutes the main binding site of many active-site directed inhibitors [63–66] (Fig. 3.3).

The catalytic domain contains a zinc ion in the active site and, in all MMPs but three (MMP-21, 27, and MMP-1-like [46]), a second zinc ion with structural function [46, 61, 64]. In the active enzyme, the catalytic zinc ion is coordinated by three histidines and one water molecule hydrogen-bonded to a catalytically relevant glutamate. Two additional labile water molecules have also been observed in crystal structures of the free enzyme, but they are easily displaced in solution by the carbonyl moiety upon binding of the substrate [52]. The catalytic domain binds from one to three calcium ions, with different affinities, that stabilize the protein folding [61, 64]. The hemopexin-like domain has the same structural features in all members of the family. The protein domain is constituted of four β-sheets composed of four antiparallel β-strands, folded in an approximately symmetric four-blade propeller around a deep channel [53, 67]. The four β-sheets are connected by loops, where a short α-helical segment can be present. The four-blade propeller is stabilized by a disulfide bridge that

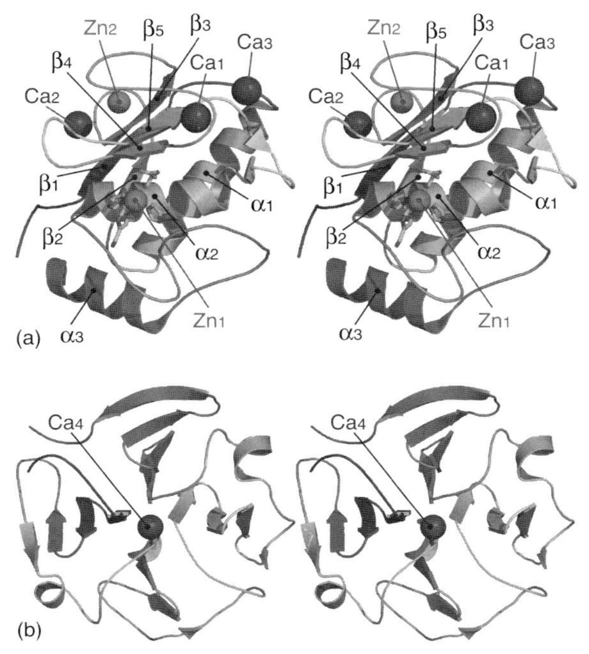

Figure 3.3 Stereo view of the catalytic (a) and hemopexin-like (b) domains of MMP-12 represented as ribbons. In the catalytic domain α-helices, β-strands, and loops are organized in a L1-β1-L2-α1-L3-β2-L4-β3-L5-β4-L6-β5-L7-α2-L8-α3 topology. The catalytic (Zn1) and the structural (Zn2) zinc ions are shown as magenta spheres of arbitrary radius. The first (Ca1), the second (Ca2), the third (Ca3) calcium ions and the calcium ion in the hemopexin-like domain are shown as blue spheres. The three histidines that bind the catalytic zinc and the catalytically relevant glutamate are represented as cyan sticks. Strands and helices are labeled with numbers and greek letters. The hemopexin-like domain is constituted by four β-sheets of four antiparallel β-strands that folds in a symmetric four-blade propeller [53, 67]. The central deep tunnel filled by water molecules is closed by a calcium ion (Ca4) at the bottom. (*See insert for color representation of this figure.*)

links the first and the fourth blades [67]. A proline-rich linker of variable length connects the catalytic and hemopexin-like domains of MMPs [46]. The linker is relatively short in collagenases, but can be long enough, as in the case MMP-9, to constitute a domain by itself [68].

3.3 Catalytic mechanism

The sequence of events occurring at the active site of MMPs while functioning has been largely revealed by investigating the hydrolysis of peptide models [51]. It is now widely accepted that the metal-coordinated water molecule, hydrogen-bonded to a conserved active-site glutamate, performs a nucleophilic attack on the peptide carbonyl group of the substrate. In the resulting intermediate, the zinc ion is coordinated in a bidentate manner by the two oxygen atoms of the gem-diol group of the transition state.

During the hydrolysis, the substrate is maintained in the active site crevice by a pool of hydrogen bonds and by the interaction of the lipophilic side chain of the residue at

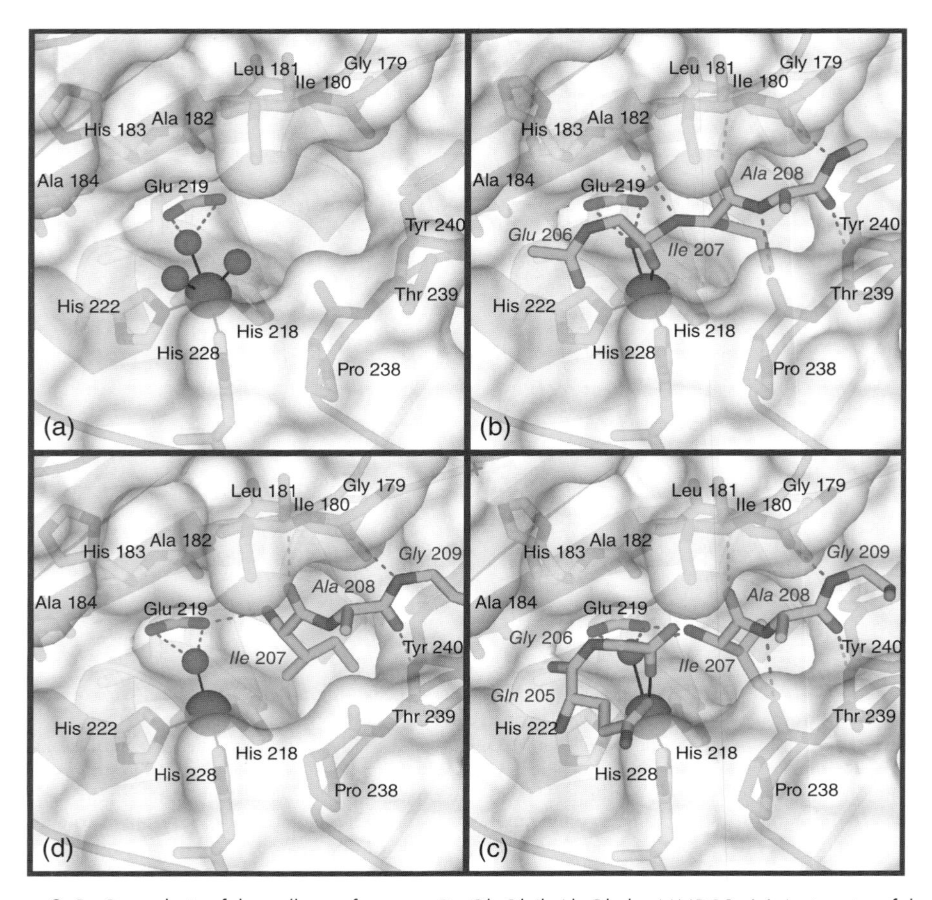

Figure 3.4 Proteolysis of the collagen fragment ProGlnGlyIleAlaGly by MMP-12. (a) Active site of the free enzyme before the interaction with the substrate. (b) Calculated model of the gemdiol intermediate. (c) X-ray structure of the two-peptide intermediate obtained by soaking the active uninhibited MMP-12 crystals with the collagen peptide. (d) Adduct of MMP-12 with the peptide fragment IleAlaGly after the release of the C-terminal fragment. (*See insert for color representation of this figure.*)

the N-terminal of the cleaved fragment. The arrangement of the two fragments in the active site, immediately after the peptide hydrolysis, has been revealed by an X-ray study on MMP-12 (Fig. 3.4) [52]. The C-terminal peptide coordinates the catalytic zinc ion by its carboxylate end, without establishing other significant interactions with the protein. The zinc ion is further coordinated by the three histidines and the water molecule, hydrogen bonded to the catalytic Glu, and maintains the pentacoordinated geometry of the transition state. Conversely, the N-terminal peptide fragment is maintained in place by the same pool of hydrogen bonds observed in the transition state, by hydrophobic interactions involving the S_1' cavity, and by new interactions established with the oxygen of the metal-coordinated water and with the carboxylic oxygen of the catalytic glutamate.

Another crystal structure sheds light on the subsequent step of the catalytic mechanism when the C-terminal peptide fragment is released from the active site and the N-terminal fragment moves away from the zinc that remains tetra-coordinated by the three histidines and the water molecule. The active site undergoes a rearrangement with the opening of the S_1' cavity and a deeper penetration of the sidechain of the N-terminal residue. Then, the N-terminal peptide fragment is also released and the protein is ready to hydrolyze a new peptide chain.

3.4 Intra- and inter-domain flexibility

The conformational rearrangement of the active site, suggested by the slight differences among the crystal structures, along the steps of the catalytic mechanism, accounts for a relatively large internal mobility of the catalytic domain in MMP-12. The internal flexibility of this protein has been well characterized by integrating the information collected from crystal structures with the NMR data obtained in solution [69].

In the X-ray structures of the protein complexed to different inhibitors, several loop regions show different conformations. The conformational heterogeneity is markedly evident for the long loop L8 that forms the S1′ cavity and establishes interactions with substrates and inhibitors. The NMR analysis clearly shows that the same regions are disordered also in solution. In the protein, several residues experience motions on the milliseconds–microseconds time scales, as revealed by ^{1}H-^{15}N relaxation, $R_{1\rho}$ and CPMG measurements. For other residues, mobility on all time scales faster than, or of the order of, milliseconds was revealed by measuring residual dipolar couplings. In addition, in several loops, conformational heterogeneity with time scales of the order of, or longer than, milliseconds has been demonstrated by the presence of resonance splittings of the amide signals in the 2D ^{1}H-^{15}N HSQC NMR spectra.

A comparative analysis carried out on solution and crystal structures released on the protein data bank has shown that conformational heterogeneity and flexibility of the loop regions are a common feature of the catalytic domains in all MMPs. The presence of wide and collective motions involving the catalytic domain is one of the challenges to be addressed by researchers to design selective inhibitors. However, this extensive flexibility is essential for binding and hydrolysis of the natural substrates. No less important in substrate recognition is the role of secondary binding sites that control the enzyme specificity [70, 71].

3.5 Elastin and collagen degradation

In most cases, the events at the active site are only the last steps of a more elaborated process that allows MMPs to hydrolyze structurally complex substrates. The hydrolysis of cross-linked elastin and of type I collagen are two illustrative examples of the complexity of the enzymatic activity of MMPs. Concerning elastin degradation, it has been long reported that the catalytic domain alone represents the biologically relevant form of MMP-12. In fact, the full length protein has been found to lack the hemopexin-like domain after activation, without compromising its elastolytic activity [72]. Therefore, the role of the hemopexin-like domain in MMP-12 has been, and still is, a matter of discussion and hypotheses. Soluble elastin has been used as a representative model of the natural substrate to investigate the elastolytic activity

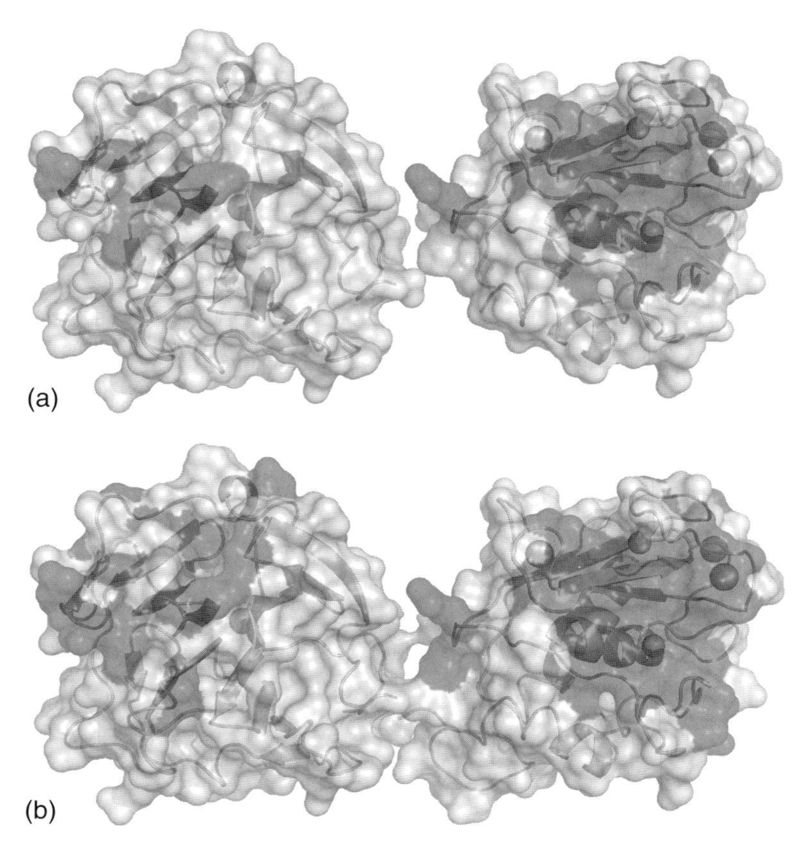

Figure 3.5 Pattern of residues interacting with elastin fragments in the isolated catalytic and hemopexin-like domains (a) and in the full length protein (b). The larger effects observed in the full length protein suggest cooperativity of the two domains in binding of elastin fragments. (*See insert for color representation of this figure.*)

of MMP-12 and to map the binding sites on the protein [72, 73]. The reported NMR analyses performed on the isolated domains show that elastin fragments extensively fit the catalytic groove and establish interactions with exosites on the catalytic domain (Fig. 3.5(a)). The binding capability of the isolated catalytic domain toward soluble elastin agrees with the retained elastolytic activity of the enzyme lacking the hemopexin-like domain. The distribution of the affected residues on the catalytic domain indicates an extended interaction of the active site crevice that involves 8–11 residues of the substrate. It is conceivable that this extended interaction may be functional to ensure an efficient binding of the enzyme to the single peptide chain of elastin, far from the cross-linked residues.

However, surprisingly, also a well-defined region of the isolated hemopexin-like domain is affected upon addition of soluble elastin (Fig. 3.5(a)) [73]. Similar but somewhat larger effects are observed in the full length MMP-12 (Fig. 3.5(b)). The larger affinity observed for the catalytic domain in the full-length protein suggests the presence of a cooperative contribution of the hemopexin-like domain to the interaction and opens new questions on its real physiological role.

The collagenolytic process by selected members of the MMP family has long been investigated and several mechanisms have been proposed. Now it is widely recognized that these MMPs utilize multiple domains to efficiently hydrolyze this highly structured substrate. The MMPs with collagenolytic activity are MMP-1, MMP-2, MMP-8, MMP-9, MMP-13, MT1-MMP, and MT2-MMP [74–77]. The presence of both the catalytic and hemopexin-like domains is required for collagen degradation by MMP-1, MMP-8, MMP-13, MT1-MMP, and MT2-MMP [78–83]. Also the linker has been demonstrated to play a role in collagenolysis, allowing the catalytic and hemopexin-like domain to adopt the correct reciprocal orientation [84–86] and/or to directly bind the substrate [87].

The collagen triple-helix is compact enough to prevent any single chain to fit the active site cavity of the catalytic domain [88]. Therefore, an active unwinding of the triple-helix by an MMP and/or the presence of a distinct conformation [88–91], or conformational flexibility of the cleavage site within collagen, have been suggested to occur during collagenolysis. In this way, one of the three peptide chains may become accessible for the binding at the active site of the catalytic domain. Alternatively, according to the "molecular tectonics" mechanism, the unwinding could be driven by specific interactions of each of the three peptide chains forming collagen with different domains [75], and by a conformational change involving the active site of the catalytic domain [92]. A further and different mechanism has been suggested by the "vulnerable" site hypothesis [93]. According to this proposed mechanism, a distinct cleavage site region within collagen could alone be responsible for collagenolysis. The occurrence of destabilization of the triple-helix and/or stabilization of an unwound triple-helix upon MMP binding has also been proposed [94].

Recently, structural studies on collagenolysis have been performed on triple helical peptides [95]. One of these, the homotrimeric α1(I)772–786 THP, is hydrolyzed by catalytic-hemopexin-like MMP-1 (hereafter FL-MMP-1) at the physiological cleavage site, and represents a suitable analog of the native collagen to investigate the enzymatic mechanism and the role of the different domains in collagenolysis [96]. The sequence of the peptide is $(GPO)_4$-GPQGIAGQRGVVGLO-$(GPO)_4$, where O is 4-hydroxyproline. The catalytic domain of MMP-8 digests THP more slowly than the analogous single-stranded peptide, while the triple helical substrate is preferred by FL- MMP-1 and MMP-8 [97]. On the other hand, deletion of the hemopexin-like domain from FL-MMP-1 greatly reduces the activity towards THPs, while it has no effects on the hydrolysis of the analogous single stranded peptides [98]. Taken altogether, these experimental data suggest that the hemopexin-like domain promotes the binding of FL-MMP-1 to the triple-helical structure. Interaction of the α1(I)772-786 THP with the FL- MMP-1 as well as with the catalytic and hemopexin-like domains has also been characterized by NMR spectroscopy. The three peptide chains forming the homotrimeric THP are chemically identical but sterically not equivalent, as also demonstrated by NMR spectroscopy [99]. Indeed, the chemical shifts of the atoms forming each of the three chains are not superimposable in the NMR spectra. These features allowed the assignment of the amino acids within THP and the determination of its solution structure. In solution, THP exhibits a rodlike structure in equilibrium with the unfolded single chain species. The NMR data collected on the isolated domains show that the hemopexin-like domain binds THP chains 1T and 2T downstream to the cleavage site, at Val23-Leu26, with low micromolar affinity. Conversely, the isolated catalytic domain binds THP with some residues forming the active site crevice but with a lower affinity and specificity than those observed for the hemopexin-like domain. However, a cooperative behavior of the catalytic and hemopexin-like domain has been observed in the FL-MMP-1, where a reciprocal reinforcement of the interaction with THP takes place. Therefore, it can be assumed that after the binding of the hemopexin-like domain that drives the interaction with THP, the second event in collagenolysis is the binding of the catalytic domain at the cleavage site. However, the simple visual inspection of the experimental structures neither allows us to understand how the hemopexin-like domain within the full length protein interacts with THP, nor does it provide any indication for the correct positioning of the catalytic domain in front of the cleavage site.

In this regard, in recent years it has been shown that several members of the MMP family are affected by a large interdomain flexibility [100–106]. This large flexibility has perhaps different functional roles in the different MMPs. It has been hypothesized that the interdomain reorientation might play a role in (i) binding and degradation of structurally unrelated substrates by permitting a variety of molecular conformations [106], (ii) ratcheting along collagen fibers during the proteolytic activity [107], and (iii) unwinding of the triple helical collagen to make the scissile bond on one of the three peptide chains accessible to cleavage [6, 88, 99].

Critical for interdomain flexibility is the length of the linker region connecting the catalytic and the hemopexin-like domain. The linker length is largely different among the members of the MMPs family (14–69 AA) (Fig. 3.2), and in MMP-1, MMP-8, MMP-12, MMP-13, is relatively short (14–16 AA) [61]. The well-defined (closed) spatial arrangement of catalytic and hemopexin-like domains in the X-ray structures of FL-MMP-1 have long induced scientists to consider all MMPs with a short linker as mainly rigid entities. Only in recent years the presence of a large inter-domain flexibility in solution has been demonstrated for FL-MMP-1 [102, 103] and MMP-12 [101]. In particular, for FL-MMP-1, the closed conformations observed in the X-ray structures are not representative of the conformations sampled by the protein in solution, where it experiences an open-closed equilibrium, spending at least one-third of the time in an extended arrangement [102]. More recently, an assessment of the most readily accessed conformations of FL-MMP-1 in solution, before the interaction with the substrate, has been obtained by paramagnetic NMR spectroscopy and small angle X-ray scattering (SAXS) through the calculation of their maximum occurrence (MO) [108]. For a given conformer, the MO is defined as the maximum weight that it can have in any ensemble that matches the averaged experimental data. In this respect, it has been amply demonstrated through several simulations that the highest MO conformations do point to the conformations with the highest weight in synthetic ensembles [109–111].

For FL-MMP-1, the conformers with the highest MO exhibit an extended arrangement (Fig. 3.6, right side), featuring an open space between the two domains and having the residues of the hemopexin-like domain that are known to bind the THP,

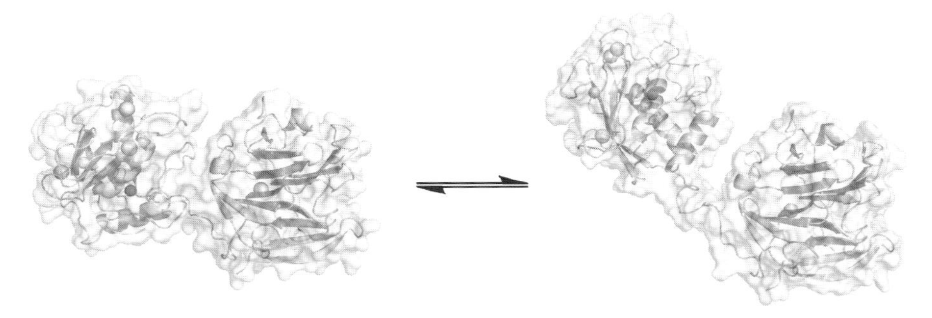

Figure 3.6 Closed (left) and open/extended (right) forms of FL-MMP-1 in equilibrium. The catalytic zinc ion is represented as a magenta sphere. (*See insert for color representation of this figure.*)

solvent exposed and available for the interaction. Thus, the NMR data have been used to develop a detailed and energetically favourable mechanism of collagenolysis well described by experimentally driven structures of the FL-MMP-1-THP complex and summarized as follows. In solution, and in the absence of collagen, FL-MMP-1 is in equilibrium between open/extended and closed structures (Fig. 3.6). When THP is available in solution, the hemopexin-like domain binds chains 1T and 2T, and then the catalytic domain finds itself facing the cleavage site of chain 1T (Fig. 3.6). The structure corresponding to the first event of collagen recognition by FL-MMP-1 suggests that a conformation resembling the closed and more energetically stable X-ray crystallographic structure can be achieved by a *back-rotation* of the catalytic and hemopexin-like domains. A set of structures reproducing the motion toward the closed form has been generated by imposing in docking calculations the same proximity observed in the X-ray structure, to the residues at the interface between the hemopexin-like and catalytic domains (Fig. 3.7, panel A). The movement and the reorientation of the two domains cause a distortion of the THP structure and drive chain 1T into the active site crevice, to establish a set of H-bonding interactions and allowing the carbonyl oxygen of the cleavage site amide bond to coordinate the catalytic zinc ion. The destabilization/unwinding of THP is consistent with the experimentally observed weakening of the interchain NOEs upon the addition of FL-MMP-1. Considering the structural similarity of MMPs at the active site, the structural models of the steps after the hydrolysis by MMP-1 (Fig. 3.7, (b)), have been derived from the X-ray crystallographic structure of the complex between the MMP-12 catalytic domain and the two fragments obtained by enzymatic cleavage of the $\alpha 1$(I) collagen model hexapeptide described above.

The analysis of the highest MO structures of FL-MMP-1 provides another piece of information on the behaviour of the protein and shows that the conformational space explored by FL-MMP-1 in solution strongly supports the mechanism described here. In fact, after the interaction of the hemopexin-like domain with the Val23-Leu26 region of the THP (Fig. 3.8, (a)) the catalytic domain can already adopt the correct positioning in front of the cleavage site (Fig. 3.8, (d)). If THP is modelled on the hemopexin-like domain of the highest MO structures, with the same binding mode determined for the first step of the above proposed mechanism, the catalytic domain is placed at the boundary between sterically overlapping and non-overlapping

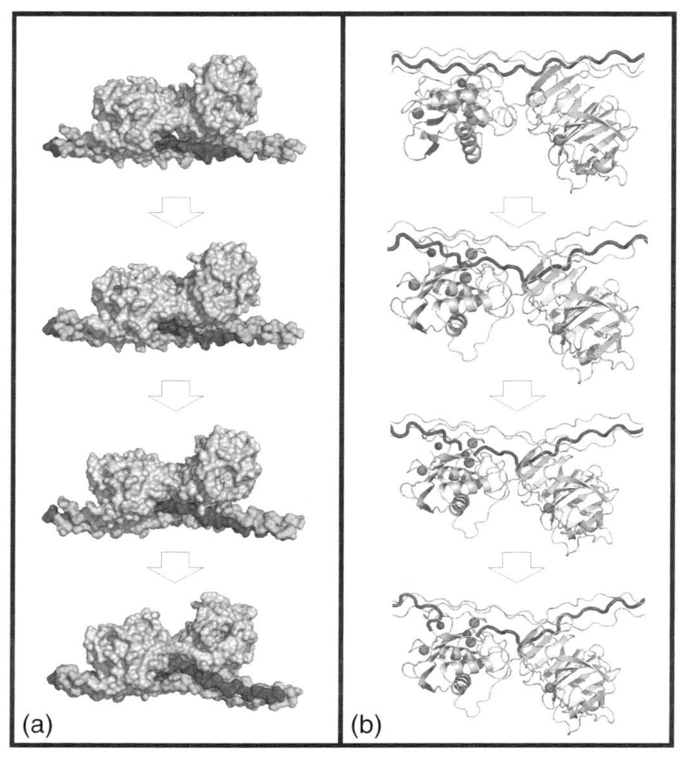

Figure 3.7 Proposed mechanism for collagenolysis. In panel (a), from the top (the experimentally-driven docked complex between FL-MMP-1 and THP) to the bottom (the unwounded THP bound to the X-ray closed conformation of FL-MMP-1) the intermediate and energetically possible structures generated by HADDOCK [112] to provide a smooth conformational transition between the initial and final states. In panel (b), starting from the experimentally-driven docked complex between FL-MMP-1 and THP (top), the closed FL-MMP-1 interacting with the released 1T chain (in red), the hydrolysis of the 1T chain with both peptide fragments still in place, and the complex with the C-terminal region of the N-terminal peptide released from the active site (bottom). (*See insert for color representation of this figure.*)

conformations, closely facing the hydrolysis site. Then, it is possible that once the hemopexin-like domain is bound to triple-helical collagen, the catalytic domain of the high MO conformations that are not colliding can easily rearrange to productively interact with the cleavage site. Therefore, conformational selection can play a role in this case, favouring the productive binding of FL-MMP-1 to collagen, and once both domains are bound to collagen, the structural conformations suggested by the experimental data would occur, essentially driven by an induced fit mechanism. This mechanism that proceeds through the "destabilization/unwinding" is also energetically possible, as shown by the energies of the calculated FL-MMP-1-THP complexes. The open forms and closed form of the FL-MMP-1 interconvert rapidly. The closed conformation is enthalpically-favored while the open forms are equivalently entropically-favored, and the initial binding of the hemopexin-like domain to the substrate does not alter this equilibrium. The subsequent interaction of the catalytic domain with the THP in front of the cleavage site in chain 1T is enthalpically-favored and entropically-disfavored. Also the back-rotation of

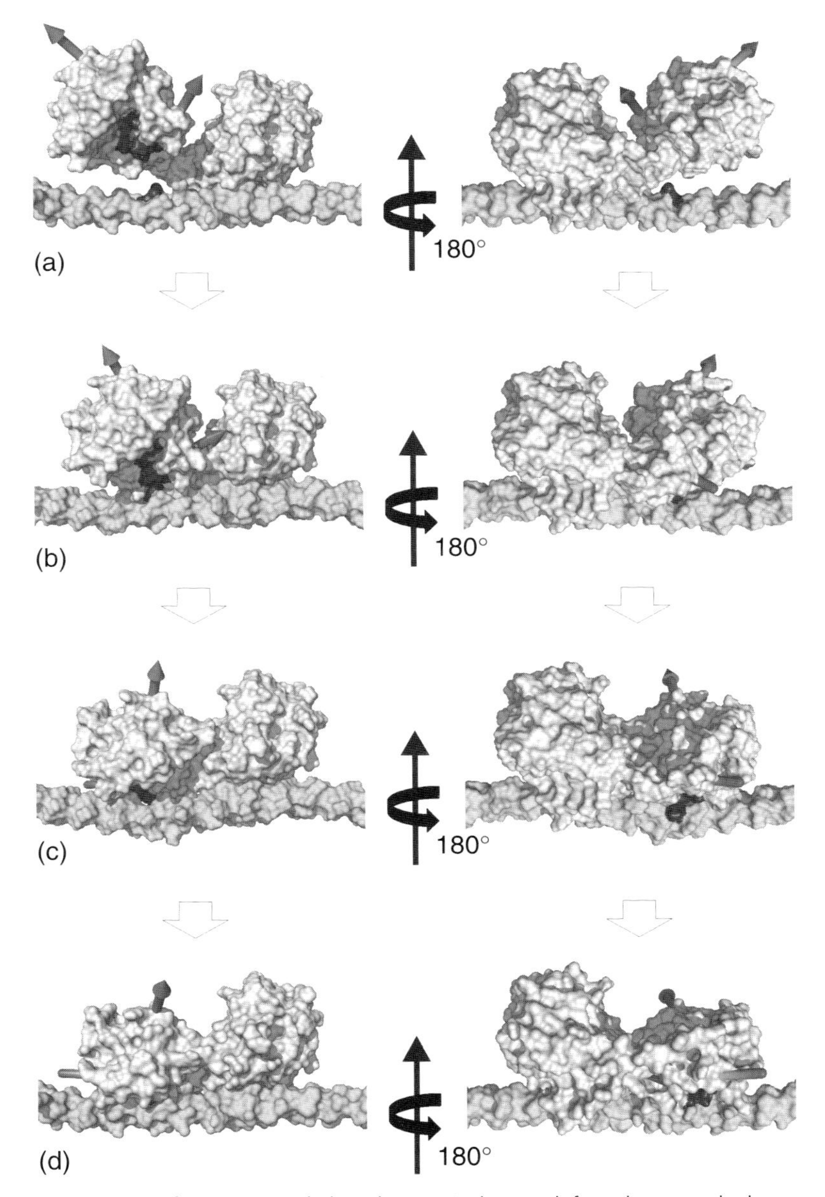

Figure 3.8 Interaction of FL-MMP-1 with the substrate. In the panel, from the top to the bottom: (a) structure with the highest MO, (b–c) two morphing intermediate steps, (d)the experimentally-driven docked complex where the hemopexin-like domain and the catalytic domain bind the triple-helical collagen. The structure with the highest MO e morphing structures were aligned to the hemopexin-like domain of the docked complex. FL-MMP-1 and THP are represented as white and yellow surfaces, respectively. In blue is the MMP consensus sequence HE*XX*H*XX*G*XX*H and the cleavage site (*Gly-Ile*) in the first chain of THP. The catalytic zinc ion is represented as an orange sphere. To facilitate visualizing the movement of the catalytic domain with respect to the hemopexin-like domain, the blue and red arrows indicate the direction of helices hA and hC of the catalytic domain defined by residues 130–141 and 250–258, respectively. (*See insert for color representation of this figure.*)

FL-MMP-1 toward the closed conformation and the associated release of chain 1T from the triple helical structure implies a balance of enthalpic gains (the new contacts between chain 1T and the active site of the catalytic domain) and losses (the breaking of the interactions between chain 1T and the other two chains of THP). The analysis shows that the calculated energy of the "closed" complex between FL-MMP-1 and THP is 300 kcal/mol compared with the initial "open" complex. At the same time, the activation energies reported for the catalyzed hydrolysis of soluble type I collagen and type I collagen fibrils by FL-MMP-1 are 26.0–49.2 and 101 kcal/mol, respectively. Collectively, the data indicate that the free energy change associated with this mechanism is more than sufficient to account for catabolism of soluble and fibrillar collagen.

Also the three populations of water associated with collagen fibrils, the first tightly bound, the second found in the interstices of microfibrils, and the third localized in the interfibrillar space, are reported to play a role in collagenolysis, facilitating the binding of the enzyme and the unwinding of the triple helical collagen [113].

A somewhat similar hypothesis on how FL-MMP-1 binds and cleaves collagen, has been proposed by Manka et al [114]. They have solved the crystallographic structure of an unproductive complex between MMP-1 and a different triple helical peptide. In the structure the catalytic domain faces a chain different from that bearing the cleavage site, and therefore a more complex rotation/translation of the triple helix is required to position the substrate in an enzymatically productive conformation. According to this hypothesis, exosites different from those previously described, should play a pivotal role in collagen binding, while a large interdomain flexibility of the protein would not be required to process the triple helix collagen.

References

1. Sela-Passwell, N., Trahtenherts, A., Kruger, A. & Sagi, I. (2011) New opportunities in drug design of metalloproteinase inhibitors: combination between structure-function experimental approaches and systems biology. *Expert Opinion on Drug Discovery*, 6(5), 527–542.
2. Shapiro, S.D. (1998) Matrix metalloproteinase degradation of extracellular matrix: biological consequences. *Current Opinion in Cell Biology*, 10, 602–608.
3. Woessner, J.F. Jr. & Nagase, H. (1999) Matrix metalloproteinases. *The Journal of Biological Chemistry*, 274, 21491–21494.
4. Page-McCaw, A., Ewald, A.J. & Werb, Z. (2007) Matrix metalloproteinases and the regulation of tissue remodelling. *Nature Reviews. Molecular Cell Biology*, 8(3), 221–233.
5. Overall, C.M. (2002) Molecular determinants of metalloproteinase substrate specificity. *Molecular Biotechnology*, 22, 51–86.
6. Visse, R. & Nagase, H. (2003) Matrix metalloproteinases and tissue inhibitors of metalloproteinases: structure, function, and biochemistry. *Circulation Research*, 92, 827–839.
7. Boire, A., Covic, L., Agarwal, A., Jacques, S., Sherifl, S. & Kuliopulos, A. (2005) PAR1 is a matrix metalloprotease-1 receptor that promotes invasion and tumorigenesis of breast cancer cells. *Cell*, 120(3), 303–313.
8. Nagase, H., Visse, R. & Murphy, G. (2006) Structure and function of matrix metalloproteinases and TIMPs. *Cardiovascular Research*, 69(3), 562–573.
9. Parks, W.C., Wilson, C.L. & Lopez-Boado, Y.S. (2004) Matrix metalloproteinases as modulators of inflammation and innate immunity. *Nature Reviews. Immunology*, 4(8), 617–629.
10. D'Alessio, S., Fibbi, G., Cinelli, M. *et al.* (2004) Matrix metalloproteinase 12-dependent cleavage of urokinase receptor in systemic sclerosis microvascular endothelial cells results in impaired angiogenesis. *Arthritis and Rheumatism*, 50(10), 3275–3285.

11. Fragai, M. & Nesi, A. (2007) Substrate Specificities of Matrix Metalloproteinase 1 in PAR-1 Exodomain Proteolysis. *ChemBioChem*, 8(12), 1367–1369.
12. Doucet, A. & Overall, C.M. (2008) Protease proteomics: revealing protease in vivo functions using systems biology approaches. *Molecular Aspects of Medicine*, 29(5), 339–358.
13. Limb, G.A., Matter, K., Murphy, G. *et al.* (2005) Matrix metalloproteinase-1 associates with intracellular organelles and confers resistance to lamin A/C degradation during apoptosis. *The American Journal of Pathology*, 166(5), 1555–1563.
14. Wang, W., Schulze, C.J., Suarez-Pinzon, W.L., Dyck, J.R., Sawicki, G. & Schulz, R. (2002) Intracellular action of matrix metalloproteinase-2 accounts for acute myocardial ischemia and reperfusion injury. *Circulation*, 106(12), 1543–1549.
15. Sawicki, G., Leon, H., Sawicka, J. *et al.* (2005) Degradation of myosin light chain in isolated rat hearts subjected to ischemia-reperfusion injury: a new intracellular target for matrix metalloproteinase-2. *Circulation*, 112(4), 544–552.
16. Kwan, J.A., Schulze, C.J., Wang, W.J. *et al.* (2004) Matrix metalloproteinase-2 (MMP-2) is present in the nucleus of cardiac myocytes and is capable of cleaving poly (ADP-ribose) polymerase (PARP) in vitro. *The FASEB Journal*, 18(2), 690–692.
17. Luo, D., Mari, B., Stoll, I. & Anglard, P. (2002) Alternative splicing and promoter usage generates an intracellular stromelysin 3 isoform directly translated as an active matrix metalloproteinase. *The Journal of Biological Chemistry*, 277(28), 25527–25536.
18. Aumailley, M. & Gayraud, B. (1998) Structure and biological activity of the extracellular matrix. *Journal of Molecular Medicine*, 76(3–4), 253–265.
19. Bosman, F.T. & Stamenkovic, I. (2003) Functional structure and composition of the extracellular matrix. *The Journal of Pathology*, 200(4), 423–428.
20. Tanzer, M.L. (2006) Current concepts of extracellular matrix. *Journal of Orthopaedic Science*, 11(3), 326–331.
21. Eble, J.A. (2009) The extracellular matrix in health and disease. *Current Pharmaceutical Design*, 15(12), 1275–1276.
22. Sternlicht, M.D. & Werb, Z. (2001) How matrix metalloproteinases regulate cell behavior. *Annual Review of Cell and Developmental Biology*, 17, 463–516.
23. Berrier, A.L. & Yamada, K.M. (2007) Cell-matrix adhesion. *Journal of Cellular Physiology*, 213(3), 565–573.
24. Gordon, M.K. & Hahn, R.A. (2010) Collagens. *Cell and Tissue Research*, 339(1), 247–257.
25. Canty, E.G. & Kadler, K.E. (2005) Procollagen trafficking, processing and fibrillogenesis. *Journal of Cell Science*, 118(Pt 7), 1341–1353.
26. Orgel, J.P., Antipova, O., Sagi, I. *et al.* (2011) Collagen fibril surface displays a constellation of sites capable of promoting fibril assembly, stability, and hemostasis. *Connective Tissue Research*, 52(1), 18–24.
27. Ricard-Blum, S. (2011) The collagen family. *Cold Spring Harbor Perspectives in Biology*, 3(1), a004978.
28. Fields, G.B. (2013) Interstitial collagen catabolism. *The Journal of Biological Chemistry*, 288, 8785–8793.
29. Muiznieks, L.D. & Keeley, F.W. (2013) Molecular assembly and mechanical properties of the extracellular matrix: A fibrous protein perspective. *Biochimica et Biophysica Acta*, 1832(7), 866–875.
30. Rosenbloom, J., Abrams, W.R. & Mecham, R. (1993) Extracellular matrix 4: the elastic fiber. *The FASEB Journal*, 7(13), 1208–1218.
31. Mithieux, S.M. & Weiss, A.S. (2005) Elastin. *Advances in Protein Chemistry*, 70, 437–461.
32. Muiznieks, L.D. & Weiss, A.S. (2007) Flexibility in the solution structure of human tropoelastin. *Biochemistry*, 46(27), 8196–8205.
33. Wise, S.G. & Weiss, A.S. (2009) Tropoelastin. *The International Journal of Biochemistry & Cell Biology*, 41(3), 494–497.
34. Muiznieks, L.D., Weiss, A.S. & Keeley, F.W. (2010) Structural disorder and dynamics of elastin. *Biochemistry and Cell Biology*, 88(2), 239–250.
35. Yeo, G.C., Keeley, F.W. & Weiss, A.S. (2011) Coacervation of tropoelastin. *Advances in Colloid and Interface Science*, 167(1–2), 94–103.
36. Yeo, G.C., Baldock, C., Tuukkanen, A. *et al.* (2012) Tropoelastin bridge region positions the cell-interactive C terminus and contributes to elastic fiber assembly. *Proceedings of the National Academy of Sciences of the United States of America*, 109(8), 2878–2883.
37. Ohuchi, E., Imai, K., Fujii, Y., Sato, H., Seiki, M. & Okada, Y. (1997) Membrane type 1 matrix metalloproteinase digests interstitial collagens and other extracellular matrix macromolecules. *Journal of Biological Chemistry*, 272(4), 2446–2451.

38. Allan, J.A., Docherty, A.J., Barker, P.J., Huskisson, N.S., Reynolds, J.J. & Murphy, G. (1995) Binding of gelatinases A and B to type-I collagen and other matrix components. *The Biochemical Journal*, 309(Pt 1), 299–306.
39. Gioia, M., Monaco, S., Fasciglione, G.F. *et al.* (2007) Characterization of the mechanisms by which gelatinase A, neutrophil collagenase, and membrane-type metalloproteinase MMP-14 recognize collagen I and enzymatically process the two alpha-chains. *Journal of Molecular Biology*, 368(4), 1101–1113.
40. Matrisian, L.M. & Bowden, G.T. (1990) Stromelysin/transin and tumor progression. *Seminars in Cancer Biology*, 1(2), 107–115.
41. Bertini, I., Calderone, V., Fragai, M., Luchinat, C., Mangani, S. & Terni, B. (2004) Crystal structure of the catalytic domain of human matrix metyalloproteinase 10. *Journal of Molecular Biology*, 336, 707–716.
42. Mecham, R.P., Broekelmann, T.J., Fliszar, C.J., Shapiro, S.D., Welgus, H.G. & Senior, R.M. (1997) Elastin degradation by matrix metalloproteinases. Cleavage site specificity and mechanisms of elastolysis. *The Journal of Biological Chemistry*, 272(29), 18071–18076.
43. Taddese, S., Weiss, A.S., Neubert, R.H. & Schmelzer, C.E. (2008) Mapping of macrophage elastase cleavage sites in insoluble human skin elastin. *Matrix Biology*, 27(5), 420–428.
44. Taddese, S., Weiss, A.S., Jahreis, G., Neubert, R.H. & Schmelzer, C.E. (2009) In vitro degradation of human tropoelastin by MMP-12 and the generation of matrikines from domain 24. *Matrix Biology*, 28(2), 84–91.
45. Overall, C.M. & Blobel, C.P. (2007) In search of partners: linking extracellular proteases to substrates. *Nature Reviews. Molecular Cell Biology*, 8(3), 245–257.
46. Andreini, C., Banci, L., Bertini, I., Luchinat, C. & Rosato, A. (2004) Bioinformatic comparison of structures and homology-models of matrix metalloproteinases. *Journal of Proteome Research*, 3, 21–31.
47. Lovejoy, B., Hassell, A.M., Luther, M.A., Weigl, D. & Jordan, S.R. (1994) Crystal structures of Recombinant 19-kDa human fibroblast collagenase complexed to itself. *Biochemistry*, 33, 8207–8217.
48. Whittaker, M., Floyd, C.D., Brown, P. & Gearing, A.J. (1999) Design and therapeutic application of matrix metalloproteinase inhibitors. *Chemistry Review*, 99, 2735–2776.
49. Cha, J. & Auld, D.S. (1997) Site-directed mutagenesis of the active site glutamate in human matrilysin: investigation of its role in catalysis. *Biochemistry*, 36, 16019–16024.
50. Lang, R., Kocourek, A., Braun, M. *et al.* (2001) Substrate specificity determinants of human macrophage elastase (MMP-12) based on the 1.1 angstrom crystal structure. *Journal of Molecular Biology*, 312(4), 731–742.
51. Bertini, I., Calderone, V., Fragai, M., Luchinat, C., Mangani, S. & Terni, B. (2003) X-ray structures of ternary enzyme-product-inhibitor complexes of MMP. *Angewandte Chemie, International Edition*, 42, 2673–2676.
52. Bertini, I., Calderone, V., Fragai, M., Luchinat, C. & Maletta, M. (2006) Snapshots of the reaction mechanism of matrix metalloproteinases. *Angewandte Chemie, International Edition*, 45, 7952–7955.
53. Piccard, H., Van den Steen, P.E. & Opdenakker, G. (2007) Hemopexin domains as multifunctional liganding modules in matrix metalloproteinases and other proteins. *Journal of Leukocyte Biology*, 81(4), 870–892.
54. Park, H.I., Ni, J., Gerkema, F.E., Liu, D., Belozerov, V.E. & Sang, Q.X. (2000) Identification and characterization of human endometase (Matrix metalloproteinase-26) from endometrial tumor. *The Journal of Biological Chemistry*, 275(27), 20540–20544.
55. Rosenblum, G., Meroueh, S., Toth, M. *et al.* (2007) Molecular structures and dynamics of the stepwise activation mechanism of a matrix metalloproteinase zymogen: Challenging the cysteine switch dogma. *Journal of the American Chemical Society*, 129(44), 13566–13574.
56. Glasheen, B.M., Kabra, A.T. & Page-McCaw, A. (2009) Distinct functions for the catalytic and hemopexin domains of a Drosophila matrix metalloproteinase. *Proceedings of the National Academy of Sciences of the United States of America*, 106(8), 2659–2664.
57. Pei, D. & Weiss, S.J. (1995) Furin-dependent intracellular activation of the human stromelysin-3 zymogen. *Nature*, 375(6528), 244–247.
58. Steffensen, B., Wallon, U.M. & Overall, C.M. (1995) Extracellular matrix binding properties of recombinant fibronectin type II-like modules of human 72-kDa gelatinase/type IV collagenase. High affinity binding to native type I collagen but not native type IV collagen. *The Journal of Biological Chemistry*, 270(19), 11555–11566.
59. Strongin, A.Y., Collier, I.E., Krasnov, P.A., Genrich, L.T., Marmer, B.L. & Goldberg, G.I. (1993) Human 92 kDa type IV collagenase: functional analysis of fibronectin and carboxyl-end domains. *Kidney International*, 43(1), 158–162.
60. Elkins, P.A., Ho, Y.S., Smith, W.W. *et al.* (2002) Structure of the C-terminally truncated human ProMMP9, a gelatin-binding matrix metalloproteinase. *Acta Crystallographica. Section D, Biological Crystallography*, 58(Pt 7), 1182–1192.

61. Bertini, I., Fragai, M. & Luchinat, C. (2009) Intra- and interdomain flexibility in matrix metalloproteinases: functional aspects and drug design. *Current Pharmaceutical Design*, 15, 3592–3605.
62. Maskos, K. & Bode, W. (2003) Structural basis of matrix metalloproteinases and tissue inhibitors of metalloproteinases. *Molecular Biotechnology*, 25(3), 241–266.
63. Lovejoy, B., Welch, A.R., Carr, S. *et al.* (1999) Crystal structures of MMP-1 and -13 reveal the structural basis for selectivity of collagenase inhibitors. *Nature Structural Biology*, 6(3), 217–221.
64. Bode, W. & Maskos, K. (2003) Structural basis of the matrix metalloproteinases and their physiological inhibitors, the tissue inhibitors of metalloproteinases. *Biological Chemistry*, 384(6), 863–872.
65. Bertini, I., Fragai, M., Giachetti, A. *et al.* (2005) Combining in silico tools and NMR data to validate protein-ligand structural models: application to matrix metalloproteinases. *Journal of Medicinal Chemistry*, 48, 7544–7559.
66. Czarny, B., Stura, E.A., Devel, L. *et al.* (2013) Molecular determinants of a selective matrix metalloprotease-12 inhibitor: insights from crystallography and thermodynamic studies. *Journal of Medicinal Chemistry*, 56(3), 1149–1159.
67. Li, J., Brick, P., Ohare, M.C. *et al.* (1995) Structure of full-length porcine synovial collagenase reveals a C-terminal domain-containing a calcium-linked, 4-bladed beta-propeller. *Structure*, 3(6), 541–549.
68. Van den Steen, P.E., Van Aelst, I., Hvidberg, V. *et al.* (2006) The hemopexin and O-glycosylated domains tune gelatinase B/MMP-9 bioavailability via inhibition and binding to cargo receptors. *The Journal of Biological Chemistry*, 281(27), 18626–18637.
69. Bertini, I., Calderone, V., Cosenza, M. *et al.* (2005) Conformational variability of MMPs: beyond a single 3D structure. *Proceedings of the National Academy of Sciences of the United States of America*, 102, 5334–5339.
70. Udi, Y., Fragai, M., Grossman, M. *et al.* (2013) Unraveling hidden regulatory sites in structurally homologous metalloproteases. *Journal of Molecular Biology*, 425(13), 2330–2346.
71. Robichaud, T.K., Steffensen, B. & Fields, G.B. (2011) Exosite interactions impact matrix metalloproteinase collagen specificities. *The Journal of Biological Chemistry*, 286(43), 37535–37542.
72. Palmier, M.O., Fulcher, Y.G., Bhaskaran, R., Duong, V.Q., Fields, G.B. & Van Doren, S.R. (2010) NMR and bioinformatics discovery of exosites that tune metalloelastase specificity for solubilized elastin and collagen triple helices. *The Journal of Biological Chemistry*, 285(40), 30918–30930.
73. Bertini, I., Fragai, M., Luchinat, C., Melikian, M. & Venturi, C. (2009) Characterization of the MMP-12-elastin adduct. *Chemistry - A European Journal*, 15, 7842–7845.
74. McCawley, L.J. & Matrisian, L.M. (2000) Matrix metalloproteinases: multifunctional contributors to tumor progression. *Molecular Medicine Today*, 6(4), 149–156.
75. Overall, C.M. (2002) Molecular determinants of metalloproteinase substrate specificity - Matrix metalloproteinase substrate binding domains, modules, and exosites. *Molecular Biotechnology*, 22(1), 51–86.
76. Morrison, C.J., Butler, G.S., Rodriguez, D. & Overall, C.M. (2009) Matrix metalloproteinase proteomics: substrates, targets, and therapy. *Current Opinion in Cell Biology*, 21(5), 645–653.
77. Bigg, H.F., Rowan, A.D., Barker, M.D. & Cawston, T.E. (2007) Activity of matrix metalloproteinase-9 against native collagen types I and III. *The FEBS Journal*, 274(5), 1246–1255.
78. Clark, I.M. & Cawston, T.E. (1989) Fragments of human fibroblast collagenase. Purification and characterization. *The Biochemical Journal*, 263(1), 201–206.
79. Knauper, V., Cowell, S., Smith, B. *et al.* (1997) The role of the C-terminal domain of human collagenase-3 (MMP-13) in the activation of procollagenase-3, substrate specificity, and tissue inhibitor of metalloproteinase interaction. *The Journal of Biological Chemistry*, 272(12), 7608–7616.
80. Knauper, V., Wilhelm, S.M., Seperack, P.K. *et al.* (1993) Direct activation of human neutrophil procollagenase by recombinant stromelysin. *The Biochemical Journal*, 295(Pt 2), 581–586.
81. Murphy, G., Allan, J.A., Willenbrock, F., Cockett, M.I., O'Connell, J.P. & Docherty, A.J. (1992) The role of the C-terminal domain in collagenase and stromelysin specificity. *The Journal of Biological Chemistry*, 267(14), 9612–9618.
82. Ohuchi, E., Imai, K., Fujii, Y., Sato, H., Seiki, M. & Okada, Y. (1997) Membrane type 1 matrix metalloproteinase digests interstitial collagens and other extracellular matrix macromolecules. *The Journal of Biological Chemistry*, 272(4), 2446–2451.
83. Hurst, D.R., Schwartz, M.A., Ghaffari, M.A. *et al.* (2004) Catalytic- and ecto-domains of membrane type 1-matrix metalloproteinase have similar inhibition profiles but distinct endopeptidase activities. *The Biochemical Journal*, 377(Pt 3), 775–779.
84. Hirose, T., Patterson, C., Pourmotabbed, T., Mainardi, C.L. & Hasty, K.A. (1993) Structure-function relationship of human neutrophil collagenase: identification of regions responsible for substrate specificity and general proteinase activity. *Proceedings of the National Academy of Sciences of the United States of America*, 90(7), 2569–2573.

85. Chung, L., Shimokawa, K., Dinakarpandian, D., Grams, F., Fields, G.B. & Nagase, H. (2000) Identification of the (183)RWTNNFREY(191) region as a critical segment of matrix metalloproteinase 1 for the expression of collagenolytic activity. *The Journal of Biological Chemistry*, 275(38), 29610–29617.
86. Iyer, S., Visse, R., Nagase, H. & Acharya, K.R. (2006) Crystal structure of an active form of human MMP-1. *Journal of Molecular Biology*, 362(1), 78–88.
87. de Souza, S.J., Pereira, H.M., Jacchieri, S. & Brentani, R.R. (1996) Collagen/collagenase interaction: does the enzyme mimic the conformation of its own substrate? *The FASEB Journal*, 10(8), 927–930.
88. Chung, L.D., Dinakarpandian, D., Yoshida, N. *et al.* (2004) Collagenase unwinds triple-helical collagen prior to peptide bond hydrolysis. *The EMBO Journal*, 23(15), 3020–3030.
89. Tam, E.M., Moore, T.R., Butler, G.S. & Overall, C.M. (2004) Characterization of the distinct collagen binding, helicase and cleavage mechanisms of matrix metalloproteinase 2 and 14 (gelatinase A and MT1-MMP) - The differential roles of the MMP hemopexin C domains and the MMP-2 fibronectin type II modules in collagen triple helicase activities. *The Journal of Biological Chemistry*, 279(41), 43336–43344.
90. Adhikari, A.S., Chai, J. & Dunn, A.R. (2011) Mechanical Load Induces a 100-Fold Increase in the Rate of Collagen Proteolysis by MMP-1. *Journal of the American Chemical Society*, 133, 1686–1689.
91. Fields, G.B. (1991) A model for interstitial collagen catabolism by mammalian collagenases. *Journal of Theoretical Biology*, 153(4), 585–602.
92. O'Farrell, T.J., Guo, R., Hasegawa, H. & Pourmotabbed, T. (2006) Matrix metalloproteinase-1 takes advantage of the induced fit mechanism to cleave the triple-helical type I collagen molecule. *Biochemistry*, 45(51), 15411–15418.
93. Salsas-Escat, R., Nerenberg, P.S. & Stultz, C.M. (2010) Cleavage site specificity and conformational selection in type I collagen degradation. *Biochemistry*, 49(19), 4147–4158.
94. Han, S., Makareeva, E., Kuznetsova, N.V. *et al.* (2010) Molecular mechanism of type I collagen homotrimer resistance to mammalian collagenases. *The Journal of Biological Chemistry*, 285(29), 22276–22281.
95. Lauer-Fields, J.L., Tuzinski, K.A., Shimokawa, K., Nagase, H. & Fields, G.B. (2000) Hydrolysis of triple-helical collagen peptide models by matrix metalloproteinases. *Journal of Biological Chemistry*, 275(18), 13282–13290.
96. Lauer-Fields, J.L., Tuzinski, K.A., Shimokawa, K.-I., Nagase, H. & Fields, G.B. (2000) Hydrolysis of triple-helical collagen peptide models by matrix metalloproteinases. *The Journal of Biological Chemistry*, 275, 13282–13290.
97. Ottl, J., Gabriel, D., Murphy, G. *et al.* (2000) Recognition and catabolism of synthetic heterotrimeric collagen peptides by matrix metalloproteinases. *Chemical Biology*, 7(2), 119–132.
98. Docquier, J.D., Benvenuti, M., Calderone, V. *et al.* (2010) High-resolution crystal strucutre of the subclass B3 metallo-beta-lactamase BJP-1: rational basis for substrate specificity and interaction with sulfonamides. *Antimicrobial Agents and Chemotherapy*, 54, 4343–4351.
99. Bertini, I., Fragai, M., Luchinat, C. *et al.* (2012) Structural basis for matrix metalloproteinase 1-catalyzed collagenolysis. *Journal of the American Chemical Society*, 134, 2100–2110.
100. Rosenblum, G., Van den Steen, P.E., Cohen, S.R. *et al.* (2007) Insights into the structure and domain flexibility of full-length pro-matrix metalloproteinase-9/gelatinase B. *Structure*, 15, 1227–1236.
101. Bertini, I., Calderone, V., Fragai, M. *et al.* (2008) Evidence of reciprocal reorientation of the catalytic and hemopexin-like domains of full-length MMP-12. *Journal of the American Chemical Society*, 130, 7011–7021.
102. Bertini, I., Fragai, M., Luchinat, C. *et al.* (2009) Interdomain flexibility in full-length matrix metalloproteinase-1 (MMP-1). *The Journal of Biological Chemistry*, 284(19), 12821–12828.
103. Jozic, D., Bourenkov, G., Lim, N.H. *et al.* (2005) X-ray structure of human proMMP-1 - New insights into procollagenase activation and collagen binding. *The Journal of Biological Chemistry*, 280(10), 9578–9585.
104. Arnold, L.H., Butt, L.E., Prior, S.H., Read, C.M., Fields, G.B. & Pickford, A.R. (2011) The interface between catalytic and hemopexin domains in matrix metalloproteinase-1 conceals a collagen binding exosite. *Journal of Biological Chemistry*, 286(52), 45073–45082.
105. Diaz, N., Suarez, D. & Valdes, H. (2008) From the X-ray compact structure to the elongated form of the full-length MMP-2 enzyme in solution: a molecular dynamics study. *Journal of the American Chemical Society*, 130(43), 14070–14071.
106. Overall, C.M. & Butler, G.S. (2007) Protease yoga: Extreme matrix metalloproteinase. *Structure*, 15(10), 1159–1161.
107. Saffarian, S., Collier, I.E., Marmer, B.L., Elson, E.L. & Goldberg, G. (2004) Interstitial collagenase is a Brownian ratchet driven by proteolysis of collagen. *Science*, 306(5693), 108–111.

108. Cerofolini, L., Fields, G.B., Fragai, M. *et al.* (2013) Examination of Matrix Metalloproteinase-1 (MMP-1) in solution: a preference for the pre-collagenolysis state. *The Journal of Biological Chemistry*, 288, 30659–30671.
109. Bertini, I., Gupta, Y.K., Luchinat, C. *et al.* (2007) Paramagnetism-based NMR restraints provide maximum allowed probabilities for the different conformations of partially independent protein domains. *Journal of the American Chemical Society*, 129, 12786–12794.
110. Bertini, I., Giachetti, A., Luchinat, C. *et al.* (2010) Conformational space of flexible biological macromolecules from average data. *Journal of the American Chemical Society*, 132, 13553–13558.
111. Bertini, I., Luchinat, C., Nagulapalli, M., Parigi, G. & Ravera, E. (2012) Paramagnetic relaxation enhancements for the characterization of the conformational heterogeneity in two-domain proteins. *Physical Chemistry Chemical Physics*, 14, 9149–9156.
112. Dominguez, C., Boelens, R. & Bonvin, A.M. (2003) HADDOCK: a protein-protein docking approach based on biochemical or biophysical information. *Journal of the American Chemical Society*, 125(7), 1731–1737.
113. Grossman, M., Born, B., Heyden, M. *et al.* (2011) Correlated structural kinetics and retarded solvent dynamics at the metalloprotease active site. *Nature Structural and Molecular Biology*, 18, 1102–1108.
114. Manka, S.W., Carafoli, F., Visse, R. *et al.* (2012) Structural insights into triple-helical collagen cleavage by matrix metalloproteinase 1. *Proceedings of the National Academy of Sciences of the United States of America*, 109(31), 12461–12466.

4 Metzincin Modulators

Dmitriy Minond

Cancer Research, Torrey Pines Research Institute for Molecular Studies, Port St. Lucie, USA

Here we address the most recent developments in MMP and ADAM protease inhibitors. Multiple reviews focus on past medicinal chemistry efforts in the development of metzincin inhibitors. In the past, there was a significant effort based on the rational design of small molecule metzincin inhibitors which involved building selectivity around a zinc-binding moiety. Multiple scaffolds featuring zinc-binding moieties have been pursued as drug candidates for various therapeutic indications. However, none have made it to the clinic. It is believed that the lack of selectivity and efficacy, combined with the metabolic liabilities of zinc-binding functionalities, are at the heart of the problem. Therefore, alternative approaches not involving zinc-binding moieties and targeting active sites are needed. Readers can refer to the most comprehensive reviews on collagenase inhibitors [1], TACE inhibitors [2], metzincin inhibitors [3], and metalloenzyme inhibitors [4, 5].

In this chapter we focus on advances that were made using approaches alternative to zinc-binding in the MMPs' and ADAMs' active sites with an emphasis on rational design. This chapter is not intended to be an exhaustive review of all activities in the field of metzincin modulator development, but rather a review of various rational and combinatorial approaches that can lead to it.

4.1 Inhibitors

4.1.1 Antibodies: targeting beyond the active site

Recent exciting approaches to development of MMP and ADAM inhibitors include monoclonal anti-ADAM17 antibodies reported by several groups. One of the approaches, known as bispecific T-cell engager antibodies (BiTE), was utilized by Yamamoto et al. [6]. In an attempt to preserve some important proteolytic activities of ADAM17, a monoclonal antibody against the membrane-proximal cysteine-rich domain of human ADAM17 (A300E) was raised, rather than an antibody targeting the catalytic domain. The membrane proximal domain of ADAM17 was shown to be

Matrix Metalloproteinase Biology, First Edition. Edited by Irit Sagi and Jean P. Gaffney.

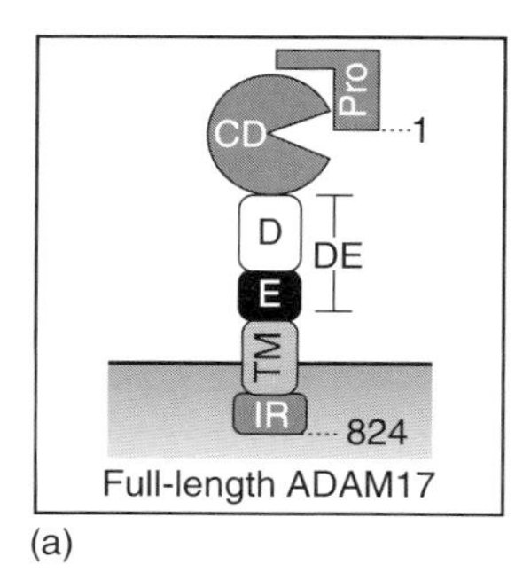

(a)

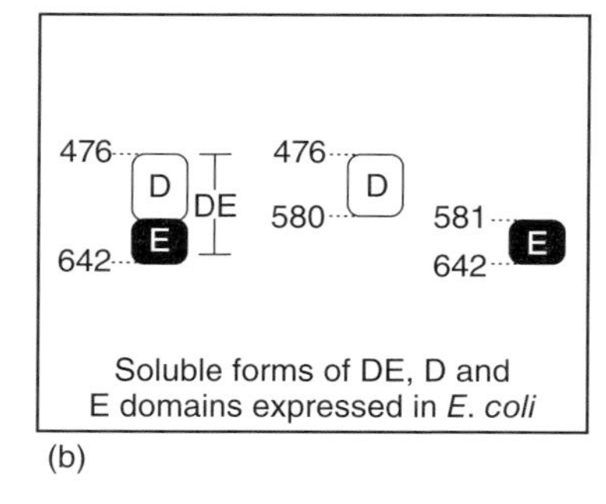

(b)

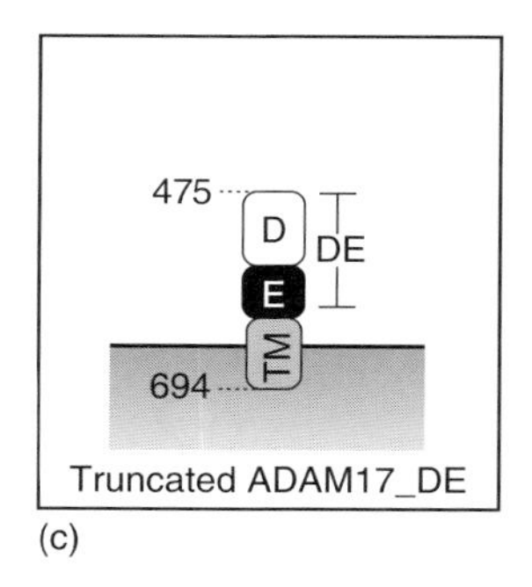

(c)

Figure 4.1 Schematic representation of the ADAM17 proteins used in the present study. (a) Full-length ADAM17 (amino acids 1–824). (b) Soluble forms of the cysteine-rich domain (DE, amino acids 476–642), the disintegrin-like domain (D, amino acids 476–580) and the membrane proximal cysteine-rich extension (E, amino acids 581–642) expressed in *E. coli*. (c) ADAM17_DE consisting of the cysteine-rich domain (DE) followed by the transmembrane region (TM) of human ADAM17 (amino acids 475–694). Pro, pro-domain; CD, catalytic domain; TM, transmembrane region; IR, intracellular region. (Reproduced with permission from Yamamoto K., Trad A., Baumgart A., Huske L., Lorenzen I., Chalaris A., Grotzinger J., Dechow T., Scheller J., and Rose-John S. (2012) A novel bispecific single-chain antibody for ADAM17 and CD3 induces T-cell-mediated lysis of prostate cancer cells. *Biochem J*, 445 (1) 135-144. © the Biochemical Society).

important for substrate binding [7] and oligomerization [8]. Therefore, targeting it represents a viable strategy.

Western blot analysis confirmed A300E bound only the membrane-proximal cysteine-rich domain and not the disintegrin domain. The constructs used included a variety of deletions: the disintegrin-like domain, membrane-proximal cysteine -rich domain, and a dual disintegrin-like domain + membrane-proximal cysteine-rich domain confirmed that A300E binds only to membrane-proximal cysteine-rich domain and not to the neighboring disintegrin-like domain (Fig. 4.1).

Additionally, A300E precipitated ADAM17 from HEK-293 lysates and was able to selectively deliver conjugated doxorubicin to breast cancer cells overexpressing ADAM17 [9]. This highly specific to ADAM17 membrane-proximal cysteine-rich domain antibody was combined with CD3 recognition sequence to produce recombinant bi-specific antibody recognizing ADAM17 on tumor cells and CD3 on T-cells (Fig. 4.2, A300E-BiTE). The authors show that the specificity of A300E-BiTE to ADAM17 membrane-proximal cysteine-rich domain is similar to that of A300E. A300-BiTE recognized ADAM17 on the surface of several human cancer cells (COLO357, Panc89, MDA-MB-231, and PC3) and CD3 on the surface of Jurkat and PBMC cells. A300E-BiTE efficacy testing demonstrated that CHO cells transfected with ADAM17 dual disintegrin-like domain + membrane-proximal cysteine-rich domain construct were lysed in the presence of PBMCs, while CHO cells without ADAM17 were spared, suggesting that A300E-BiTE can selectively recognize ADAM17 on cell surface and recruit effector cells to the cells expressing target protein. Finally, A300E-BiTE enabled lysis of several cell lines, including cancerous ones (PBMCs, MDA-MB-231, HCT116, SW620), in the presence of freshly isolated T-cells. The authors did not discuss whether ADAM17 on the surface of healthy cells retains activity in the presence of A300E-BiTE. ADAM17 cleaves a variety of cell surface proteins, some of which might have protective roles in cancer; therefore, it is

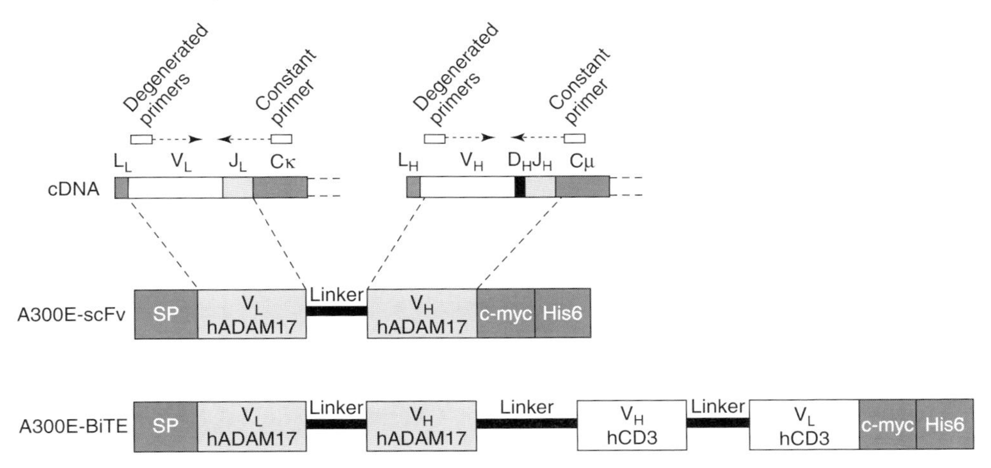

Figure 4.2 Design and expression of ADAM17-specific A300E-BiTE. Schematic representation of generation of A300E-BiTE. To identify cDNA sequences of V_H and V_L of mouse monoclonal antibody, ten primer sets and seven primer sets were used to amplify V_H and V_L cDNA. After analysis of DNA sequences from V_H and V_L fragments the construct of A300E-scFv and BiTE were introduced into pET23a and pcDNA3.1 vectors respectively. Linker indicates flexible linker (Gly4Ser). The c-Myc and His6 tags are fused for detection and purification respectively. (Reproduced with permission from Yamamoto K., Trad A., Baumgart A., Huske L., Lorenzen I., Chalaris A., Grotzinger J., Dechow T., Scheller J., and Rose-John S. (2012) A novel bispecific single-chain antibody for ADAM17 and CD3 induces T-cell-mediated lysis of prostate cancer cells. *Biochem J*, *445* (1) 135–144. © the Biochemical Society).

important to know the repertoire of ADAM17 substrates cleaved in the presence of A300E-BiTE to predict potential side effects due to ADAM17 inhibition.

Another ADAM17-specific inhibitory antibody was developed by targeting both catalytic and non-catalytic domains [10]. The authors hypothesized that by targeting non-catalytic domains in addition to the catalytic domain, the specificity of the resulting antibody would be increased. To verify that the antibody does not interact with the catalytic site residues, a small zinc-binding ADAM17 inhibitor CT1746 was used to block the active site. The blocked ectodomain of ADAM17 was then exposed to an antibody phage-display library. The resulting antibody (D1) bound the ectodomain of ADAM17, but not the catalytic domain-only construct. This confirms binding to non-catalytic domains. In order to introduce catalytic domain-binding capability, the non-catalytic domain-binding portion of D1 (Fig. 4.3, D1-V_H) was cloned into a phage-display library. The resulting library was selected against the ectodomain of ADAM17 without CT1746 to allow access to catalytic cleft. The resulting antibodies were screened against both the ectodomain of ADAM17 and the catalytic domain-only constructs, yielding a series of variants capable of binding either construct independently. The A12 (D1(A12)) demonstrated the highest affinity to both constructs and therefore was considered the lead candidate. A surface plasmon resonance (SPR) study showed that D1(A12) had greater than 10-fold affinity for binding of the ADAM17 ectodomain than the catalytic domain-only ($K_D = 0.46 \pm 0.7$ nM versus 5.21 ± 0.1 nM for ADAM17 ectodomain and catalytic domain-only, respectively). This suggests that D1(A12) engages both catalytic and non-catalytic domains. D1(A12) did not inhibit ADAM10 in a biochemical dose

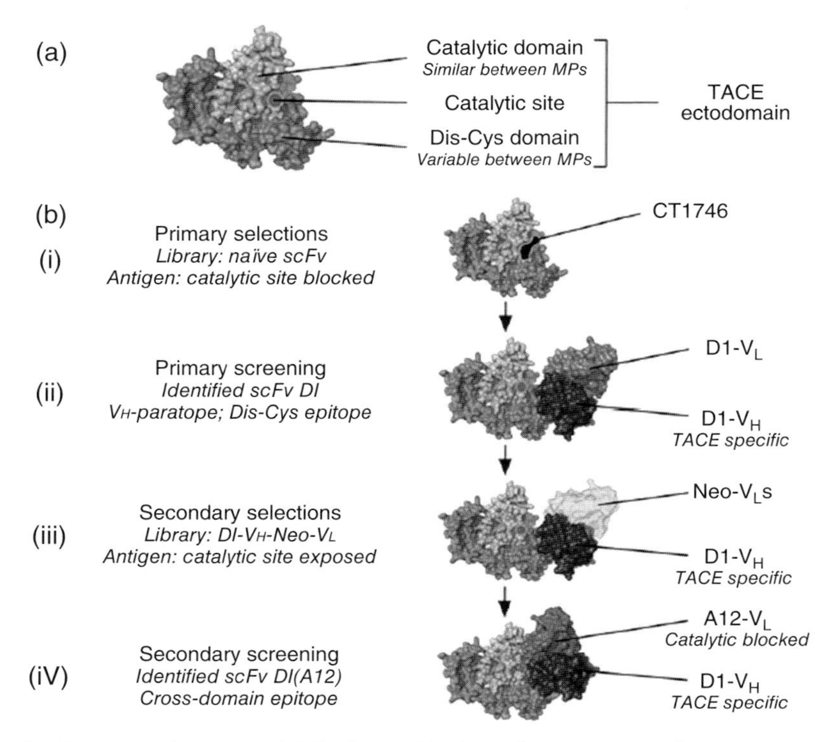

Figure 4.3 Experimental overview. (a) The human TACE ectodomain consists of an amino-terminal metalloprotease catalytic domain (light red) and a carboxyl-terminal noncatalytic Dis-Cys domain (light blue) (I-TASSER model). We exploited this multidomain topology to develop a truly specific ADAM inhibitor using two-step antibody phage display. (b) (i) First, the catalytic site of TACE ectodomain was blocked during primary antibody phage-display selections using the small-molecule inhibitor CT1746. This prevented the selection of antibodies with catalytic-cleft epitopes that could cross-react with non-target metalloproteases. (ii) Primary screening revealed the inhibitory scFv antibody clone D1. This scFv bound specifically to the TACE Dis-Cys domain through its variable heavy (V_H) domain. (iii) A D1-V_H-bias antibody phage display library was produced to introduce new variable light (neo-V_L) chains while maintaining the TACE specificity provided by the D1-V_H. Secondary selections were performed in the absence of CT1746 in order to provide the neo-V_L chains with uninterrupted access to the TACE catalytic site. (iv) Secondary screening identified several neo-VL scFvs capable of binding the isolated TACE catalytic domain. Due to Dis-Cys domain binding through the D1-V_H these "cross-domain" antibodies maintained their strict specificity for TACE. D1-V_H-neo-V_L scFv clone A12 (D1(A12)) exhibited the highest affinity for the TACE ectodomain and is the most selectively potent cell-surface ADAM inhibitor ever described. (Reproduced with permission from Tape, C. J., Willems, S. H., Dombernowsky, S. L., Stanley, P. L., Fogarasi, M., Ouwehand, W., McCafferty, J., and Murphy, G. (2011) Cross-domain inhibition of TACE ectodomain *Proc Natl Acad Sci* U S A 108, 5578–5583). (*See insert for color representation of this figure.*)

response experiment (highest concentration tested 1000 nM). In a MCF7 cell-based assay, D1(A12) inhibited PMA-induced, but not ionomycin-induced, shedding of alkaline phosphatase (AP)-tagged HB-EGF, suggesting selectivity for ADAM17 over ADAM10. Additional cell-based testing in multiple cancer cell lines, including TOV21G, HeLa, IGROV-1, and PC3, demonstrated that D1(A12) is more efficacious in inhibition of shedding known ADAM17 substrates such as TNFα, HB-EGF, TGFα, and AREG than N-TIMP-3. Shedding of each of the above-mentioned substrates was inhibited by D1(A12) with low nanomolar IC_{50} values in cell-based assay, suggesting potential *in vivo* applicability. Indeed, in the follow-up study [11]

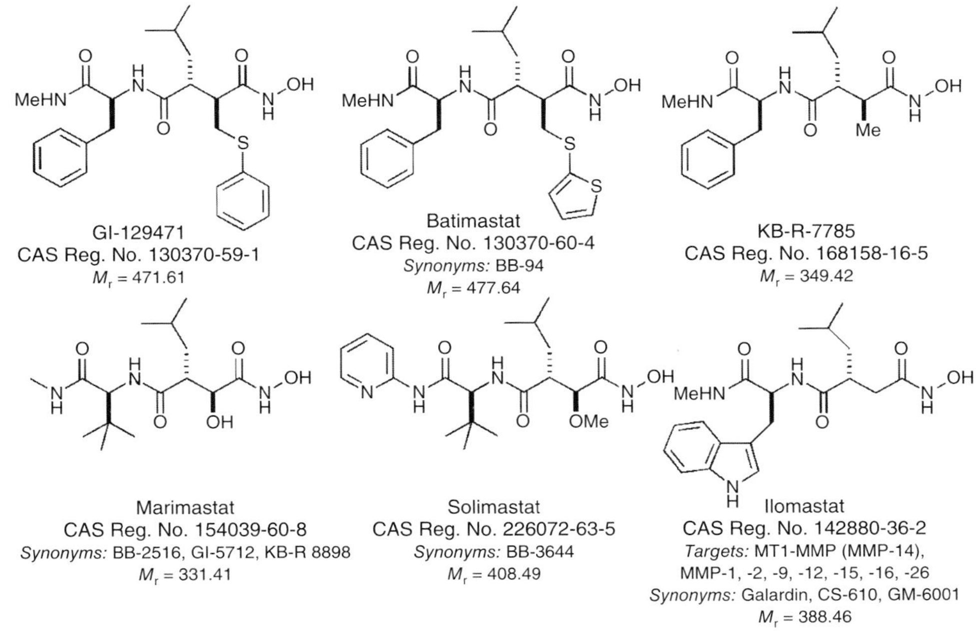

GI-129471
CAS Reg. No. 130370-59-1
M_r = 471.61

Batimastat
CAS Reg. No. 130370-60-4
Synonyms: BB-94
M_r = 477.64

KB-R-7785
CAS Reg. No. 168158-16-5
M_r = 349.42

Marimastat
CAS Reg. No. 154039-60-8
Synonyms: BB-2516, GI-5712, KB-R 8898
M_r = 331.41

Solimastat
CAS Reg. No. 226072-63-5
Synonyms: BB-3644
M_r = 408.49

Ilomastat
CAS Reg. No. 142880-36-2
Targets: MT1-MMP (MMP-14), MMP-1, -2, -9, -12, -15, -16, -26
Synonyms: Galardin, CS-610, GM-6001
M_r = 388.46

Figure 4.4 Collagen-based, peptidomimetic hydroxamates. (Reproduced with permission from Fisher, J. F., and Mobashery, S. (2006) Recent advances in MMP inhibitor design. *Cancer Metastasis Rev* 25, 115–136. Copyright © 2006, Springer).

D1(A12) demonstrated suitable pharmacokinetics at 10 mg/kg i.p. dosing (non-tumor bearing mice: $C_{max} = 523 \pm 58$ nM, $T_{max} = 2$ days, $t_{1/2} = 8.6$ days and IGROV1-Luc tumor-bearing mice: $C_{max} = 425 \pm 51$ nM). To measure *in vivo* efficacy, D1(A12) was dosed at 10 mg/kg i.p. in mice bearing IGROV1-Luc tumor ($n = 11$), which reduced average tumor burden by 44% as compared to vehicle control. Analysis of concentrations of ADAM17 substrates in plasma and ascites revealed 4.4-fold decrease of soluble TNFR1-α, 5.4-fold decrease of soluble AREG and 15-fold decrease of soluble TGFα in ascites, suggesting that smaller size of tumors in mice treated with D1(A12) is possibly due to inhibition of EGFR signaling and increase of TNFR signaling.

4.1.2 Peptide-based inhibitors

The design of first generation peptide-based inhibitors of MMPs typically relied on the development of short peptides coupled to a zinc-binding moiety. Good examples of such inhibitors are the class known as "peptidomimetics," which were developed to target MMPs in cancer (Fig. 4.4). While being potent inhibitors of MMPs, they suffered from the lack of selectivity as the relatively short amino acid portion of the molecule could only provide interactions in the vicinity of the active site.

To gain selectivity, the Fields laboratory incorporated a zinc-binding moiety (phosphinate) into collagen V-based sequence [12, 13] that spans 8 residues which interacted with several secondary substrate binding sites in the structure of the enzyme (Fig. 4.5, P_4–P_4' subsites).

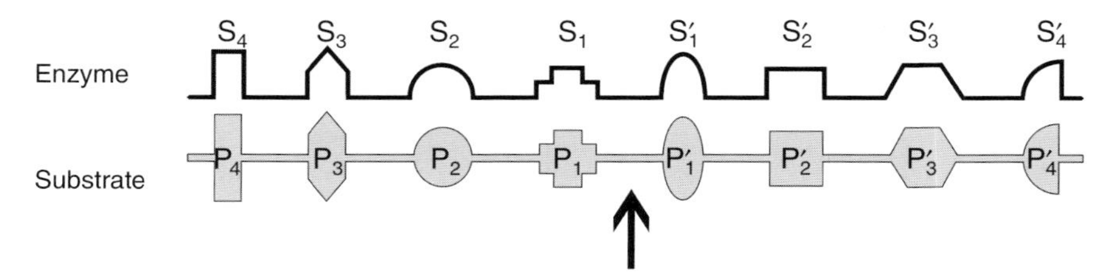

Figure 4.5 Nomenclature used for enzyme and substrate subsites. The arrow marks the site of protease hydrolysis. (Reproduced with permission from Lauer-Fields, J., Brew, K., Whitehead, J. K., Li, S., Hammer, R. P., and Fields, G. B. (2007) Triple-helical transition state analogues: a new class of selective matrix metalloproteinase inhibitors. *J Am Chem Soc* 129, 10408–10417).

$(CH_3(CH_2)_4CO_2H)$-(Gly-Pro-Hyp)$_4$-Gly-Pro-Pro-Gly$\Psi$$\{PO_2HCH_2\}$

(*R,S*)Val-Val-Gly-Glu-Gln-Gly-Glu-Gln-Gly-Pro-Pro-(Gly-Pro-Hyp)$_4$-NH_2)

Figure 4.6 Sequence of triple-helical peptide containing phosphinate group.

The sequence was based on α1 chain of collagen V (Gly-Pro-Pro-Gly$_{439}$ ~Val$_{440}$-Val-Gly-Glu-Gln), which is hydrolyzed at the Gly$_{439}$-Val$_{440}$ bond efficiently by MMP-2 and MMP-9 (Fig. 4.6) but not MMP-1, MMP-3, or MT1-MMP [14]. To induce triple-helical conformation and thus increase selectivity toward collagenolytic MMPs, four Gly-Pro-Hyp repeats were added on both termini of the peptide to create peptides named f1 and f2 (f1 had Val in *S* configuration, while f2 had Val in *R* configuration). A portion of the resulting triple-helical peptides were lipidated on the N-termini with hexanoic acid ($CH_3(CH_2)_4CO_2H$) to create peptide-amphiphiles with increased thermal stability (Fig. 4.6, f3 and f4).

Unlipidated peptides f1 and f2 exhibited weak triple-helical signatures when examined by circular dichroism, while lipidated f3 and f4 were more pronounced triple-helices. Melting temperatures (T_m) of f1-f4 constructs were determined by monitoring molar ellipticity at $l = 225$ nm while increasing temperature from 5 to 85°C. f1 and f2 exhibited T_m of approximately 17.5 C while f3 and f4 had T_m of approximately 25°C. Size exclusion chromatography (SEC) showed that f1-f3 were monomeric at 37°C while f4 was trimeric; therefore, initial inhibition studies were conducted at 10°C where f1–f4 were triple-helical. f3 and f4 exhibited low nanomolar K_i values at 10°C (1–6 nM), while at 37°C both f3 and f4 had significantly increased K_i values for the inhibition of MMP-2, but not MMP-9 (f3 MMP-2 $K_i^{10°C} = 4.14 \pm 0.47$ nM versus $K_i^{37°C} = 19.23 \pm 0.647$ nM; f4 MMP-2 $K_i^{10°C} = 5.48 \pm 0.00$ nM versus $K_i^{37°C} = 38.32 \pm 27.85$ nM; f3 MMP-9 $K_i^{10°C} = 1.76 \pm 0.05$ nM versus $K_i^{37°C} = 1.29 \pm 0.00$ nM; f4 MMP-9 $K_i^{10°C} = 2.20 \pm 0.347$ nM versus $K_i^{37°C} = 2.34 \pm 0.23$ nM). These data suggested that triple-helicity can modulate MMP-2, but not MMP-9, inhibition. MMP-1, MMP-3, MT-1-MMP were not inhibited by f3 or f4 (tested up to 25 μM), while MMP-8 and MMP-13 were inhibited weakly in the micromolar range, suggesting a good selectivity profile of collagen V-derived inhibitors.

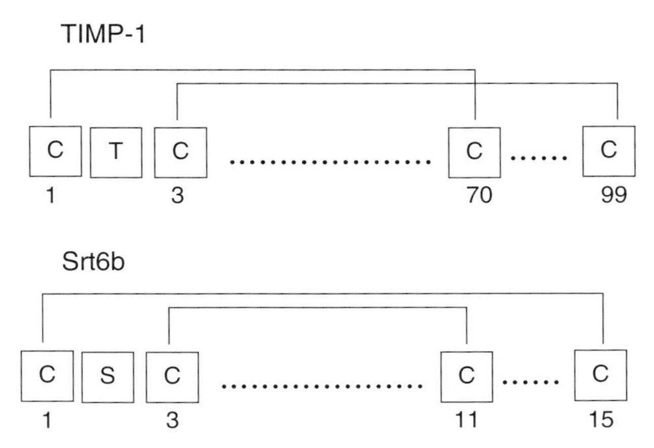

Figure 4.7 A comparison of disulfide topology and sequences of human N-TIMP-1 and sarafotoxin 6b. (Adapted from Lauer-Fields, J. L., Cudic, M., Wei, S., Mari, F., Fields, G. B., and Brew, K. (2007) Engineered sarafotoxins as tissue inhibitor of metalloproteinases-like matrix metalloproteinase inhibitors. *J Biol Chem* 282, 26948–26955. Rights holder: AMERICAN SOC FOR BIOCHEMISTRY & MOLECULAR BIOLOGY).

Another example of rational design of peptide-derived MMP inhibitors was based on the similarity of novel sarafotoxin peptides to the well-known endogenous MMP inhibitors – TIMPs [15]. Sarafotoxins are derived from the venom of snakes belonging to the genus *Atractaspis*. The authors noticed a topological similarity of one of the members (Srt6b) of sarafotoxin peptide family with TIMPs (Fig. 4.7). Sarafotoxins are significantly smaller than TIMPs (21-25 residues) and, therefore, are more amenable to the engineering effort. An important feature of sarafotoxins is a well-defined conformation with two disulfide bonds, which lends itself to peptide chemistry methods. In initial studies, Srt6b was found to be a low micromolar inhibitor of MMP-2 and MMP-3. As the Val-Ile-Trp sequence was found to be responsible for most of the affinity to endothelin receptors, potentially leading to unwanted side effects, (-)Val-Ile-Trp deletion mutant of Srt6b was synthesized (Table 4.1, STX). STX exhibited a complete loss of inhibitory activity towards all tested metzincins (Table 4.2). Using TIMPs as a template, the authors introduced Ala in position 4 and Val in position 16 (Table 4.1, STX-A4 and STX-V16). STX-V16, much like STX, did not inhibit any of the tested enzymes, while STX-A4 exhibited 20-50 μM K_i values towards MMP-1, MMP-2, and MMP-9 (Table 4.2). Substitution of Ala for Lys in position 4 did not affect inhibitory activity or selectivity profile (Table 4.2, STX-S4). Examination of the MMP3/TIMP-1 complex suggested that additional interactions can be gained using the C-terminal region of sarafotoxin derivatives. Sequence information from 13 different human and non-human TIMPs was used to create a combinatorial pool of residues encountered in these positions (Table 4.3). The resulting combinatorial library was applied to an affinity column with MMP-1 and MMP-3 bound on the solid support and eluted with isopropanol. Peptide STX-S4-CT (Table 4.1), which was bound by both enzymes, was tested for inhibitory activity. STX-S4-CT exhibited 20–30-fold improvement of affinity toward MMP-9 and 4-fold toward MMP-1, while its affinity for MMP-2 remained relatively

Table 4.1 Sequences of sarafotoxin analogs.

Name	Sequence
Srt6b	Cys Ser Cys Lys Asp Met Thr Asp Lys Glu Cys Leu Tyr Phe Cys His Gln Asp Val Ile Trp
STX	Cys Ser Cys Lys Asp Met Thr Asp Lys Glu Cys Leu Tyr Phe Cys His Gln Asp
STX-V16	Cys Ser Cys Lys Asp Met Thr Asp Lys Glu Cys Leu Tyr Phe Cys Val Gln Asp
STX-A4	Cys Ser Cys Ala Asp Met Thr Asp Lys Glu Cys Leu Tyr Phe Cys His Gln Asp
STX-S4	Cys Ser Cys Ser Asp Met Thr Asp Lys Glu Cys Leu Tyr Phe Cys His Gln Asp
STX-CT	Cys Ser Cys Lys Asp Met Thr Asp Lys Glu Cys Leu Tyr Phe Cys Met Ser Glu Met Ser
STX-S4-CT	Cys Ser Cys Ser Asp Met Thr Asp Lys Glu Cys Leu Tyr Phe Cys Met Ser Glu Met Se

[a]Reproduced with permission, from Lauer-Fields, J. L., Cudic, M., Wei, S., Mari, F., Fields, G. B., and Brew, K. (2007) Engineered sarafotoxins as tissue inhibitor of metalloproteinases-like matrix metalloproteinase inhibitors *J Biol Chem* 282, 26948-26955. Rights holder: AMERICAN SOC FOR BIOCHEMISTRY & MOLECULAR BIOLOGY.

Table 4.2 Apparent Ki values of Srt variants for different MMPs (μM).

Name	MMP-1	MMP-2	MMP-3	MMP-9	MT1-MMP	ADAM17	ADAMTS-4
STX	NA	NA	>100	NA	>100	>100	>100
STX-V16	NA	NA	>100	NA	>100	>100	>100
STX-A4	21.5 ± 0.6	41.8 ± 3.5	>100	24.5 ± 0.9	>100	>100	>100
STX-S4	22.0 ± 2.2	35.0 ± 4.4	>100	29.3 ± 0.0	>100	>100	>100
STX-CT	>100	>100	>100	25.3 ± 3.5	>100	>100	>100
STX-S4-CT	4.5 ± 0.0	21.6 ± 2.2	>100	1.0 ± 0.1	>100	>100	>100

NA = not applicable
[a]Reproduced with permission, from Lauer-Fields, J. L., Cudic, M., Wei, S., Mari, F., Fields, G. B., and Brew, K. (2007) Engineered sarafotoxins as tissue inhibitor of metalloproteinases-like matrix metalloproteinase inhibitors *J Biol Chem* 282, 26948-26955. Rights holder: AMERICAN SOC FOR BIOCHEMISTRY & MOLECULAR BIOLOGY.

Table 4.3 Amino acid residues encountered in positions 16–20 of TIMPs used to create a combinatorial pool for sarafotoxin S4 engineering.

Position 16	Position 17	Position 18	Position 19	Position 20
Val	Ser	Glu	Met	Ala
Leu	Ala	Ser	Ser	Ser
Met	–	Asp	Asp	–
Ala	–	–	Glu	–

[a]Adapted from from Lauer-Fields, J. L., Cudic, M., Wei, S., Mari, F., Fields, G. B., and Brew, K. (2007) Engineered sarafotoxins as tissue inhibitor of metalloproteinases-like matrix metalloproteinase inhibitors *J Biol Chem* 282, 26948-26955. Rights holder: AMERICAN SOC FOR BIOCHEMISTRY & MOLECULAR BIOLOGY.

unaffected. This study demonstrated that rational peptide chemistry approaches can be used to achieve selectivity and potency toward a metzincin of interest.

4.1.3 Small molecules: non-zinc binding exosite inhibitors

In this section we focus on small molecule inhibitors that do not target zinc of metzincin active site. As discussed in previous sections, a rational approach to the design of metzincin zinc-binding inhibitors involved inclusion of a zinc-binding moiety and optimization of a scaffold to achieve selectivity. This rational approach was based on the existing knowledge of metzincins' catalytic machinery involving zinc. In this section we review rational approaches to the design and discovery of small molecules that do not bind zinc of an active site.

Our first case study demonstrates the benefits of a rational design of an exosite-binding substrate that led to a discovery of a novel exosite-binding inhibitor. The sequence of this exosite-binding substrate was based on consensus sequence from collagen type I, II, and III that is recognized and cleaved by collagenolytic MMPs (MMP-1, MMP-8, MMP-13). As discussed in the previous section (Peptide-based inhibitors) collagenolytic MMPs possess multiple secondary substrate-binding sites (Fig. 4.5) that collagen-based substrate can interact with. Therefore, the authors reasoned, the small molecules that interact with these secondary collagen-binding sites can potentially inhibit its hydrolysis [16] which can be detected by the assay utilizing such substrates. Collagen I–III sequence was modified to include Mca/Dnp fluorophore-quencher pair (((7-methoxycoumarin-4-yl)acetyl) and 2,4-dinitrophenyl) attached to a Lys side chain. Fluorophore and quencher were incorporated in P_5 and P_5' positions of the substrate, which allowed for optimal quenching of Mca fluorescence. Additionally, five repeats of Gly-Pro-Hyp sequence were added on both termini of the peptide to induce triple-helical conformation, characteristic of fibrillar type I–III collagens. Resulting fluorogenic triple-helical substrate fTHP-15, [(Gly-Pro-Hyp)$_5$-Gly-Pro-Lys(Mca)-Gly-Pro-Gln-Gly~ Leu-Arg-Gly-Gln-Lys(Dnp) -Gly-Val-Arg-(Gly-Pro-Hyp)$_5$-NH_2] has been used to develop high-throughput screening assay for inhibition of MMP-13 in 1536 well plate format. HTS campaign against the NIH small molecule library yielded 54 compounds that exhibited inhibitory activity against MMP-13-mediated hydrolysis of fTHP-15 when tested in single concentration at 4 μM. 34 out of 54 primary hit compounds confirmed their activity in the dose response assay. 30 out of 34 confirmed hits were tested in RP-HPLC-based assay for inhibition of triple-helical exosite-binding substrate (fTHP-15) and linear active site-only ([Mca-Lys-Pro-Leu-Lys(Dnp)-Ala-Arg-NH_2], fSSP, a.k.a. Knight substrate) substrate. As can be seen from Fig. 4.8, only one compound (compound Q) exhibited statistically significant preferential inhibition of fTHP over fSSP, suggesting a unique mode of inhibition of MMP-13.

As fSSP interacts only with an active site and fTHP interacts with additional secondary binding sites, it stands to reason that the compound that preferentially inhibits fTHP can potentially do so via interacting with one of the secondary binding sites. Structure-activity relationship studies yielded two derivatives of compound Q (compounds Q, Q1 and Q2 are referred to as compounds **4**, **20** and **24** in [17], respectively) which were more potent toward MMP-13 and exhibited improved selectivity profile (Table 4.4).

Compound **20** (Q1) did not inhibit MMP-1 or MMP-8, and showed low levels of inhibition of MMP-2, MMP-9, and MMP-14. This is different from compound **4** (Q), which inhibited MMP-8. Compound **24** (Q2) was less selective than compound **20**, inhibiting MMP-8 and showing greater levels of inhibition of MMP-14. Compounds Q1 and Q2 were characterized in kinetic assays with MMP-13 using fTHP-15 as a substrate. As evidenced by Fig. 4.9 lines of best fit crossing on the *x* axis indicated a

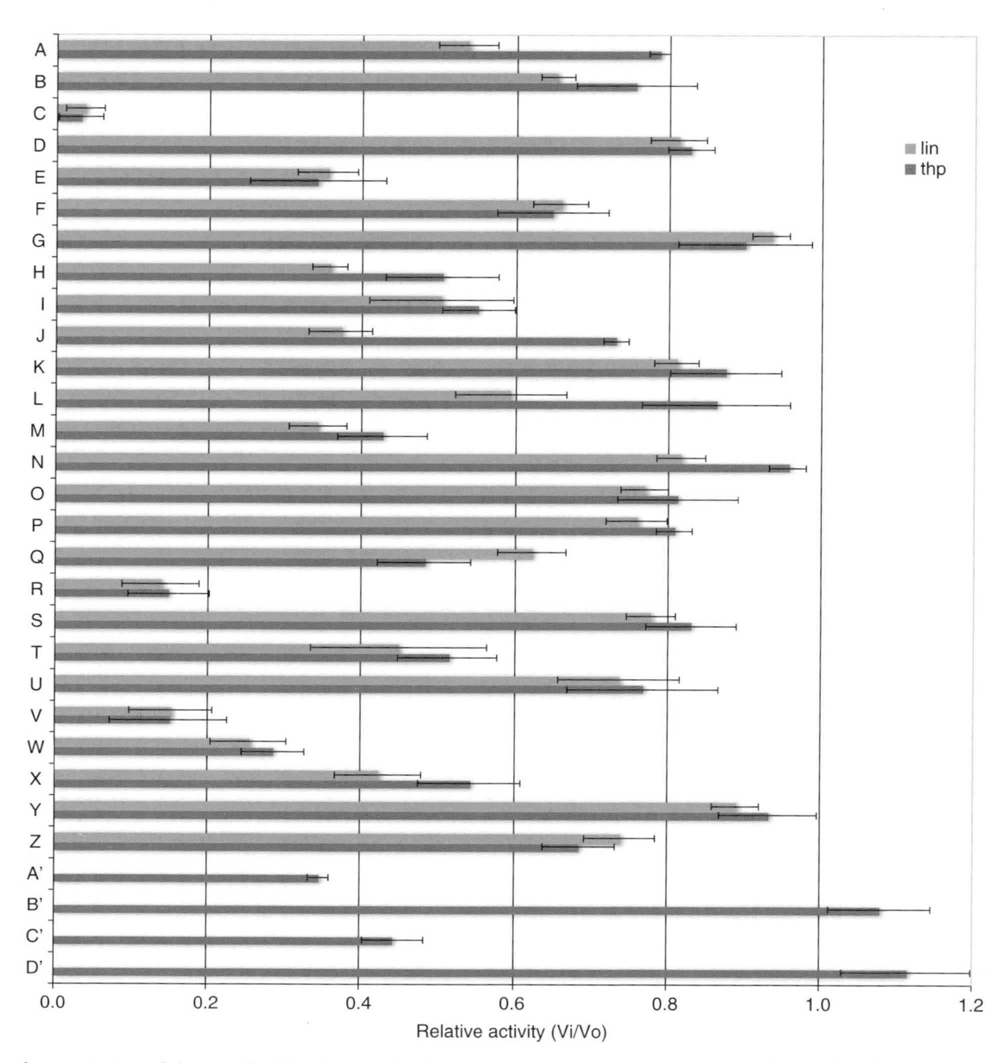

Figure 4.8 Inhibition of MMP-13 by 30 different compounds, as monitored by RP-HPLC and fluorescence spectroscopy. The change in RP-HPLC peak areas or relative fluorescence units for 10 nM MMP-13 hydrolysis of 10 μM fTHP-15 or 5 μM Knight fSSP was monitored at an inhibitor concentration of 100 μM. Assays were performed in triplicate. (Reproduced with permission from Lauer-Fields, J. L., Minond, D., Chase, P. S., Baillargeon, P. E., Saldanha, S. A., Stawikowska, R., Hodder, P., and Fields, G. B. (2009) High throughput screening of potentially selective MMP-13 exosite inhibitors utilizing a triple-helical FRET substrate. *Bioorg Med Chem* 17, 990–1005. © PERGAMON). (*See insert for color representation of this figure.*)

non-competitive mode of inhibition of MMP-13-mediated hydrolysis of the fTHP-15 by compound **20**. The K_i value was calculated to be 824 ± 171 nM. Lines of best fit crossing on the x axis also indicated a non-competitive mode of inhibition of MMP-13-mediated hydrolysis of the fTHP-15 by compound **24**. The K_i value was calculated to be 1,526 ± 260 nM. The K_i value for compound **4** was 3,800 nM (Table 4.4). These results suggested that compounds of Q series do not bind to the active site of MMP-13. Dual inhibition kinetic assays demonstrated non-mutually

Table 4.4 Inhibition of MMP-1, MMP-2, MMP-8, MMP-9, MMP-13, and MMP-14 activity by compounds 4, 20, and 24.

Target	Inhibition (%) at [inhibitor] = 40 μM (K_i, μM)		
	4 (Q)	20 (Q1)	24 (Q2)
—			
MMP-1	0	0	0
MMP-2	25 ± 11	19 ± 2.1	26 ± 9.2
MMP-8	17 ± 3.0	0	8.0 ± 5.0
MMP-9	10 ± 4.0	7.0 ± 3.0	16 ± 4.0
MMP-13	51 ± 0 (3.8)	67 ± 2.0 (0.8)	61 ± 1.0 (1.5)
MMP-14	22 ± 5.0	20 ± 6.0	31 ± 9.0

Assays were performed using 40 μM of each compound and fTHP-15 as substrate as described [16]. Enzyme concentrations were 10 nM for MMP-2, MMP-9, and MMP-14, 2.3 nM for MMP-13, 4 nM for MMP-1, and 2 nM for MMP-8.

[a]Adapted from Roth, J., Minond, D., Darout, E., Liu, Q., Lauer, J., Hodder, P., Fields, G. B., and Roush, W. R. (2011) Identification of novel, exosite-binding matrix metalloproteinase-13 inhibitor scaffolds *Bioorg Med Chem Lett* 21, 7180-7184

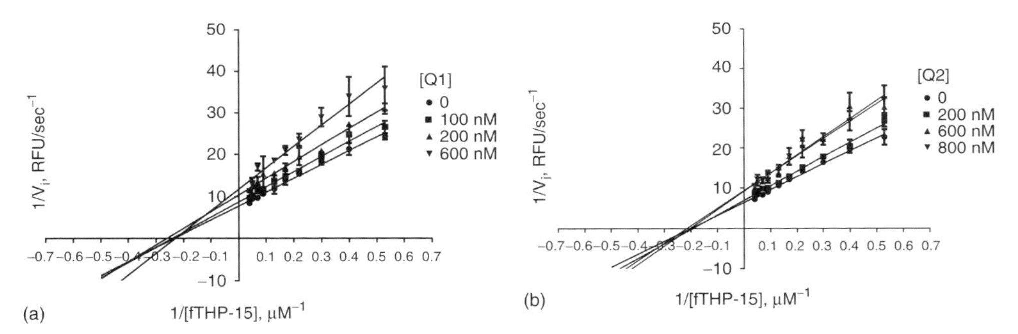

Figure 4.9 Lineweaver–Burk plot of MMP-13 inhibition of fTHP-15 hydrolysis by compound 20 (a) or 24 (b). (Reproduced with permission from Roth, J., Minond, D., Darout, E., Liu, Q., Lauer, J., Hodder, P., Fields, G. B., and Roush, W. R. (2011) Identification of novel, exosite-binding matrix metalloproteinase-13 inhibitor scaffolds. *Bioorg Med Chem Lett* 21, 7180–7184. © PERGAMON).

exclusive binding between compound **4** and other known exosite inhibitors of MMP-13 (Timothy P. Spicer, Jianwen Jiang, Alexander B. Taylor, Jun Yong Choi, P. John Hart, William R. Roush, Gregg B. Fields, Peter S. Hodder, and Dmitriy Minond. Characterization of Selective Inhibitors of Matrix Metalloproteinase 13 That Prevent Articular Cartilage Degradation *In Vitro*. *J Med Chem*. 2014 Nov 26; 57(22):9598-611. PMID:25330343) suggesting either a novel mode of binding to a known exosite or binding to a novel region within MMP-13 structure. Studies are currently under way to structurally characterize the compounds of Q series and further develop them into *in vivo* probes for biological systems where MMP-13 can potentially play a role.

Matrix metalloproteinases, especially collagenolytic ones, are relatively well-studied enzymes. Multiple 3D structures are available as a result of NMR and XRD studies (reviewed in [18]). Several collagen secondary binding sites have been discovered by various research groups [19–32] which lead to the idea of utilizing collagen-like substrate for the discovery of inhibitors of collagenolysis. As a contrast to the MMP exosite inhibitor discovery, our next example of rational approach to the discovery of small non-zinc binding inhibitors of ADAM17 (TACE) describes the case where pre-existing knowledge about secondary substrate binding sites was limited. In our research group we were interested in discovering ADAM17 exosite-binding inhibitors, but very few reports described potential ADAM's secondary substrate binding sites [33, 34] and neither structural nor kinetic studies describing substrate interactions with these potential binding sites existed. We hypothesized that interactions between substrates binding to the enzyme's active and secondary binding sites should result in increased affinity as opposed to the substrate that interacts only with the active site. With that in mind we proceeded to design such a substrate. However, it was not immediately clear what substrate features would enable it to interact with hypothetical secondary binding sites. Substrate recognition by ADAM proteases is a largely unexplored area. ADAMTS and MMP substrate specificity was shown to be due to a combination of sequence features and substrate topology [35–37]. Cleavage site sequence specificity was addressed for several members of the ADAM family [38–40]; however, there were no studies of the effects of secondary structure on substrate recognition by ADAM proteases. Also, it was unknown whether post-translational modifications of ADAM substrates play a role in substrate specificity. Most ADAM substrates are cell surface-bound proteins, which are either known or predicted to be glycosylated. For example, the cleavage site of TNFα by ADAM17 is only four residues away from a glycosylated residue [41], while glycosylation occurs 14 residues away from the TGFα cleavage site [42] and more than 200 residues away from the L-selectin cleavage site [43]. Literature and database searches revealed several additional validated substrates of ADAM proteases with glycosylation close to the cleavage sites; however, in most cases the type of carbohydrate present on the ADAM substrate was never identified. Two best-characterized ADAM substrates, with respect to carbohydrate moieties present, appeared to be TNFα and IL6 receptor (IL6-R). Pro-TNFα was reported to be *O*-glycosylated at Ser 80 in human B-cell lymphoblastic leukemia cells (BALL-1) [41]. The mucin type 2,3- and 2,6-sialylated and non-sialylated core 1 structure (β-Gal-(1→3)-GalNAc), also known as T or TF antigen, were identified. The physiological cleavage site by ADAM17 is at the A^{76}~V bond, only 4 residues away from the modified Ser residue. Interleukin-6 receptor (IL6-R), which is cleaved by both ADAM10 and 17 [39, 44], was reported to be *N*-glycosylated by *N*-acetylglucosamine (GlcNAc) at four different positions [45, 46]. Glycosylated Asn^{350} is only seven residues away from Q^{357}~D scissile bond. We decided to focus on canonical substrate of ADAM17, TNFα. Immediate sequence around the cleavage site (A~V) was used ($PLAQA^{76}$~$VRSS^{80}S$). Ser80 was glycosylated with β-Gal-(1→3)-GalNAc (Fig. 4.10, TF antigen). We incorporated a fluorophore/quencher pair (EDANS/DABCYL) attached to side chains of Glu and Lys on different sides of the scissile bond to allow for continuous kinetic readout.

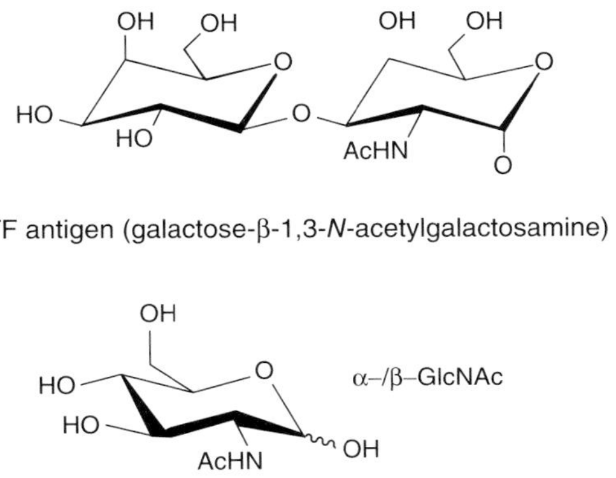

TF antigen (galactose-β-1,3-*N*-acetylgalactosamine)

α-/β-GlcNAc

Figure 4.10 β-Gal-(1→3)-GalNAc (TF antigen) and *N*-acetylglucosamine (GlcNAc) found on TNFα and IL6-R.

Non-glycosylated substrate was also synthesized for comparative studies on effects of glycosylation.

Both peptides exhibited random coil signature in 100% aqueous solution; however, at higher TFE concentrations glycosylated substrate showed lower α-helical content as compared to a non-glycosylated substrate. This suggested that TF antigen interferes with formation of the hydrogen bond pattern of an α-helix which is most likely attributable to steric hindrance. Glycosylation affected cleavage site specificity of several tested metzincins (ADAM8, ADAM12, MMP-1, MMP-2, MMP-8, MMP-9, and MMP-13) such that additional bonds were cleaved in glycosylated versus non-glycosylated substrate. Interestingly, ADAM10 and ADAM17 cleaved only at canonical $A^{76}{\sim}V$ bond. ADAM8 and 17 exhibited higher activity toward the glycosylated substrate as evidenced by 16- and 6-fold higher k_{cat}/K_M values, respectively (Table 4.5). The ADAM12 k_{cat}/K_M value was slightly increased for the glycosylated substrate. In the case of ADAM17, the increase in activity was due to the cumulative effects of changes in both K_M and k_{cat}, whereas ADAM8 improvement of activity was almost entirely due to an increase in k_{cat}. ADAM10 activity toward the glycosylated substrate was approximately three-fold lower than for the non-glycosylated substrate.

These kinetic data suggested differences in ADAM10 and 17 substrate binding site(s) corresponding to the Ser residue in position P4′ of the TNFα-based substrate. Existing structural and modeling studies reveal the presence of a large S_3' cleft in ADAM17 structure that can potentially accommodate a bulky disaccharide (βGal-1,3-αGalNAc) [47–49] suggesting that the carbohydrate moiety potentially interacts with this hypothetical exosite and, therefore, could be exploited for inhibitor discovery. In order to test this hypothesis, HTS assays were developed for ADAM10 and 17 using both glycosylated and non-glycosylated substrates. Parallel screens of mixture libraries against ADAM10 and 17 yielded one library (Fig. 4.11(b), #1344) preferentially inhibiting ADAM17 in the glycosylated substrate assay, suggesting the possibility of the presence of exosite inhibitors in the above-mentioned library mixture. Most interestingly, library 1344 was not active against ADAM17 in the non-glycosylated substrate assay and, therefore, would have been discarded from further studies if just the conventional active site-only substrate was utilized. This unusual preference toward a glycosylated, potentially exosite-binding substrate was

Table 4.5 Kinetic parameters for ADAM hydrolysis of glycosylated and non-glycosylated substrates.

Enzyme	kcat/KM ($M^{-1}s^{-1}$)		kcat (s^{-1})		KM (μM)	
	Glycosylated	Non-glycosylated	Glycosylated	Non-glycosylated	Glycosylated	Non-glycosylated
ADAM8	$1.4 \pm 0.1 \times 103$	$2.2 \pm 0.01 \times 102$	0.04 ± 0.01	0.01 ± 0.03	31 ± 11	46 ± 15
ADAM9	ND	$1.2 \pm 0.11 \times 103$	ND	0.06 ± 0.01	ND	49 ± 6
ADAM10	$0.8 \pm 0.3 \times 104$	$2.5 \pm 0.71 \times 104$	0.06 ± 0.01	0.28 ± 0.03	8.5 ± 1.3	12 ± 3
ADAM12	$1.3 \pm 0.41 \times 103$	$8.5 \pm 0.51 \times 102$	0.02 ± 0.01	0.01 ± 0.00	20 ± 11	12 ± 1
ADAM17	$7.6 \pm 1.71 \times 104$	$1.2 \pm 0.11 \times 104$	0.25 ± 0.03	0.14 ± 0.01	3.0 ± 0.5	12 ± 3

ND = not determined due to low reaction velocity

[a]Reproduced with permission, from Minond, D., Cudic, M., Bionda, N., Giulianotti, M., Maida, L., Houghten, R. A., and Fields, G. B. (2012) Discovery of novel inhibitors of a disintegrin and metalloprotease 17 (ADAM17) using glycosylated and non-glycosylated substrates *J Biol Chem* 287, 36473-36487. Rightsholder: AMERICAN SOC FOR BIOCHEMISTRY & MOLECULAR BIOLOGY.

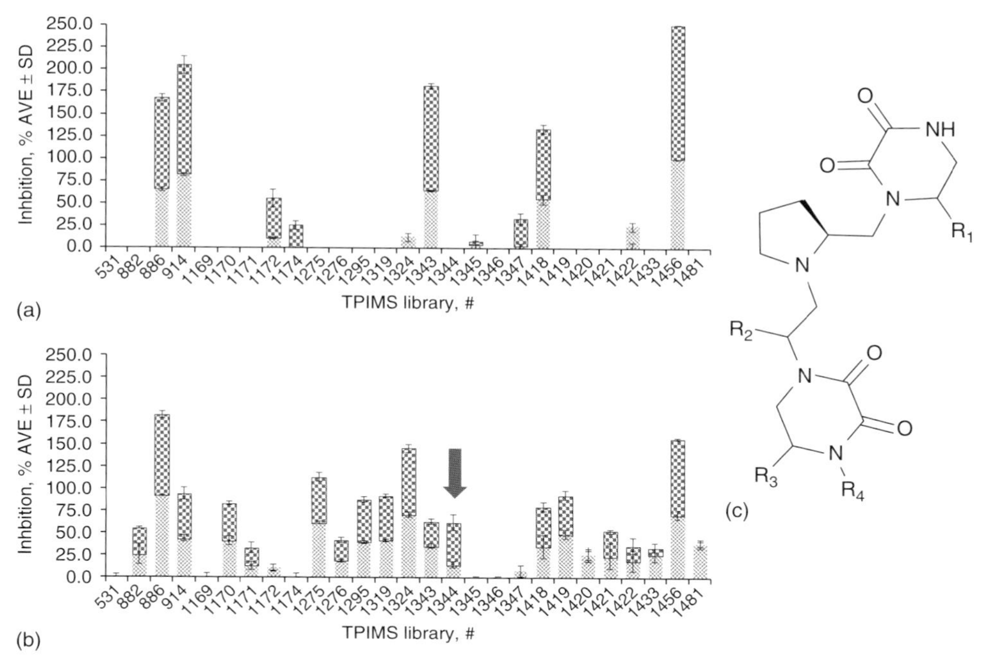

Figure 4.11 Results of the pilot "scaffold ranking" screen of TPIMS drug-like library against ADAM10 and 17. Shown is an ADAM10 (a) and ADAM17 (b) screen using glycosylated (red checked bars) and non-glycosylated substrate (blue bars). The arrow indicates library containing potential exosite inhibitors of ADAM17. All assays were performed in triplicate. Activity and selectivity of all libraries were confirmed in reversed-phase HPLC-based assays. (c), basic scaffold of library 1344. (Adapted from Minond, D., Cudic, M., Bionda, N., Giulianotti, M., Maida, L., Houghten, R. A., and Fields, G. B. (2012) Discovery of novel inhibitors of a disintegrin and metalloprotease 17 (ADAM17) using glycosylated and non-glycosylated substrates *J Biol Chem* 287, 36473–36487. Rightsholder: AMERICAN SOC FOR BIOCHEMISTRY & MOLECULAR BIOLOGY).

reminiscent of a preferential inhibition of fTHP exosite substrate hydrolysis over fSSP substrate by compound Q, reviewed in the previous section.

This prompted us to deconvolute library 1344 in order to identify individual compounds that could potentially inhibit ADAM17 via exosite binding. The basic scaffold of library 1344 (Fig. 4.11(c)), comprised of 738,192 members ($26 \times 26 \times 26 \times 42$), has four sites of diversity (R_1, R_2, R_3, and R_4) and, therefore, is made up of four separate sub-libraries, each having a single defined position (R) and three mixture positions (X). To assess which moieties imparted selectivity toward ADAM17, we screened four sets of mixtures, totaling 120 mixtures ($26\,R_1 + 26\,R_2 + 26\,R_3 + 42\,R_4$) against ADAM10 and 17. Most functional groups in positions R_1, R_2, and R_3 yielded libraries that either selectively inhibited ADAM17 or were inactive (Fig. 4.12(a–c)). Functionalities in position R_4 appear to influence the selectivity for ADAM17 to the greatest degree (Fig. 4.12(d)). Approximately 50% of all substitutions tested in this position were inhibiting both enzymes to an equal extent.

Based on the dose response experiments with *mixture libraries*, we synthesized 36 *individual compounds* ($2\,R_1 \times 3\,R_2 \times 2\,R_3 \times 3\,R_4$) containing functional groups for each defined R position that exhibited the most selectivity and potency toward ADAM17 and tested them against the target enzyme (ADAM17) and counter targets

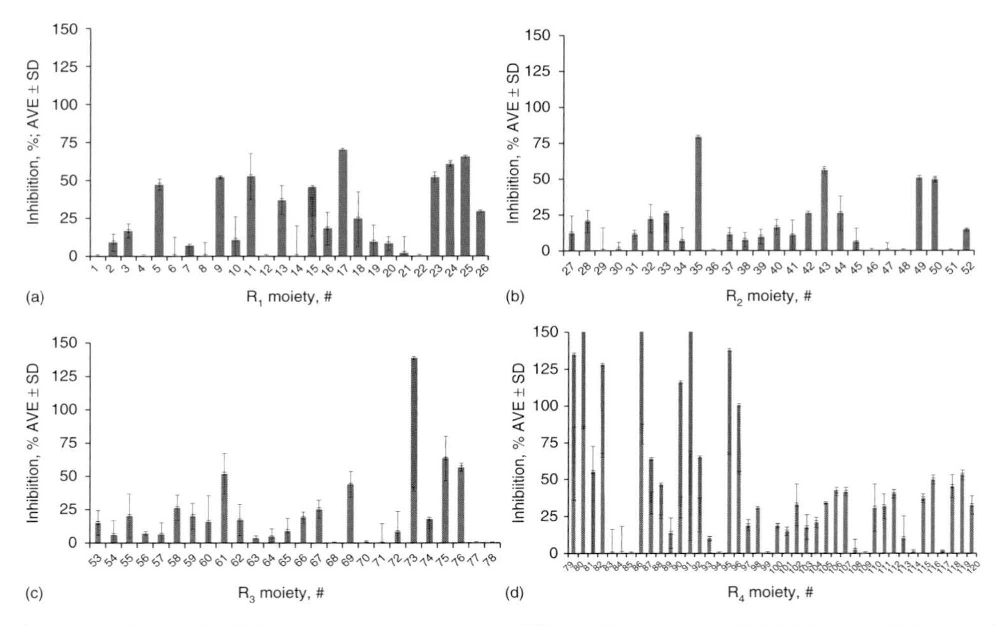

Figure 4.12 Results of the positional scan analysis of library 1344 against ADAM10 and -17. Positional scan of R_1 (a), R_2 (b), R_3 (c), and R_4 (d) defined moieties against ADAM10 (red bars) and ADAM17 (blue bars) using glycosylated substrate. (Adapted from Minond, D., Cudic, M., Bionda, N., Giulianotti, M., Maida, L., Houghten, R. A., and Fields, G. B. (2012) Discovery of novel inhibitors of a disintegrin and metalloprotease 17 (ADAM17) using glycosylated and non-glycosylated substrates. *J Biol Chem* 287, 36473–36487. Rightsholder: AMERICAN SOC FOR BIOCHEMISTRY & MOLECULAR BIOLOGY).

(ADAM10, MMP-8, and MMP-14) in a dose response experiment. Only one of the 36 tested individual compounds reached 25% inhibition at the highest tested concentration (40 μM) in the counter target assays, suggesting a highly selective nature of these compounds (Fig. 4.13).

We characterized a lead of the series, compound 15, to determine whether it indeed inhibits ADAM17 via exosite-binding. In order to rule out zinc-binding inhibition mechanism by compound 15, we performed dual inhibition kinetics using *N*-hydroxyacetamide (Fig. 4.14(a), AHA), a known zinc-binder and competitive mM-range inhibitor of many metalloproteases, in combination with inhibitor #15, following previously described methodology [50]. AHA was used in the concentration range of 0 to 2.5 mM, in combination with #15 at concentrations between 0 and 7.5 μM. When initial velocities from this experiment were organized in a Yonetani-Theorell plot they formed the series of intersecting lines of best fit (Fig. 4.14(b)). In Yonetani-Theorell plots the intersecting lines suggest simultaneous (i.e., mutually non-exclusive) binding of both inhibitors to the enzyme [51]. Since AHA is known to bind to Zn, #15 most likely acts via a non-Zn-binding mechanism and, potentially, outside of active site and beyond the catalytic domain as suggested by a recent report [10]. Single inhibition kinetics of #15 suggest non-competitive inhibition ($K_i/K_i' = 0.5 \pm 0.6$, Fig. 4.14(c)) consistent with mutually non-exclusive AHA binding. We investigated the possibility of #15 binding outside of the catalytic domain by performing dose response experiments with #15 against

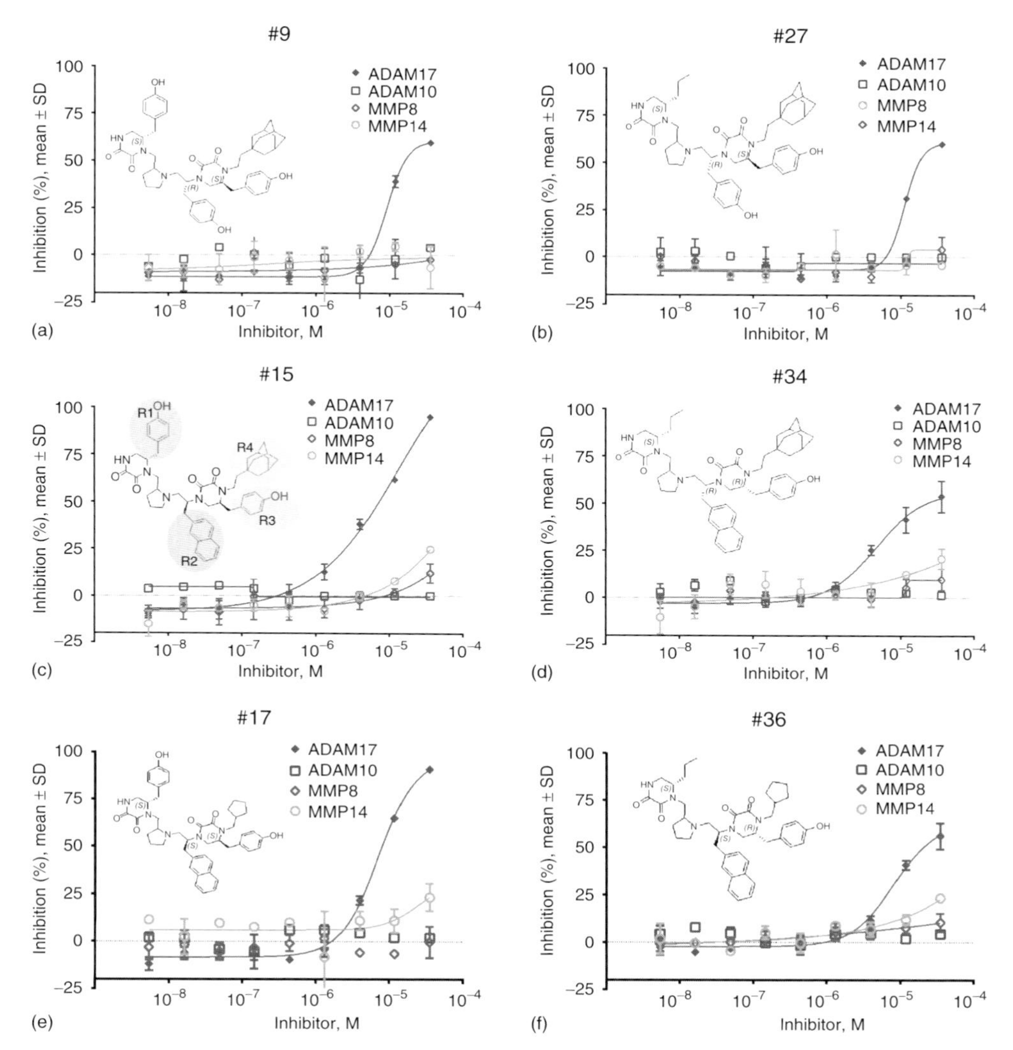

Figure 4.13 Results of dose response study of most ADAM17 selective and potent individual compounds. Structures of individual compounds are shown as inserts. (Adapted from Minond, D., Cudic, M., Bionda, N., Giulianotti, M., Maida, L., Houghten, R. A., and Fields, G. B. (2012) Discovery of novel inhibitors of a disintegrin and metalloprotease 17 (ADAM17) using glycosylated and non-glycosylated substrates. *J Biol Chem* 287, 36473–36487. Rightsholder: AMERICAN SOC FOR BIOCHEMISTRY & MOLECULAR BIOLOGY).

ADAM17 ectodomain (ECD) and catalytic domain-only (CD) constructs. AHA and MMP-9/MMP-13 inhibitor were utilized as controls. Both controls exhibited nearly identical IC_{50} values for inhibition of either construct (AHA $IC_{50} = 1.5 \pm 0.2$ mM versus 1.6 ± 0.2 mM for ADAM17 ECD and CD, respectively; MMP-9/MMP-13 inhibitor $IC_{50} = 0.3 \pm 0.03$ nM versus 0.5 ± 0.04 nM for ADAM17 ECD and CD, respectively; Fig. 4.14 (d)), whereas #15 inhibited ADAM17 ECD ten-fold more potently (#15 $IC_{50} = 4.2 \pm 0.4$ μM versus 47 ± 4 μM for ADAM17 ECD and CD, respectively; Fig. 4.14). This suggests certain cooperativity between catalytic and non-catalytic domains of ADAM17 in the binding of #15. One possibility is that

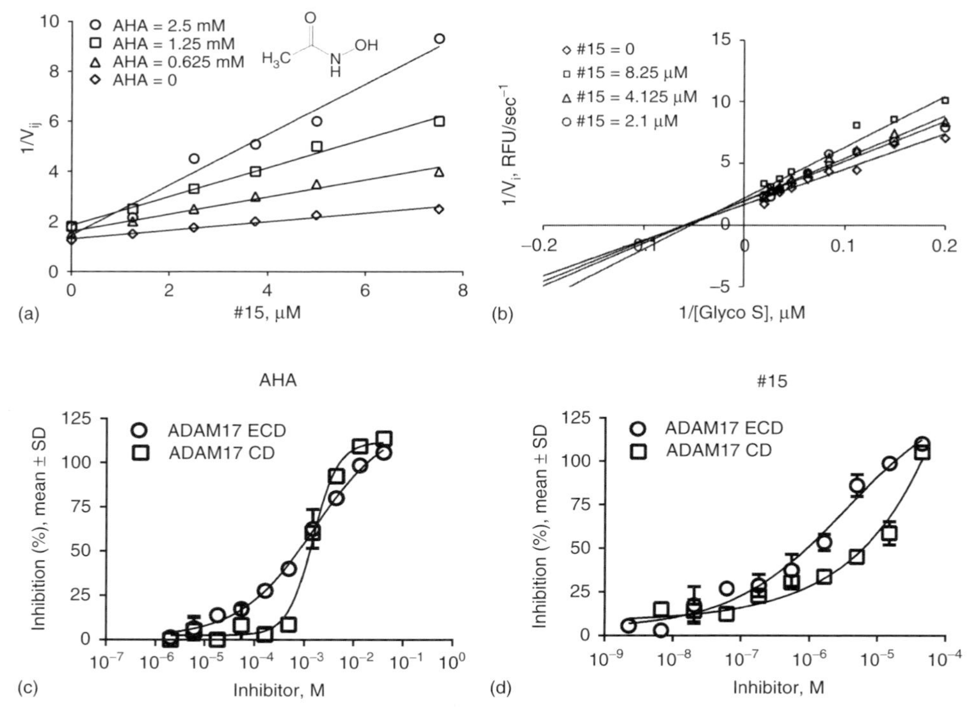

Figure 4.14 Characterization of mechanism of inhibition of ADAM17 catalytic domain and ectodomain by compound #15. (a) Yonetani-Theorell plot of glycosylated substrate hydrolysis by ADAM17 in the presence of AHA and compound #15. Note the non-parallel lines of best fit indicating mutually non-exclusive binding by two inhibitors. Structure of N-hydroxyacetamide (AHA) shown as insert. (b) Lineweaver-Burke plot of glycosylated substrate hydrolysis by ADAM17 in the presence of compound #15. Dose response study of inhibition of ADAM17 catalytic domain and ectodomain by (c) AHA and (d) compound #15. (Adapted from Minond, D., Cudic, M., Bionda, N., Giulianotti, M., Maida, L., Houghten, R. A., and Fields, G. B. (2012) Discovery of novel inhibitors of a disintegrin and metalloprotease 17 (ADAM17) using glycosylated and non-glycosylated substrates *J Biol Chem* 287, 36473–36487. Rightsholder: AMERICAN SOC FOR BIOCHEMISTRY & MOLECULAR BIOLOGY).

#15 can bind across domains due to the spatial proximity of the non-catalytic and catalytic domains, as described in the case of the inhibitory antibody reported by Tape et al., [10]. Existence of a binding pocket within the catalytic domain capable of accommodating #15, whose size is affected by the presence of a non-catalytic domain, is yet another possibility. The small size of #15, as compared to an antibody, makes the latter model more likely.

Several other exosite-binding MMP inhibitors were reported [52–58]; however, the rational design considerations were either not revealed or inhibitors were identified as a result of HTS or virtual screening.

4.1.4 Protein-based inhibitors

In this section we focus on design approaches using endogenous inhibitors of metzincins – tissue inhibitors of metalloproteinases (TIMPs). Four known mammalian TIMPs (TIMP-1–4) possess varying degrees of selectivity and potency toward

Table 4.6 IC_{50} values for phage-displayed TIMP-2 variants from the screening of libraries that mutate three regions on TIMP-2.

N-TIMP-2 variant	L1 sequence	MMP-1cd IC_{50}, nM	MMP-3cd IC_{50}, nM	Selectivity, MMP-3/MMP-1
WT	CSCSPV	13 ± 4	13 ± 3	1.0
TM1	CVCQGL	9 ± 3	54 ± 17	6.4
TM8	CDCAPV	16 ± 3	140 ± 74	8.6
TM10	CVCKTS	31 ± 9	96 ± 36	2.3
TM13	CVVCTE	16 ± 4	36 ± 9	2.3
TM14	CSCNST	15 ± 3	61 ± 13	4.0

The data were obtained by competitive phage ELISA. All mutants contained WT sequences in L2 and L3.
[a]Reproduced with permission, from Bahudhanapati, H., Zhang, Y., Sidhu, S. S., and Brew, K. (2011) Phage display of tissue inhibitor of metalloproteinases-2 (TIMP-2): identification of selective inhibitors of collagenase-1 (metalloproteinase 1 (MMP-1)). *J Biol Chem* 286, 31761-31770. Rightsholder: AMERICAN SOC FOR BIOCHEMISTRY & MOLECULAR BIOLOGY.

metzincins [59]. TIMPs interact with metzincins via multiple rather large binding sites, therefore, systematic mutagenesis might generate several million sequence variants [60]. Authors of this study employed phage display to tackle this problem. TIMP-2 was used as a template due to low nanomolar affinities toward most MMPs [61]. MMP-1 catalytic domain was used as a selection target due to pro-cancer activity, while MMP-3 catalytic domain was utilized as a counter target due to protective effects during tumorigenesis [62, 63]. Three regions within TIMP-2 that were shown to interact with MMP binding sites ((Cys^1-Val^6 (L1), $Asp^{34}-Ile^{40}$ (L2), and $Pro^{67}-Gly^{73}$ (L3) [64–68]) were mutated. After several iterative rounds of selection using MMP-1cd and MMP-3cd, several TIMP-2 mutants exhibited selectivity toward MMP-1cd as opposed to WT TIMP-2. Interestingly, all selective mutants were from TIMP-2 region L1 which is responsible for blocking access to a catalytic cleft. Clone TM8 was the most selective, showing 10-fold preference toward MMP-1cd (Table 4.6).

TM8 was chosen as a template for further optimization. Non-cysteine residues were randomized and resulting mutants were re-screened against MMP-1cd and MMP-3cd. None of the mutants showed improved selectivity over TM8. Testing of TIMP-2 mutants against an expanded panel of MMPs showed that TM8's derivative CDCS was overall the most selective inhibitor of MMP-1 [60], exhibiting approximately 10–50 fold selectivity over MMP-2, MMP-7, MMP-8, MMP-13, and MMP-14. Interestingly, it was 5-fold more selective toward MMP-9 over MMP-1.

Other examples of protein-based inhibitors of metzincins include pro-domains of ADAM10 [69] and ADAM17 [70]. Authors of the ADAM17 study expressed pro-domain in *E. coli* and were able to fold it correctly. Circular dichroism study showed that pro-domain has secondary structure, which was denatured by Gdn-HCl. The single transition between folded and unfolded states was observed, suggesting thermodynamically stable native state. Inhibition studies showed that pro-domain is significantly more active against catalytic domain-only construct of ADAM17 as opposed to the catalytic and disintegrin/cysteine-rich domains construct ($IC_{50} = 70$ nM versus > 2,000 nM for catalytic and disintegrin/cysteine-rich domains construct, respectively). To study the role of so-called "cysteine switch"

(PKVCGYLKVD[190]) in ADAM17 inhibition, the authors conducted alanine scan experiment. Even though all IC_{50} values for Ala mutants were within narrow range (49–132 nM) this study suggested that ADAM17 pro-domain is amenable to protein engineering effort, which can potentially produce the selective inhibitor of ADAM17.

In case of ADAM10 pro-domain, the authors expressed mouse orthologue (residues 23–213) which inhibited human ADAM10 with $K_i = 36 \pm 9$ nM. MMP-1, MMP-2, MMP-3, MMP-9, MMP-13, and MMP-14 were not inhibited by 3 μM pro-domain. Both catalytic and catalytic/disintegrin constructs of ADAM17 were only weakly inhibited at 30 μM dose, whereas ADAM9 and ADAM8 were inhibited in a low micromolar range, suggesting a highly selective inhibition profile of ADAM10 pro-domain-based inhibitor. Truncation of WT ADAM10 pro-domain resulted in a 10–30-fold loss of potency. In the cell-based experiments, the pro-domain of ADAM10 was able to dose-dependently inhibit ionophore-induced beta-cellulin shedding in HEK293 and MDCK cells. Interestingly, it did not inhibit shedding of amphiregulin in IMPE cells, suggesting certain substrate selectivity.

Summary and future directions

It is possible to achieve inhibition of metzincins via secondary binding site mechanism using rational approaches or combination of rational approaches with high-throughput ones. This was achieved by (i) raising antibodies using target enzyme-derived antigen from outside of the catalytic cleft (i.e., A300E-BiTE, D1(A12)); (ii) using cognate substrate or inhibitor sequences as starting templates to design an inhibitor (i.e., f1 triple-helical peptide, STX4 sarafotoxin, TIMP-2 mutants, ADAM10, and ADAM17 pro-domains); and (iii) using cognate-like secondary binding site substrates to guide discovery of secondary binding site inhibitors (i.e., fTHP-15 substrate in case of compound Q and glycosylated TNFα-based substrate in case of compound 15). Most of these approaches led to the discovery or development of more or less selective inhibitors of various metzincins, which, in some cases, were effective in *in vitro* systems. This suggests that these approaches or combinations thereof can be used for practical applications such as metzincin-directed drug or probe discovery. While enzyme selectivity was addressed in these studies by using counter targets, the question of substrate selectivity still remains largely untouched. It is now understood that some metzincins should not be inhibited in certain disease scenarios (i.e., MMP-8 and MMP-14 in skin and breast cancer [71–73]). However, it is much less known whether any of cognate substrates' hydrolysis should be spared from inhibition. As demonstrated in case of γ-secretase inhibitor discovery for Alzheimer's disease, inhibition of γ-secretase abrogated release of β-amyloid, but also prevented a cleavage of Notch leading to toxicity (reviewed in [74]). It is well known that metzincins have broad substrate repertoirs; therefore, it is possible that total abrogation of activity of a target enzyme might be unnecessary. The understanding of which substrate's hydrolysis should be spared in the particular disease will come from the future proteomic studies. Development of rational approaches to discovery and development of substrate-selective inhibitors presents, in our opinion, an interesting new direction for future metzincin inhibition studies.

Another interesting future direction is activators or potentiators of metzincins. The most obvious and immediate benefit from potentiation of metzincin activity could potentially be in the case of Alzheimer's disease. As discussed in several recent reviews [75–77] an α-secretase activity diminishes release of insoluble β-amyloid and increases release of soluble and neuro-protective α-amyloid and, therefore, represents a viable target for a therapy. It is believed that α-secretase activity is most likely attributable to ADAM10 [75]. Another possibility is to attempt to potentiate activity of metzincins that have protective effects in cancer. As discussed above, MMP-8 and MMP-14 should be spared from inhibition in certain types of cancer. Therefore, it would be interesting to see whether the potentiation of activity of these enzymes can lead to the increase of their protective effect.

In conclusion, metzincin inhibition still remains a viable therapeutic approach. Selective inhibitors of metzincins can be used as tools to probe their biological and pathophysiological roles and help validate them as therapeutic targets. There are, however, very few selective inhibitors of metzincins available. It is our hope that more of such probes will become available to the researchers studying metzincin biology.

References

1. Yadav, M.R., Murumkar, P.R. & Zambre, V.P. (2012) Advances in studies on collagenase inhibitors. *EXS*, 103, 83–135.
2. DasGupta, S., Murumkar, P.R., Giridhar, R. & Yadav, M.R. (2009) Current perspective of TACE inhibitors: a review. *Bioorganic and Medicinal Chemistry*, 17, 444–459.
3. Georgiadis, D. & Yiotakis, A. (2008) Specific targeting of metzincin family members with small-molecule inhibitors: progress toward a multifarious challenge. *Bioorganic and Medicinal Chemistry*, 16, 8781–8794.
4. Fisher, J.F. & Mobashery, S. (2006) Recent advances in MMP inhibitor design. *Cancer Metastasis Reviews*, 25, 115–136.
5. Yiotakis, A. & Dive, V. (2008) Synthetic active site-directed inhibitors of metzincins: achievement and perspectives. *Molecular Aspects of Medicine*, 29, 329–338.
6. Yamamoto, K., Trad, A., Baumgart, A. *et al.* (2012) A novel bispecific single-chain antibody for ADAM17 and CD3 induces T-cell-mediated lysis of prostate cancer cells. *The Biochemical Journal*, 445, 135–144.
7. Lorenzen, I., Lokau, J., Dusterhoft, S. *et al.* (2012) The membrane-proximal domain of A Disintegrin and Metalloprotease 17 (ADAM17) is responsible for recognition of the interleukin-6 receptor and interleukin-1 receptor II. *FEBS Letters*, 586, 1093–1100.
8. Lorenzen, I., Trad, A. & Grotzinger, J. (2011) Multimerisation of A disintegrin and metalloprotease protein-17 (ADAM17) is mediated by its EGF-like domain. *Biochemical and Biophysical Research Communications*, 415, 330–336.
9. Trad, A., Hansen, H.P., Shomali, M. *et al.* (2013) ADAM17-overexpressing breast cancer cells selectively targeted by antibody-toxin conjugates. *Cancer Immunology, Immunotherapy*, 62, 411–421.
10. Tape, C.J., Willems, S.H., Dombernowsky, S.L. *et al.* (2011) Cross-domain inhibition of TACE ectodomain. *Proceedings of the National Academy of Sciences of the United States of America*, 108, 5578–5583.
11. Richards, F.M., Tape, C.J., Jodrell, D.I. & Murphy, G. (2012) Anti-tumour effects of a specific anti-ADAM17 antibody in an ovarian cancer model in vivo. *PLoS ONE*, 7, e40597.
12. Lauer-Fields, J., Brew, K., Whitehead, J.K., Li, S., Hammer, R.P. & Fields, G.B. (2007a) Triple-helical transition state analogues: a new class of selective matrix metalloproteinase inhibitors. *Journal of the American Chemical Society*, 129, 10408–10417.
13. Lauer-Fields, J.L., Whitehead, J.K., Li, S., Hammer, R.P., Brew, K. & Fields, G.B. (2008) Selective modulation of matrix metalloproteinase 9 (MMP-9) functions via exosite inhibition. *The Journal of Biological Chemistry*, 283, 20087–20095.
14. Lauer-Fields, J.L., Sritharan, T., Stack, M.S., Nagase, H. & Fields, G.B. (2003) Selective hydrolysis of triple-helical substrates by matrix metalloproteinase-2 and -9. *The Journal of Biological Chemistry*, 278, 18140–18145.

15. Lauer-Fields, J.L., Cudic, M., Wei, S., Mari, F., Fields, G.B. & Brew, K. (2007b) Engineered sarafotoxins as tissue inhibitor of metalloproteinases-like matrix metalloproteinase inhibitors. *The Journal of Biological Chemistry*, 282, 26948–26955.
16. Lauer-Fields, J.L., Minond, D., Chase, P.S. *et al.* (2009) High throughput screening of potentially selective MMP-13 exosite inhibitors utilizing a triple-helical FRET substrate. *Bioorganic and Medicinal Chemistry*, 17, 990–1005.
17. Roth, J., Minond, D., Darout, E. *et al.* (2011) Identification of novel, exosite-binding matrix metalloproteinase-13 inhibitor scaffolds. *Bioorganic and Medicinal Chemistry Letters*, 21, 7180–7184.
18. Maskos, K. & Bode, W. (2003) Structural basis of matrix metalloproteinases and tissue inhibitors of metalloproteinases. *Molecular Biotechnology*, 25, 241–266.
19. Clark, I.M. & Cawston, T.E. (1989) Fragments of human fibroblast collagenase. Purification and characterization. *The Biochemical Journal*, 263, 201–206.
20. Murphy, G., Allan, J.A., Willenbrock, F., Cockett, M.I., O'Connell, J.P. & Docherty, A.J. (1992) The role of the C-terminal domain in collagenase and stromelysin specificity. *The Journal of Biological Chemistry*, 267, 9612–9618.
21. Knauper, V., Osthues, A., DeClerck, Y.A., Langley, K.E., Blaser, J. & Tschesche, H. (1993) Fragmentation of human polymorphonuclear-leucocyte collagenase. *The Biochemical Journal*, 291 (Pt 3), 847–854.
22. De Souza, S.J., Pereira, H.M., Jacchieri, S. & Brentani, R.R. (1996) Collagen/collagenase interaction: does the enzyme mimic the conformation of its own substrate? *The FASEB Journal*, 10, 927–930.
23. Knauper, V., Cowell, S., Smith, B. *et al.* (1997a) The role of the C-terminal domain of human collagenase-3 (MMP-13) in the activation of procollagenase-3, substrate specificity, and tissue inhibitor of metalloproteinase interaction. *The Journal of Biological Chemistry*, 272, 7608–7616.
24. Knauper, V., Docherty, A.J., Smith, B., Tschesche, H. & Murphy, G. (1997b) Analysis of the contribution of the hinge region of human neutrophil collagenase (HNC, MMP-8) to stability and collagenolytic activity by alanine scanning mutagenesis. *FEBS Letters*, 405, 60–64.
25. Chung, L., Shimokawa, K., Dinakarpandian, D., Grams, F., Fields, G.B. & Nagase, H. (2000) Identification of the (183)RWTNNFREY(191) region as a critical segment of matrix metalloproteinase 1 for the expression of collagenolytic activity. *The Journal of Biological Chemistry*, 275, 29610–29617.
26. Brandstetter, H., Grams, F., Glitz, D. *et al.* (2001) The 1.8-A crystal structure of a matrix metalloproteinase 8-barbiturate inhibitor complex reveals a previously unobserved mechanism for collagenase substrate recognition. *The Journal of Biological Chemistry*, 276, 17405–17412.
27. Knauper, V., Patterson, M.L., Gomis-Ruth, F.X. *et al.* (2001) The role of exon 5 in fibroblast collagenase (MMP-1) substrate specificity and inhibitor selectivity. *European Journal of Biochemistry*, 268, 1888–1896.
28. Tam, E.M., Wu, Y.I., Butler, G.S., Stack, M.S. & Overall, C.M. (2002) Collagen binding properties of the membrane type-1 matrix metalloproteinase (MT1-MMP) hemopexin C domain. The ectodomain of the 44-kDa autocatalytic product of MT1-MMP inhibits cell invasion by disrupting native type I collagen cleavage. *The Journal of Biological Chemistry*, 277, 39005–39014.
29. Tsukada, H. & Pourmotabbed, T. (2002) Unexpected crucial role of residue 272 in substrate specificity of fibroblast collagenase. *The Journal of Biological Chemistry*, 277, 27378–27384.
30. Chung, L., Dinakarpandian, D., Yoshida, N. *et al.* (2004) Collagenase unwinds triple-helical collagen prior to peptide bond hydrolysis. *The EMBO Journal*, 23, 3020–3030.
31. Jozic, D., Bourenkov, G., Lim, N.H. *et al.* (2005) X-ray structure of human proMMP-1: new insights into procollagenase activation and collagen binding. *The Journal of Biological Chemistry*, 280, 9578–9585.
32. Pelman, G.R., Morrison, C.J. & Overall, C.M. (2005) Pivotal molecular determinants of peptidic and collagen triple helicase activities reside in the S3' subsite of matrix metalloproteinase 8 (MMP-8): the role of hydrogen bonding potential of ASN188 and TYR189 and the connecting cis bond. *The Journal of Biological Chemistry*, 280, 2370–2377.
33. Takeda, S., Igarashi, T. & Mori, H. (2007) Crystal structure of RVV-X: an example of evolutionary gain of specificity by ADAM proteinases. *FEBS Letters*, 581, 5859–5864.
34. Hall, T., Pegg, L.E., Pauley, A.M., Fischer, H.D., Tomasselli, A.G. & Zack, M.D. (2009) ADAM8 substrate specificity: influence of pH on pre-processing and proteoglycan degradation. *Archives of Biochemistry and Biophysics*, 491, 106–111.
35. Minond, D., Lauer-Fields, J.L., Nagase, H. & Fields, G.B. (2004) Matrix metalloproteinase triple-helical peptidase activities are differentially regulated by substrate stability. *Biochemistry*, 43, 11474–11481.
36. Minond, D., Lauer-Fields, J.L., Cudic, M. *et al.* (2006) The roles of substrate thermal stability and P2 and P1' subsite identity on matrix metalloproteinase triple-helical peptidase activity and collagen specificity. *The Journal of Biological Chemistry*, 281, 38302–38313.

37. Lauer-Fields, J.L., Minond, D., Sritharan, T., Kashiwagi, M., Nagase, H. & Fields, G.B. (2007c) Substrate conformation modulates aggrecanase (ADAMTS-4) affinity and sequence specificity. Suggestion of a common topological specificity for functionally diverse proteases. *The Journal of Biological Chemistry*, 282, 142–150.
38. Moss, M.L. & Rasmussen, F.H. (2007) Fluorescent substrates for the proteinases ADAM17, ADAM10, ADAM8, and ADAM12 useful for high-throughput inhibitor screening. *Analytical Biochemistry*, 366, 144–148.
39. Caescu, C.I., Jeschke, G.R. & Turk, B.E. (2009) Active-site determinants of substrate recognition by the metalloproteinases TACE and ADAM10. *The Biochemical Journal*, 424, 79–88.
40. Moss, M.L., Rasmussen, F.H., Nudelman, R., Dempsey, P.J. & Williams, J. (2010) Fluorescent substrates useful as high-throughput screening tools for ADAM9. *Combinatorial Chemistry & High Throughput Screening*, 13, 358–365.
41. Takakura-Yamamoto, R., Yamamoto, S., Fukuda, S. & Kurimoto, M. (1996) O-glycosylated species of natural human tumor-necrosis factor-alpha. *European Journal of Biochemistry*, 235, 431–437.
42. Bringman, T.S., Lindquist, P.B. & Derynck, R. (1987) Different transforming growth factor-alpha species are derived from a glycosylated and palmitoylated transmembrane precursor. *Cell*, 48, 429–440.
43. Wollscheid, B., Bausch-Fluck, D., Henderson, C. *et al.* (2009) Mass-spectrometric identification and relative quantification of N-linked cell surface glycoproteins. *Nature Biotechnology*, 27, 378–386.
44. Chalaris, A., Garbers, C., Rabe, B., Rose-John, S. & Scheller, J. (2011) The soluble Interleukin 6 receptor: generation and role in inflammation and cancer. *European Journal of Cell Biology*, 90, 484–494.
45. Mullberg, J., Oberthur, W., Lottspeich, F. *et al.* (1994) The soluble human IL-6 receptor. Mutational characterization of the proteolytic cleavage site. *The Journal of Immunology*, 152, 4958–4968.
46. Cole, A.R., Hall, N.E., Treutlein, H.R. *et al.* (1999) Disulfide bond structure and N-glycosylation sites of the extracellular domain of the human interleukin-6 receptor. *The Journal of Biological Chemistry*, 274, 7207–7215.
47. Huang, A., Joseph-McCarthy, D., Lovering, F. *et al.* (2007) Structure-based design of TACE selective inhibitors: manipulations in the S1'-S3' pocket. *Bioorganic and Medicinal Chemistry*, 15, 6170–6181.
48. Bahia, M.S. & Silakari, O. (2010) Tumor necrosis factor alpha converting enzyme: an encouraging target for various inflammatory disorders. *Chemical Biology and Drug Design*, 75, 415–443.
49. Healy, E.F., Romano, P., Mejia, M. & Lindfors, G. 3rd (2010) Acetylenic inhibitors of ADAM10 and ADAM17: in silico analysis of potency and selectivity. *Journal of Molecular Graphics and Modelling*, 29, 436–442.
50. Gooljarsingh, L.T., Lakdawala, A., Coppo, F. *et al.* (2008) Characterization of an exosite binding inhibitor of matrix metalloproteinase 13. *Protein Sciences*, 17, 66–71.
51. Martinez-Irujo, J.J., Villahermosa, M.L., Mercapide, J., Cabodevilla, J.F. & Santiago, E. (1998) Analysis of the combined effect of two linear inhibitors on a single enzyme. *The Biochemical Journal*, 329 (Pt 3), 689–698.
52. Engel, C.K., Pirard, B., Schimanski, S. *et al.* (2005) Structural basis for the highly selective inhibition of MMP-13. *Chemistry and Biology*, 12, 181–189.
53. Johnson, A.R., Pavlovsky, A.G., Ortwine, D.F. *et al.* (2007) Discovery and characterization of a novel inhibitor of matrix metalloprotease-13 that reduces cartilage damage in vivo without joint fibroplasia side effects. *The Journal of Biological Chemistry*, 282, 27781–27791.
54. Heim-Riether, A., Taylor, S.J., Liang, S. *et al.* (2009) Improving potency and selectivity of a new class of non-Zn-chelating MMP-13 inhibitors. *Bioorganic and Medicinal Chemistry Letters*, 19, 5321–5324.
55. Piecha, D., Weik, J., Kheil, H. *et al.* (2009) Novel selective MMP-13 inhibitors reduce collagen degradation in bovine articular and human osteoarthritis cartilage explants. *Inflammation Research*, 59, 379–389.
56. Pochetti, G., Montanari, R., Gege, C., Chevrier, C., Taveras, A.G. & Mazza, F. (2009) Extra binding region induced by non-zinc chelating inhibitors into the S1' subsite of matrix metalloproteinase 8 (MMP-8). *Journal of Medicinal Chemistry*, 52, 1040–1049.
57. Gao, D.A., Xiong, Z., Heim-Riether, A. *et al.* (2010) SAR studies of non-zinc-chelating MMP-13 inhibitors: improving selectivity and metabolic stability. *Bioorganic and Medicinal Chemistry Letters*, 20, 5039–5043.
58. Taylor, S.J., Abeywardane, A., Liang, S. *et al.* (2011) Fragment-based discovery of indole inhibitors of matrix metalloproteinase-13. *Journal of Medicinal Chemistry*, 54, 8174–8187.
59. Murphy, G. (2011) Tissue inhibitors of metalloproteinases. *Genome Biology*, 12, 233.
60. Bahudhanapati, H., Zhang, Y., Sidhu, S.S. & Brew, K. (2011) Phage display of tissue inhibitor of metalloproteinases-2 (TIMP-2): identification of selective inhibitors of collagenase-1 (metalloproteinase 1 (MMP-1)). *The Journal of Biological Chemistry*, 286, 31761–31770.

61. Olson, M.W., Bernardo, M.M., Pietila, M. *et al.* (2000) Characterization of the monomeric and dimeric forms of latent and active matrix metalloproteinase-9. Differential rates for activation by stromelysin 1. *The Journal of Biological Chemistry*, 275, 2661–2668.
62. Witty, J.P., Lempka, T., Coffey, R.J. Jr. & Matrisian, L.M. (1995) Decreased tumor formation in 7,12-dimethylbenzanthracene-treated stromelysin-1 transgenic mice is associated with alterations in mammary epithelial cell apoptosis. *Cancer Research*, 55, 1401–1406.
63. McCawley, L.J., Crawford, H.C., King, L.E. Jr., Mudgett, J. & Matrisian, L.M. (2004) A protective role for matrix metalloproteinase-3 in squamous cell carcinoma. *Cancer Research*, 64, 6965–6972.
64. Gomis-Ruth, F.X., Maskos, K., Betz, M. *et al.* (1997) Mechanism of inhibition of the human matrix metalloproteinase stromelysin-1 by TIMP-1. *Nature*, 389, 77–81.
65. Fernandez-Catalan, C., Bode, W., Huber, R. *et al.* (1998) Crystal structure of the complex formed by the membrane type 1-matrix metalloproteinase with the tissue inhibitor of metalloproteinases-2, the soluble progelatinase A receptor. *The EMBO Journal*, 17, 5238–5248.
66. Iyer, S., Wei, S., Brew, K. & Acharya, K.R. (2007) Crystal structure of the catalytic domain of matrix metalloproteinase-1 in complex with the inhibitory domain of tissue inhibitor of metalloproteinase-1. *The Journal of Biological Chemistry*, 282, 364–371.
67. Maskos, K., Lang, R., Tschesche, H. & Bode, W. (2007) Flexibility and variability of TIMP binding: X-ray structure of the complex between collagenase-3/MMP-13 and TIMP-2. *Journal of Molecular Biology*, 366, 1222–1231.
68. Grossman, M., Tworowski, D., Dym, O. *et al.* (2010) The intrinsic protein flexibility of endogenous protease inhibitor TIMP-1 controls its binding interface and affects its function. *Biochemistry*, 49, 6184–6192.
69. Moss, M.L., Bomar, M., Liu, Q. *et al.* (2007) The ADAM10 prodomain is a specific inhibitor of ADAM10 proteolytic activity and inhibits cellular shedding events. *The Journal of Biological Chemistry*, 282, 35712–35721.
70. Gonzales, P.E., Solomon, A., Miller, A.B., Leesnitzer, M.A., Sagi, I. & Milla, M.E. (2004) Inhibition of the tumor necrosis factor-alpha-converting enzyme by its pro domain. *The Journal of Biological Chemistry*, 279, 31638–31645.
71. Balbin, M., Fueyo, A., Tester, A.M. *et al.* (2003) Loss of collagenase-2 confers increased skin tumor susceptibility to male mice. *Nature Genetics*, 35, 252–257.
72. Szabova, L., Chrysovergis, K., Yamada, S.S. & Holmbeck, K. (2008) MT1-MMP is required for efficient tumor dissemination in experimental metastatic disease. *Oncogene*, 27, 3274–3281.
73. Palavalli, L.H., Prickett, T.D., Wunderlich, J.R. *et al.* (2009) Analysis of the matrix metalloproteinase family reveals that MMP8 is often mutated in melanoma. *Nature Genetics*, 41, 518–520.
74. Wolfe, M.S. (2012) Gamma-Secretase inhibitors and modulators for Alzheimer's disease. *Journal of Neurochemistry*, 120 (Suppl 1), 89–98.
75. Fahrenholz, F. (2007) Alpha-secretase as a therapeutic target. *Current Alzheimer Research*, 4, 412–417.
76. Endres, K. & Fahrenholz, F. (2012) Regulation of alpha-secretase ADAM10 expression and activity. *Experimental Brain Research*, 217, 343–352.
77. Postina, R. (2012) Activation of alpha-secretase cleavage. *Journal of Neurochemistry*, 120 (Suppl 1), 46–54.

5 Therapeutics Targeting Matrix Metalloproteinases

Jillian Cathcart[1], Ashleigh Pulkoski-Gross[1], Stanley Zucker[2], and Jian Cao[1]

[1] *Department of Medicine, Stony Brook University, Stony Brook, USA*
[2] *VA Medical Center, Northport, USA*

5.1 Introduction

Matrix metalloproteinases (MMPs) are directly implicated in almost every biological process involving remodeling of the extracellular matrix, including basement membrane, from embryo implantation to tissue necrosis. As a consequence, altered expression and activity levels of MMPs have also been implicated in a diverse set of pathologies [1]. For example, MMPs (especially MMP-2 and MMP-9) have been shown to be overexpressed in a wide range of malignant tumors and this overexpression is directly correlated with tumor aggressiveness, stage, and prognosis [2]. Overexpression of MMP-3 (stromelysin) has been implicated in arthritis [3]. MMP-12 has been observed to be overexpressed in patients with pulmonary disorders, including emphysema and chronic obstructive pulmonary disorder (COPD) [4]. The list of MMPs in pathology is as diverse as it is long, and will likely continue to grow as our understanding of disease progression advances. Because of this, efforts are being taken to identify or design inhibitors targeting MMPs.

Despite their significance in pathology, to date only one MMP inhibitor (MMPI), doxycycline, has been approved by the Food and Drug Administration (FDA) for adult periodontitis and rosacea. MMPs are in fact quite pleiotropic, and due to being expressed in all tissue types and having significant structural similarities development of efficacious inhibitors which do not cause significant adverse events has proved challenging [5]. To complicate things further, MMPs are not only highly similar within the family but also share functional and structural similarities with proteases belonging to the family of a disintegrin and metalloproteinase (ADAM) with thrombospondin motif (ADAM-TS). *In vitro* tests have demonstrated that many broad-spectrum MMPIs will also bind to and inhibit ADAMs.

The majority of MMPIs tested clinically bind within the catalytic site and/or the nearby S1′ specificity pocket. The catalytic sites are largely conserved and are similar amongst MMPs and ADAMs; however, the S1′ pocket is a hydrophobic cavity unique to each of the MMPs and exhibits considerable variation in dimensions and

Matrix Metalloproteinase Biology, First Edition. Edited by Irit Sagi and Jean P. Gaffney.

residues lining the pocket. The S1′ sites also vary by depth; they are considered shallow (MMP-1 and MMP-7), intermediate (MMP-2, MMP-8, and MMP-9), or deep (MMP-3, MMP-11, MMP-12, MMP-13, and MMP-14) [6, 7]. Careful design and thorough assessment of a drug's specificity is of utmost importance in order to achieve high potency and efficacy while minimizing off-target effects.

The means for development of therapeutics is evolving quickly as biotechnology advances, due largely to improvements in methods for drug screening and specific drug design. In the mid-1990s, the design of the first peptide-based potent MMPIs ushered in an optimistic (albeit brief) era in which these enzymes and their potential inhibitors received intense scrutiny. As discussed later, lessons learned from these clinical trials have underscored the significance of the natural role of MMPs and the importance of specificity.

5.2 Peptidomimetic MMP inhibitors

The term 'peptidomimetic' encompasses any pseudopeptide derivative. In the context of peptidomimetic inhibitors, these compounds bind in an enzymatic catalytic site as the natural substrate normally would. The compounds are derived from the amino acids of the protease's endogenous ligand and are based on the sequences at the substrate cleavage site. For example, MMP-1 cleaves interstitial collagen (types I, II, and III) at a specific site approximately three-fourths of the length of the molecule from the N terminus. The triple helical fragments resulting from this cleavage will denature within the body and become prone to nonspecific proteolysis. This normal activity can be manipulated for therapeutic use with drugs that mimic endogenous substrates by binding to the catalytic site to inhibit enzyme function.

Often the cleavage sites of a peptidomimetic are optimized via residue or backbone substitutions to improve binding of the drug [5, 8]. Features of most peptidomimetic MMPIs, which bind reversibly in a stereospecific manner, include (i) chelation of the catalytic Zn^{2+}, (ii) occupancy of subsites along the extended binding site (S, S′), and (iii) quintessential occupancy of the S1′ site [9]. Generally, the scissile bond around which the peptidomimetic is built is chemically modified in a manner that prevents hydrolysis of the substrate, thereby rendering the enzyme inactive as long as the compound remains bound [10]. The prototype for substrate-based compounds targeting collagenases (MMP-1, MMP-8, and MMP-13) was built around the glycine-isoleucine and glycine-leucine cleavage sites in collagen [11]. This peptidomimetic, and the generation following, chelated the catalytic Zn^{2+} ion via a hydroxamic acid functional group. The compound binds as an anion and forms bidentate interactions with the target molecule by acting as a Lewis base to coordinate the Zn^{2+} ion. Other important contacts include formation of a hydrogen bond with the essential glutamic acid adjacent to the first histidine, which coordinates the Zn^{2+} ion in the unbound enzyme. An amino group on the inhibitor also forms hydrogen bonds with the backbone carbonyl of a proximate alanine. The global attraction between the MMP and the hydroxamate-containing inhibitor is a complex synergy lacking unfavorable interactions, a prospect that tantalized the pharmaceutical industry for nearly a decade [12].

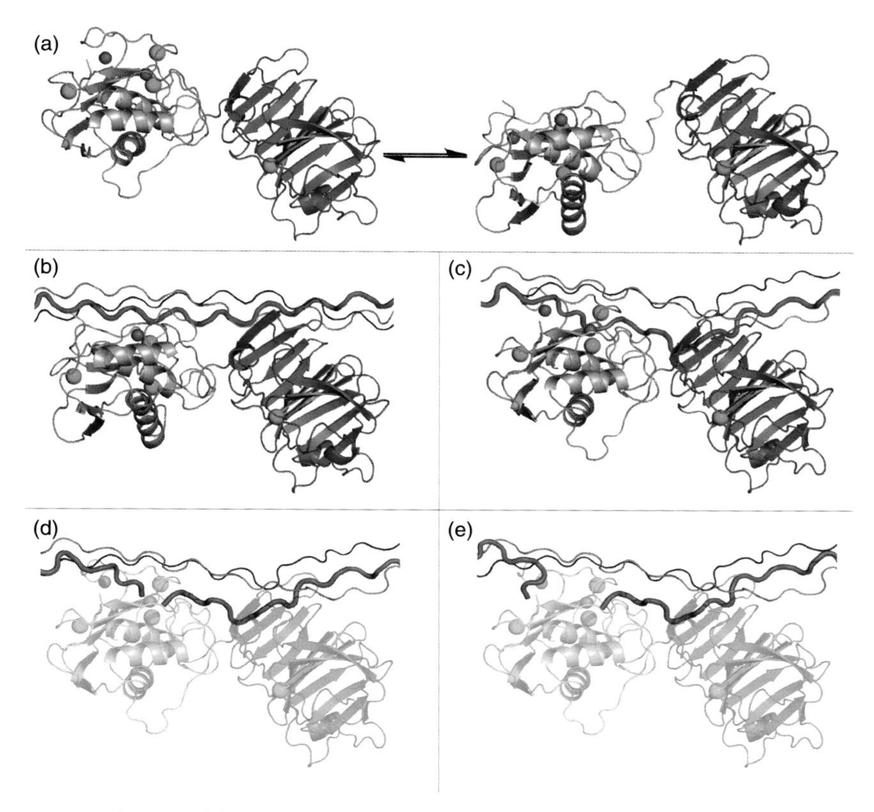

Figure 1.8 Mechanism of the initial steps of collagenolysis. (a) Closed (left) and open/extended (right) forms of MMP-1 in equilibrium. (b) The extended protein binds THP chains 1T-2T at Val23-Leu26 with the HPX domain and the residues around the cleavage site with the CAT domain. The THP is still in a compact conformation. (c) Closed FL-MMP-1 interacting with the released 1T chain (in magenta). (d) After hydrolysis, both peptide fragments (C- and N-terminal) are initially bound to the active site. (e) The C-terminal region of the N-terminal peptide fragment is released. (Reprinted with permission from [16]. Copyright (2012) American Chemical Society).

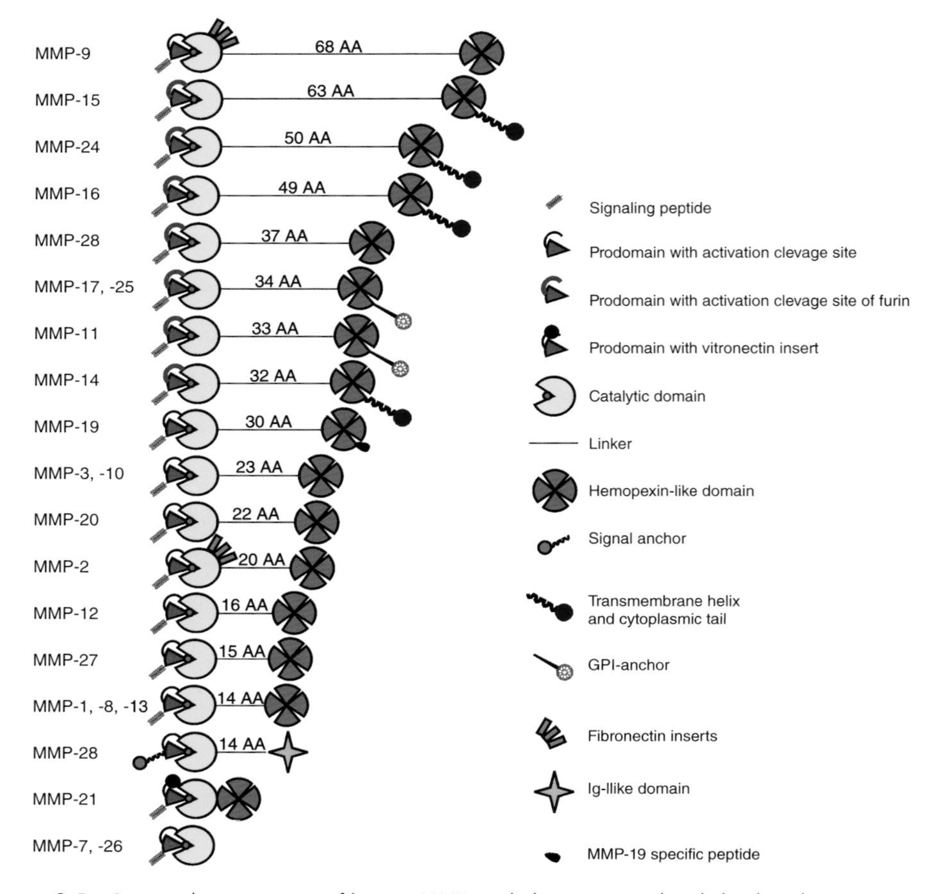

Figure 3.1 Structural organization of human MMPs with the corresponding linker length.

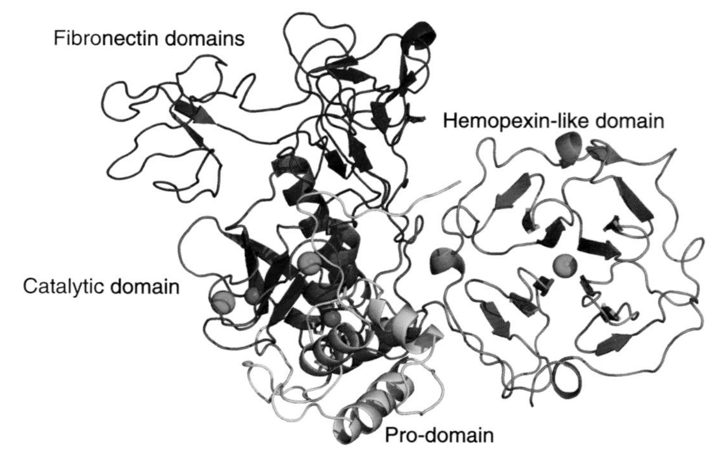

Figure 3.2 Ribbon representation of the inactive human proMMP-2. The prodomain, catalytic domain, fibronectin domains, and hemopexin domain are shown in yellow, red, blue, and orange, respectively. The catalytic and the structural zinc ions are represented as magenta spheres and calcium ions as green spheres.

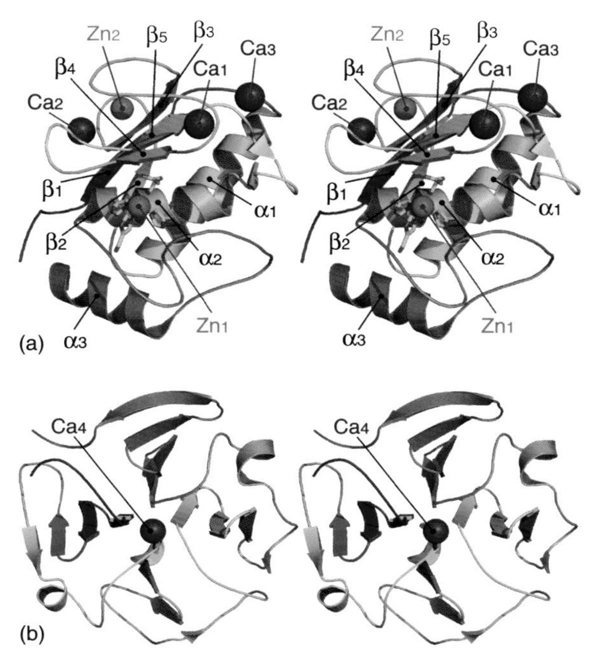

Figure 3.3 Stereo view of the catalytic (a) and hemopexin-like (b) domains of MMP-12 represented as ribbons. In the catalytic domain α-helices, β-strands, and loops are organized in a L1-β1-L2-α1-L3-β2-L4-β3-L5-β4-L6-β5-L7-α2-L8-α3 topology. The catalytic (Zn1) and the structural (Zn2) zinc ions are shown as magenta spheres of arbitrary radius. The first (Ca1), the second (Ca2), the third (Ca3) calcium ions and the calcium ion in the hemopexin-like domain are shown as blue spheres. The three histidines that bind the catalytic zinc and the catalytically relevant glutamate are represented as cyan sticks. Strands and helices are labeled with numbers and greek letters. The hemopexin-like domains is constituted by four β-sheets of four antiparallel β-strands that folds in a symmetric four-blade propeller [53, 67]. The central deep tunnel filled by water molecules is closed by a calcium ion (Ca4) at the bottom.

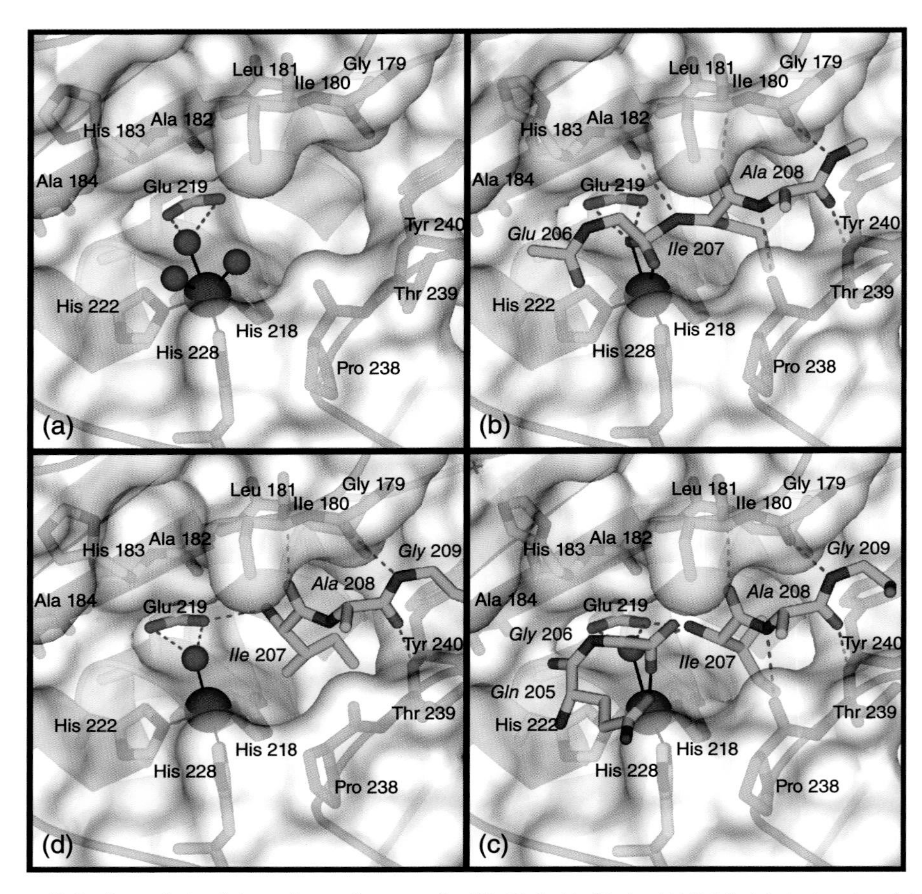

Figure 3.4 Proteolysis of the collagen fragment ProGlnGlyIleAlaGly by MMP-12. (a) Active site of the free enzyme before the interaction with the substrate. (b) Calculated model of the gemdiol intermediate. (c) X-ray structure of the two-peptide intermediate obtained by soaking the active uninhibited MMP-12 crystals with the collagen peptide. (d) Adduct of MMP-12 with the peptide fragment IleAlaGly after the release of the C-terminal fragment.

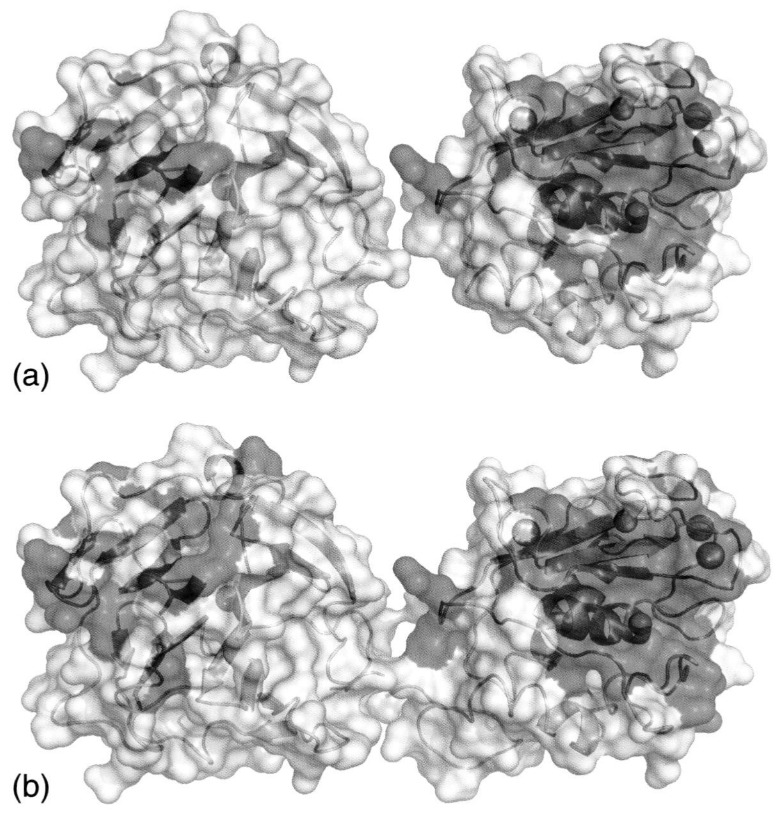

Figure 3.5 Pattern of residues interacting with elastin fragments in the isolated catalytic and hemopexin-like domains (a) and in the full length protein (b). The larger effects observed in the full length protein suggest cooperativity of the two domains in binding of elastin fragments.

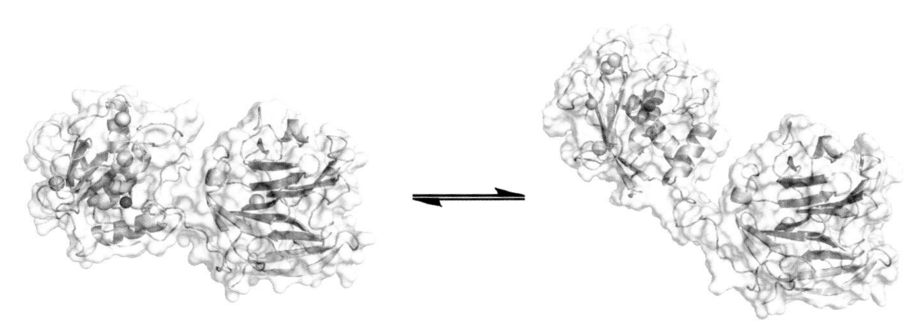

Figure 3.6 Closed (left) and open/extended (right) forms of FL-MMP-1 in equilibrium. The catalytic zinc ion is represented as a magenta sphere.

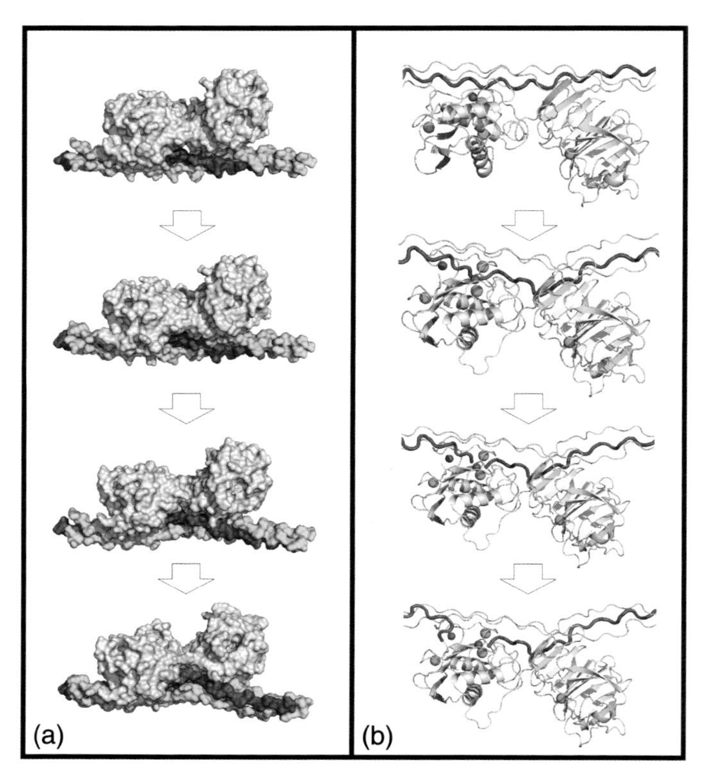

Figure 3.7 Proposed mechanism for collagenolysis. In panel (a), from the top (the experimentally-driven docked complex between FL-MMP-1 and THP) to the bottom (the unwounded THP bound to the X-ray closed conformation of FL-MMP-1) the intermediate and energetically possible structures generated by HADDOCK [112] to provide a smooth conformational transition between the initial and final states. In panel (b), starting from the experimentally-driven docked complex between FL-MMP-1 and THP (top), the closed FL-MMP-1 interacting with the released 1T chain (in red), the hydrolysis of the 1T chain with both peptide fragments still in place, and the complex with the C-terminal region of the N-terminal peptide released from the active site (bottom).

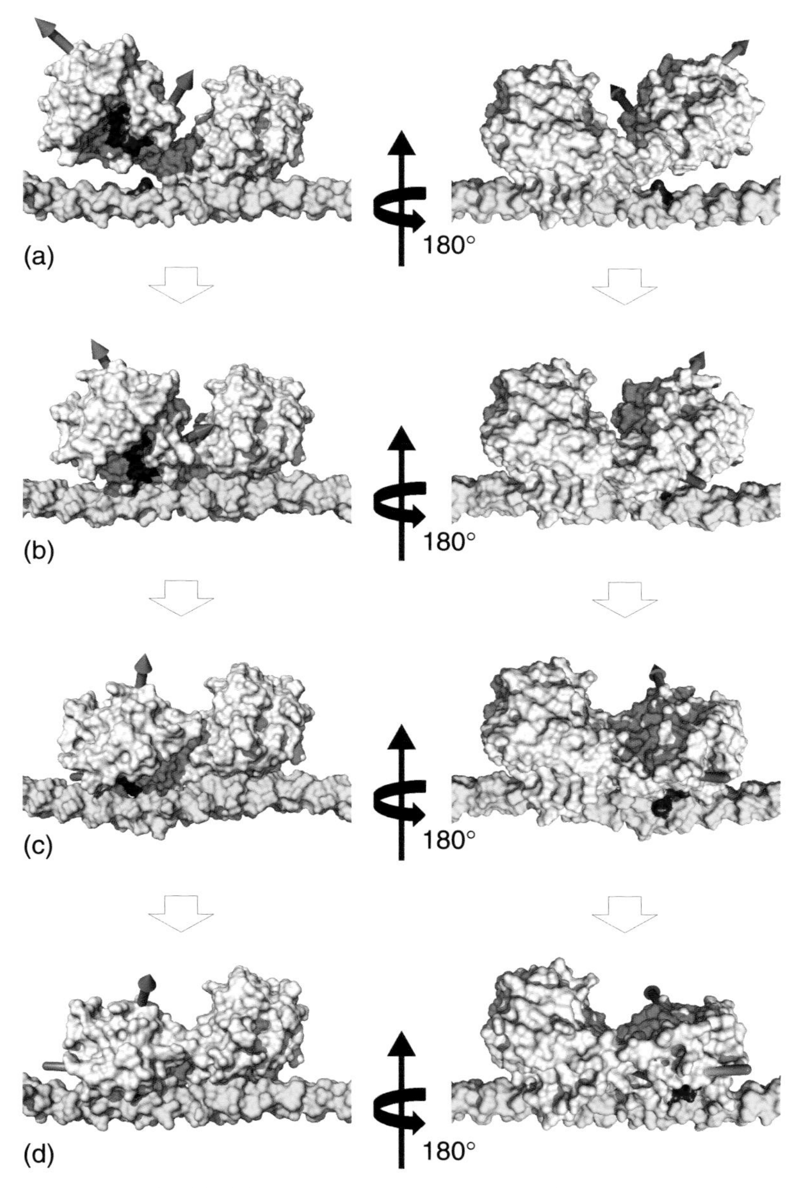

Figure 3.8 Interaction of FL-MMP-1 with the substrate. In the panel, from the top to the bottom: (a) structure with the highest MO, (b–c) two morphing intermediate steps, (d)the experimentally-driven docked complex where the hemopexin-like domain and the catalytic domain bind the triple-helical collagen. The structure with the highest MO e morphing structures were aligned to the hemopexin-like domain of the docked complex. FL-MMP-1 and THP are represented as white and yellow surfaces, respectively. In blue is the MMP consensus sequence HE*XX*H*XX*G*XX*H and the cleavage site (*Gly-Ile*) in the first chain of THP. The catalytic zinc ion is represented as an orange sphere. To facilitate visualizing the movement of the catalytic domain with respect to the hemopexin-like domain, the blue and red arrows indicate the direction of helices hA and hC of the catalytic domain defined by residues 130–141 and 250–258, respectively.

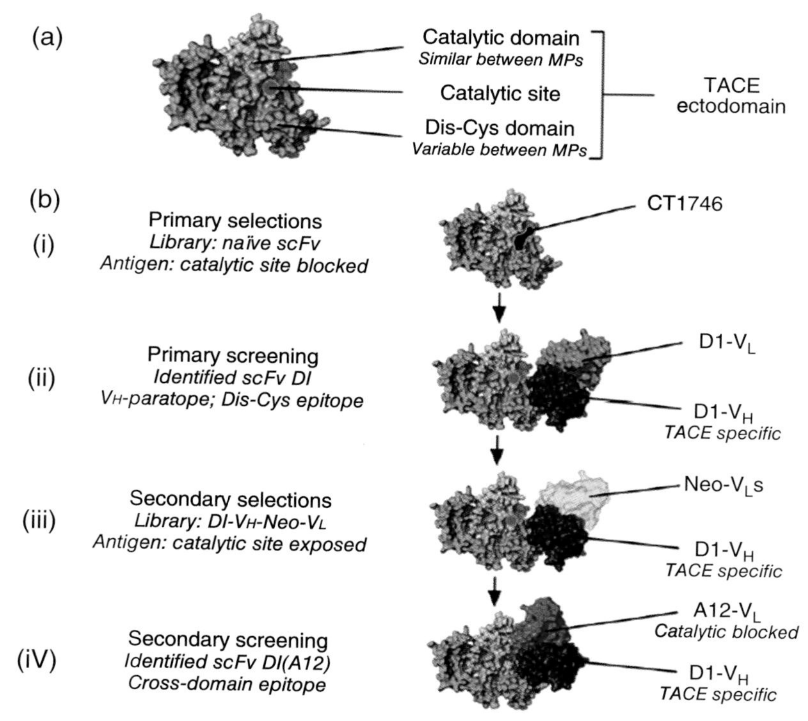

Figure 4.3 Experimental overview. (a) The human TACE ectodomain consists of an amino-terminal metalloprotease catalytic domain (light red) and a carboxyl-terminal noncatalytic Dis-Cys domain (light blue) (I-TASSER model). We exploited this multidomain topology to develop a truly specific ADAM inhibitor using two-step antibody phage display. (b) (i) First, the catalytic site of TACE ectodomain was blocked during primary antibody phage-display selections using the small-molecule inhibitor CT1746. This prevented the selection of antibodies with catalytic-cleft epitopes that could cross-react with non-target metalloproteases. (ii) Primary screening revealed the inhibitory scFv antibody clone D1. This scFv bound specifically to the TACE Dis-Cys domain through its variable heavy (V_H) domain. (iii) A D1-V_H-bias antibody phage display library was produced to introduce new variable light (neo-V_L) chains while maintaining the TACE specificity provided by the D1-V_H. Secondary selections were performed in the absence of CT1746 in order to provide the neo-V_L chains with uninterrupted access to the TACE catalytic site. (iv) Secondary screening identified several neo-VL scFvs capable of binding the isolated TACE catalytic domain. Due to Dis-Cys domain binding through the D1-V_H these "cross-domain" antibodies maintained their strict specificity for TACE. D1-V_H-neo-V_L scFv clone A12 (D1(A12)) exhibited the highest affinity for the TACE ectodomain and is the most selectively potent cell-surface ADAM inhibitor ever described. (Reproduced with permission from Tape, C. J., Willems, S. H., Dombernowsky, S. L., Stanley, P. L., Fogarasi, M., Ouwehand, W., McCafferty, J., and Murphy, G. (2011) Cross-domain inhibition of TACE ectodomain *Proc Natl Acad Sci* U S A 108, 5578–5583).

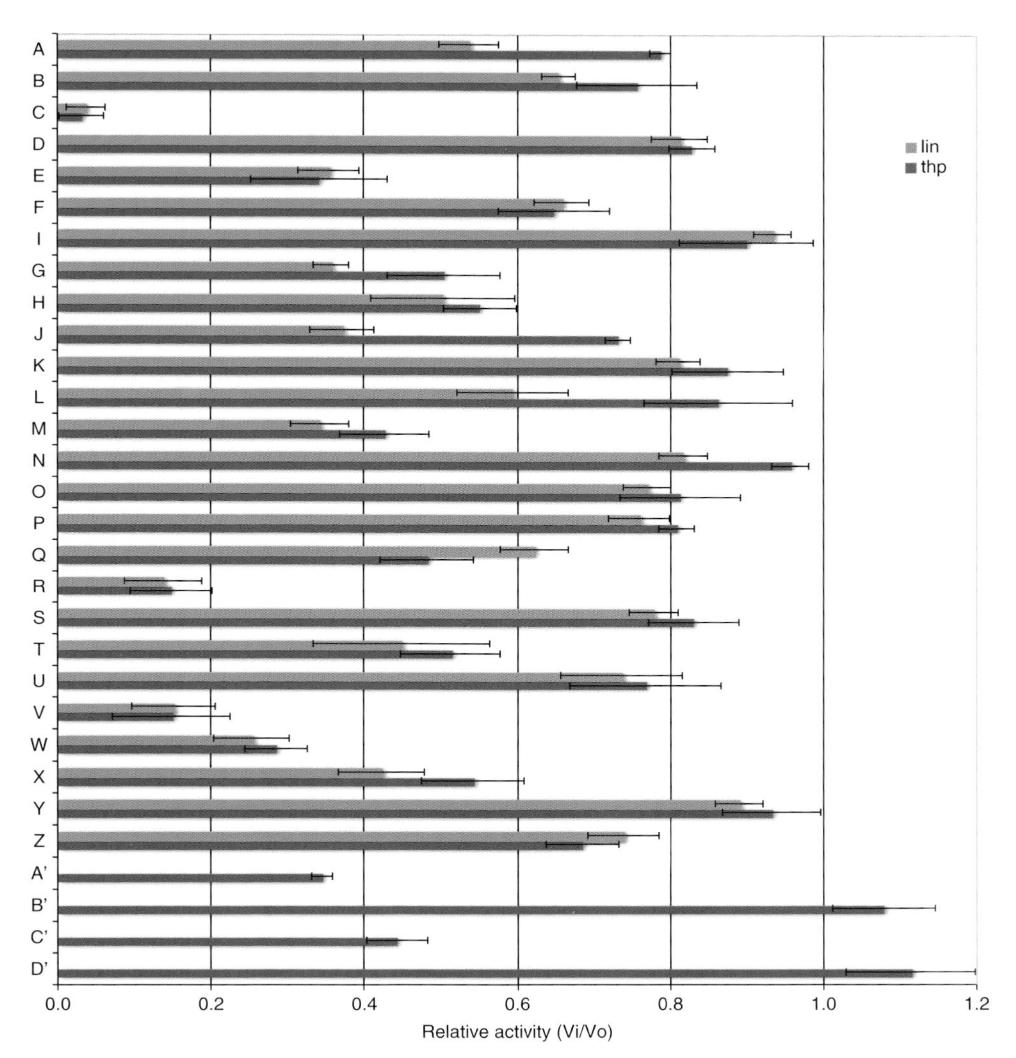

Figure 4.8 Inhibition of MMP-13 by 30 different compounds, as monitored by RP-HPLC and fluorescence spectroscopy. The change in RP-HPLC peak areas or relative fluorescence units for 10 nM MMP-13 hydrolysis of 10 μM fTHP-15 or 5 μM Knight fSSP was monitored at an inhibitor concentration of 100 μM. Assays were performed in triplicate. (Reproduced with permission from Lauer-Fields, J. L., Minond, D., Chase, P. S., Baillargeon, P. E., Saldanha, S. A., Stawikowska, R., Hodder, P., and Fields, G. B. (2009) High throughput screening of potentially selective MMP-13 exosite inhibitors utilizing a triple-helical FRET substrate *Bioorg Med Chem* 17, 990–1005. ©PERGAMON).

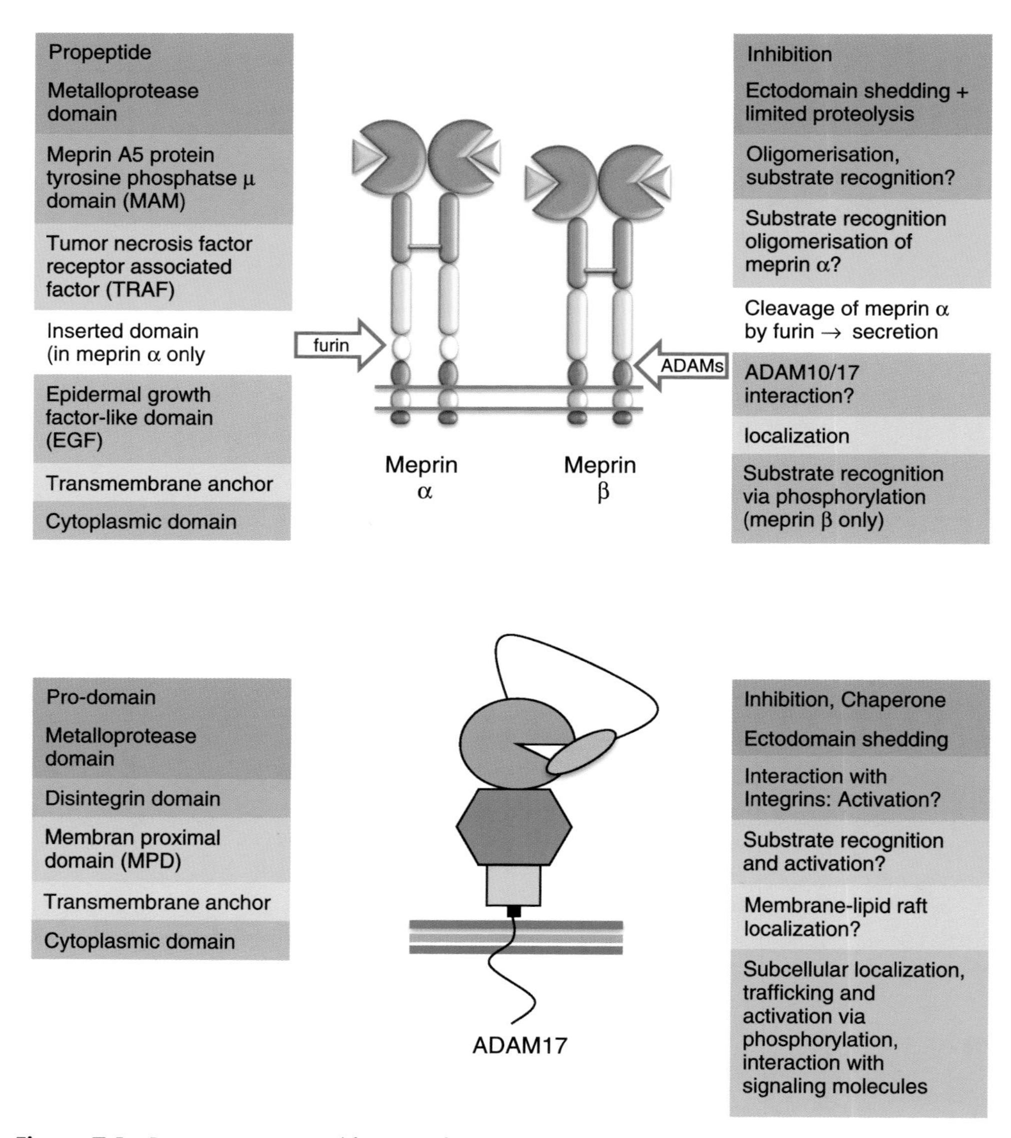

Figure 7.1 Domain structure and function of meprin α, meprin β and ADAM17. The functions of the domains are indicated in the figure.

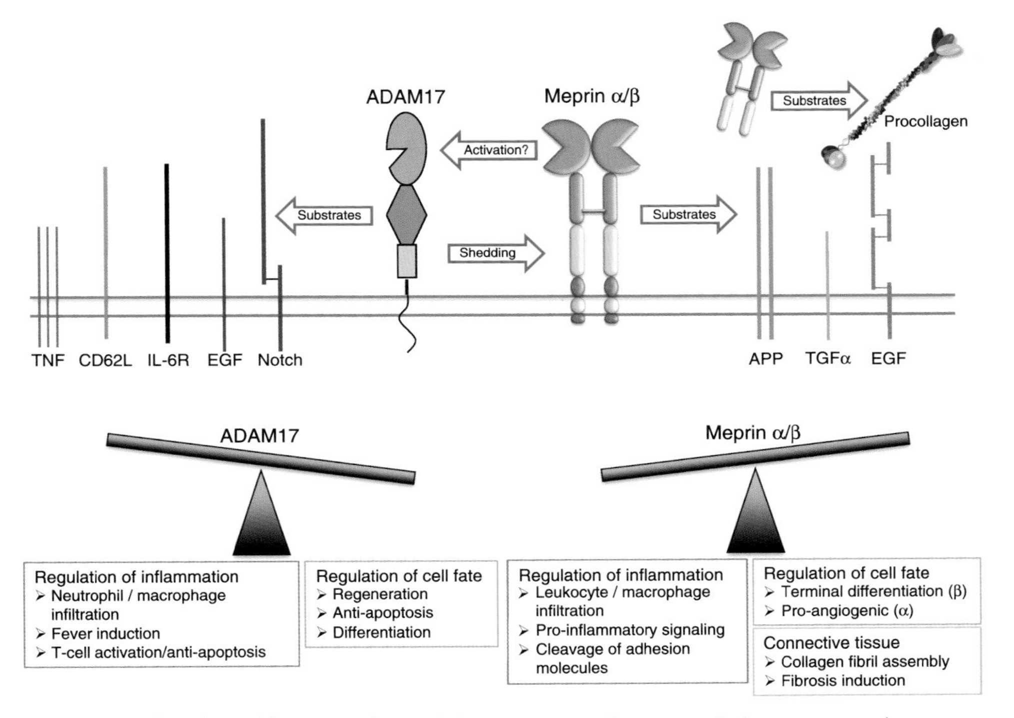

Figure 7.2 Physiological functions of ADAM17, meprin α, and meprin β. Both proteases orchestrate different processes in development and during the activation of the immune system.

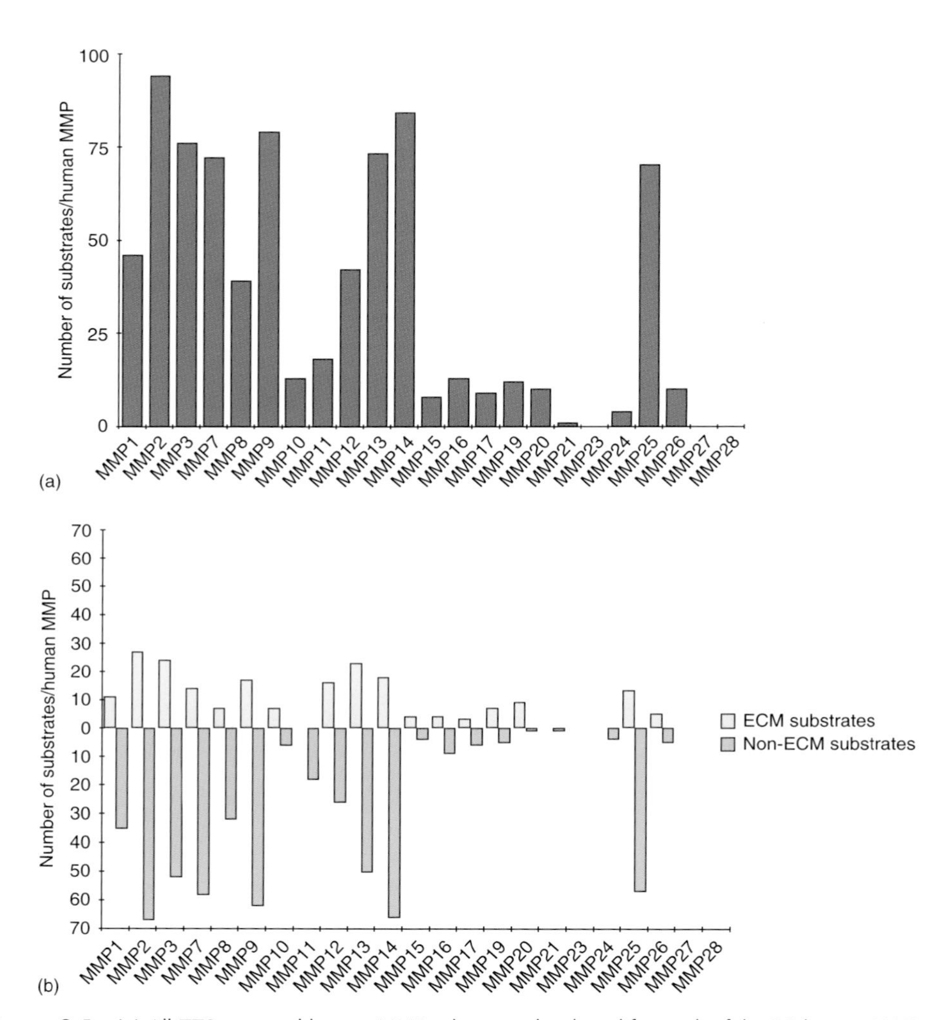

Figure 8.1 (a) All 773 reported human MMP substrates distributed for each of the 23 human MMPs. (b) All 773 reported human MMP substrates: the ECM substrates are shown in blue and the non-ECM substrates are shown in green.

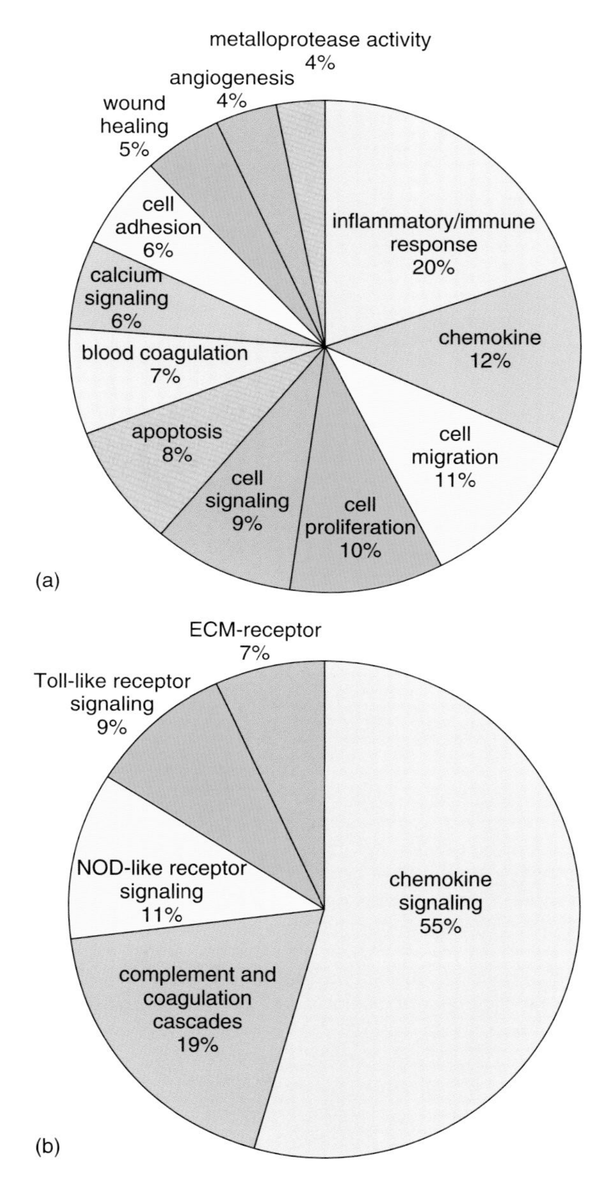

Figure 8.2 (a) Gene Ontology (GO) terms enrichment of all 246 reported non-ECM human MMP substrates. (b) Pathway enrichment analysis of the 246 reported non-ECM human MMP substrates.

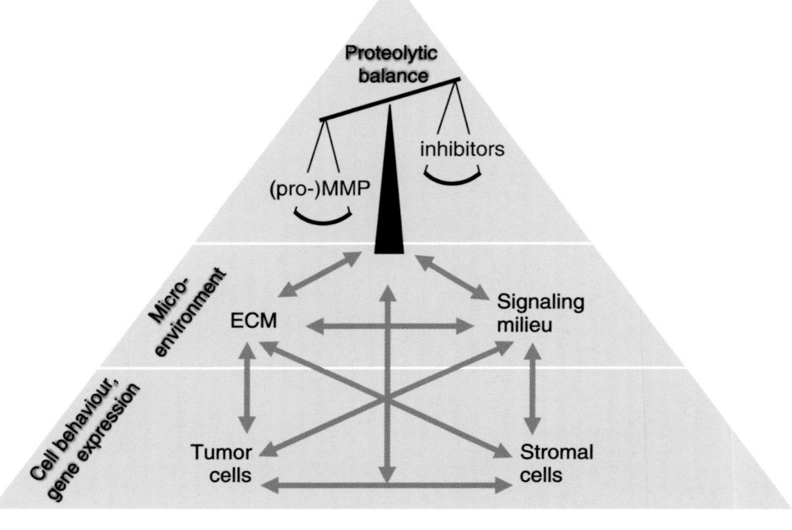

Figure 9.1 Biological activity of an individual MMP within a local tumor microenvironment. MMPs are central regulators of tumor extracellular environment in terms of both extracellular matrix (ECM) turnover and the signaling milieu controlling cell function. Proteolytic balance is tightly controlled at the protein level by activation of individual MMPs from inactive zymogens (proMMPs) and by the binding of inhibitors. Upon activation, each MMP mediates specific effects on the local microenvironment, dependent on its substrate repertoire. These effects derive either down-stream of the MMPs individual ECM substrates or via activation and/or inactivation of signaling molecules, such as cytokines and growth factors. In consequence, the proteolytic balance influences gene expression and behavior of cancer as well as stromal cells, which in turn are major determinants of the proteolytic balance, the local ECM composition and signaling milieu.

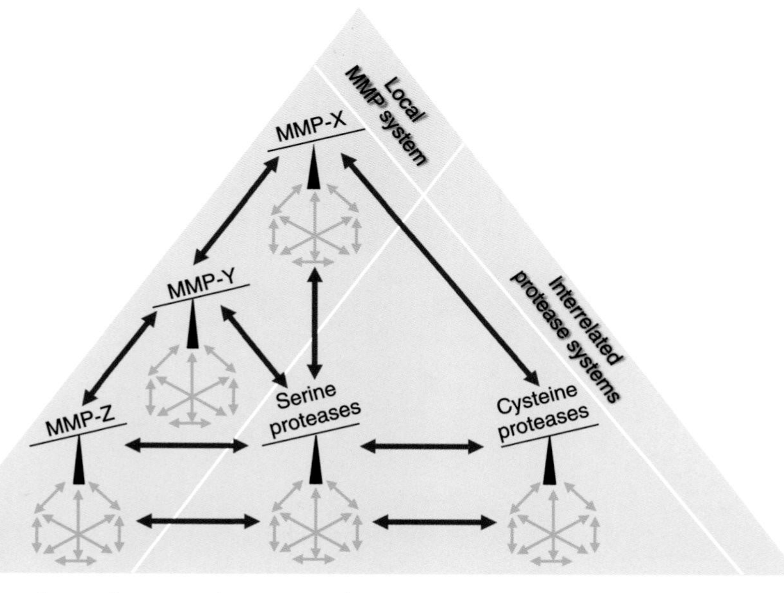

Figure 9.2 Local proteolytic network consisting of interrelated protease systems. The biological effects of an individual MMP (as depicted in Fig. 9.1.) are embedded in the interaction with other MMPs that exhibit partially over-lapping but also distinct substrate specificities, forming the local MMP system. Individual proteolytic systems are interrelated, mutually influencing each other in the modulation of protease activity, substrate availability, and action within a tissue.

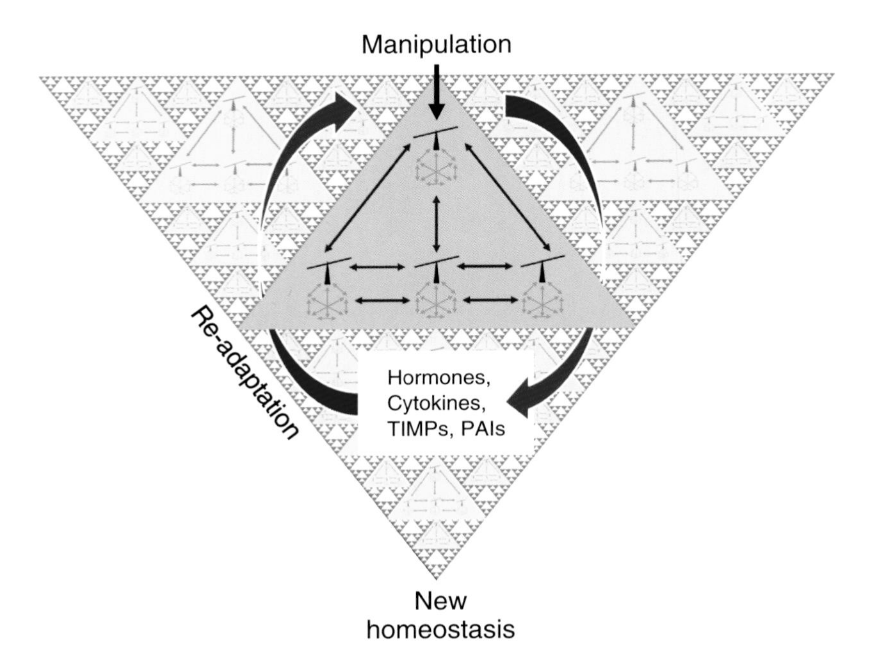

Figure 9.3 Interconnectivity of local proteolytic networks within an organism. Local proteolytic tissue networks (as depicted in Fig. 9.2.) within an organism communicate with each other over a distance via the circulatory system, forming the proteolytic internet. Information is transmitted systemically via up-regulation or down-regulation of soluble factors such as cytokines, hormones, as well as secreted protease inhibitors such as TIMP-1 and PAI-1. The status of homeostasis in the regional proteolytic network of an organ is thereby reported to other tissues in the body. Accordingly, any manipulation of a single member of the proteolytic network results in a re-adaptation. This process is subject to a multitude of net effects that altogether impact on the formation of a new homeostasis, which determines the susceptibility of the organism to disease.

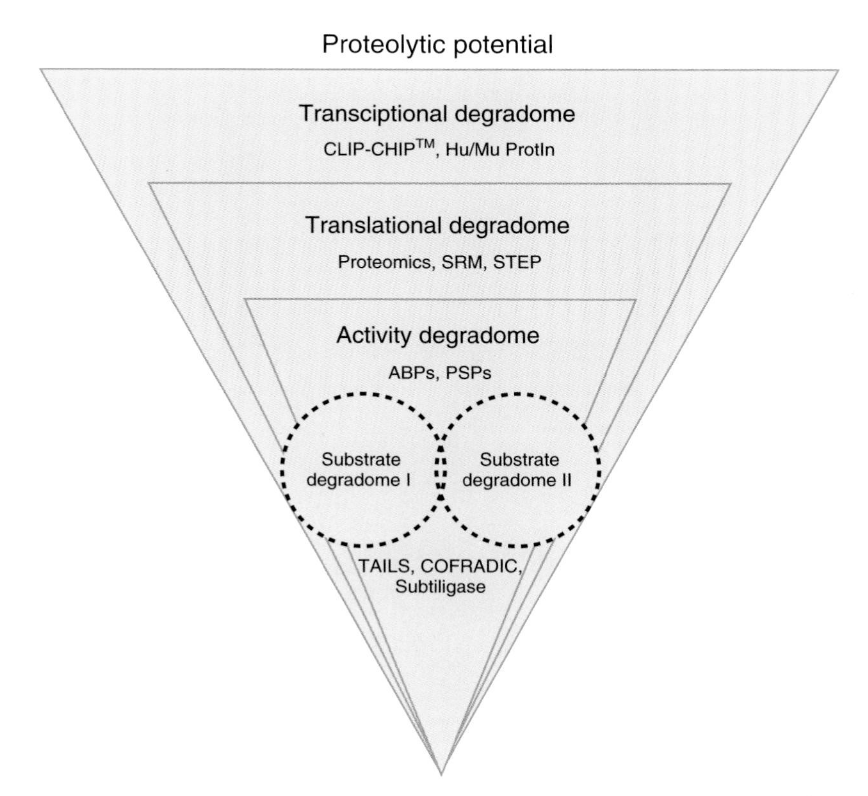

Figure 9.4 Degradomes and degradomics approaches. The transcriptional degradome defines the translational degradome, of which the activity degradome represents the active proteases. Individual active proteases give rise to partial overlapping substrate degradomes. All together, they define the proteolytic potential of a system. For each level of complexity powerful degradomics techniques have been developed. CLIP-CHIP™, Hu/Mu ProtIn, dedicated protease microarrays; SRM, selected reaction monitoring; STEP, STandard of Expressed Protein peptides; ABPs, activity-based probes; PSPs, proteolytic signature peptides; TAILS, terminal amine isotopic labeling of substrates; COFRADIC, combined diagonal fractional chromatography; Subtiligase, engineered peptide ligase for modification of protein N termini.

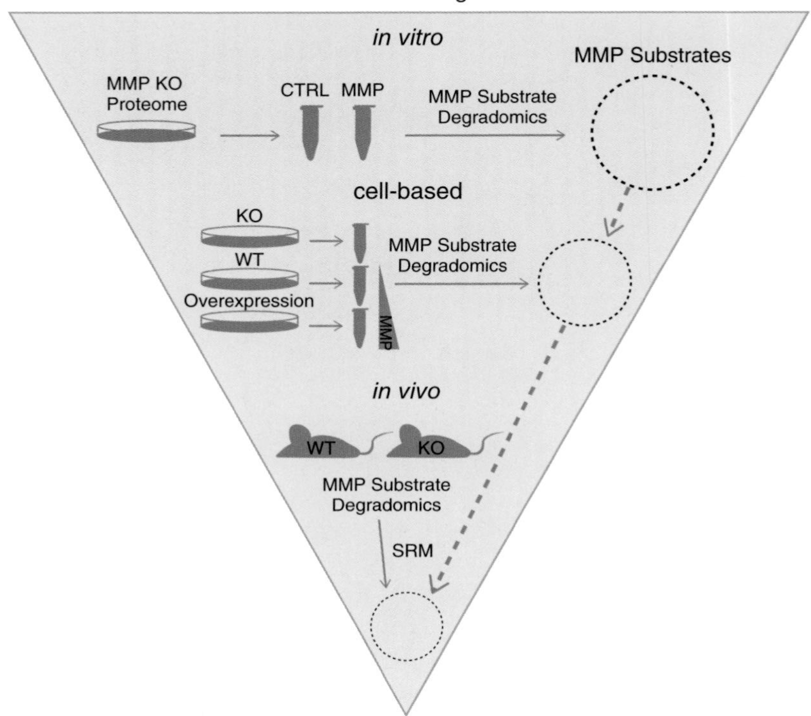

Figure 9.5 Integrated strategy to elucidate physiological MMP substrates. Multiple candidate substrates from unbiased *in vitro* and cell-based experiments serve as templates for the development of targeted SRM assays that are applied in appropriate *in vivo* models. KO, knockout; WT, wild-type; SRM, selected reaction monitoring.

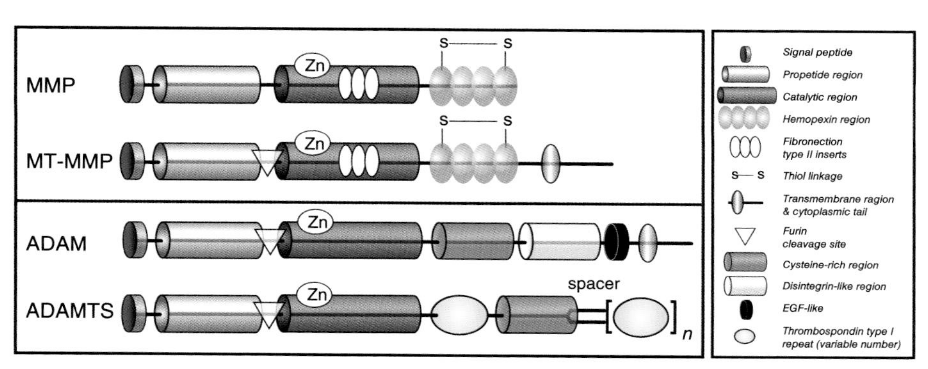

Figure 10.1 Basic domain structure of MMP and ADAM family members. The characteristic domain structure of MMPs includes (i) the signal peptide domain, which guides the enzyme into the rough endoplasmic reticulum during synthesis, (ii) the propeptide domain, which sustains the latency of these enzymes until it is removed or disrupted, (iii) the catalytic domain, which houses the highly conserved Zn^{2+} binding region and is responsible for enzyme activity, (iv) the hemopexin domain, which determines the substrate specificity of MMPs, and (v) a small hinge region, which enables the hemopexin region to present substrate to the active core of the catalytic domain. The subfamily of membrane-type MMPs (MT-MMPs) possesses an additional transmembrane domain and an intracellular domain. MMPs are produced in a latent form and most are activated by extracellular proteolytic cleavage of the propeptide. MT-MMPs also contain a cleavage site for furin proteases, providing the basis for furin-dependent activation of latent MT-MMPs prior to secretion. ADAMs are multidomain proteins composed of propeptide, metalloprotease, disintegrin-like, cysteine-rich, and epidermal growth factor-like domains. Membrane-anchored ADAMs contain a transmembrane and cytoplasmic domain. ADAMTSs have at least one Thrombospondin type I Sequence Repeat (TSR) motif [1]. (Reprinted with permission ©(2009) American Society of Clinical Oncology. All rights reserved).

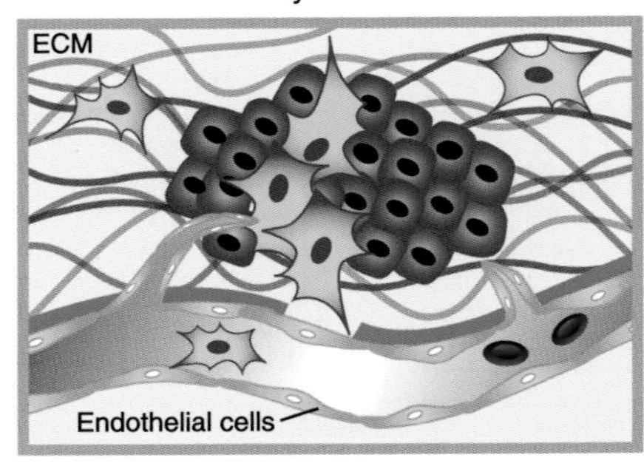

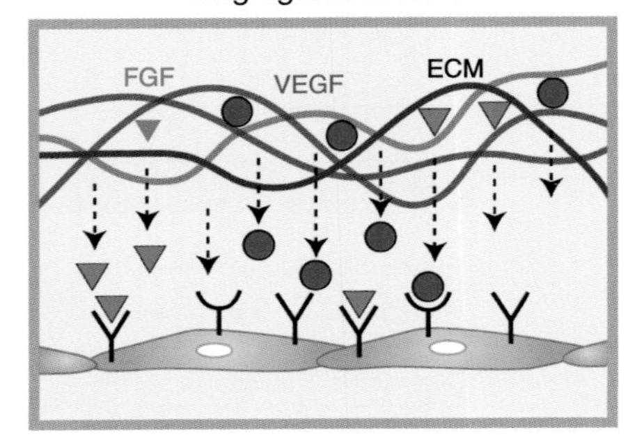

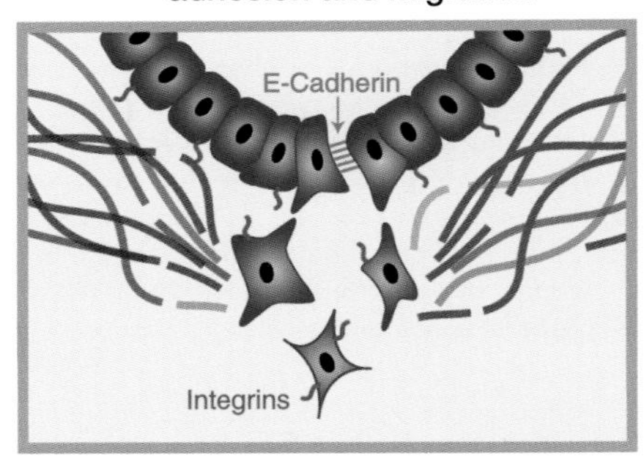

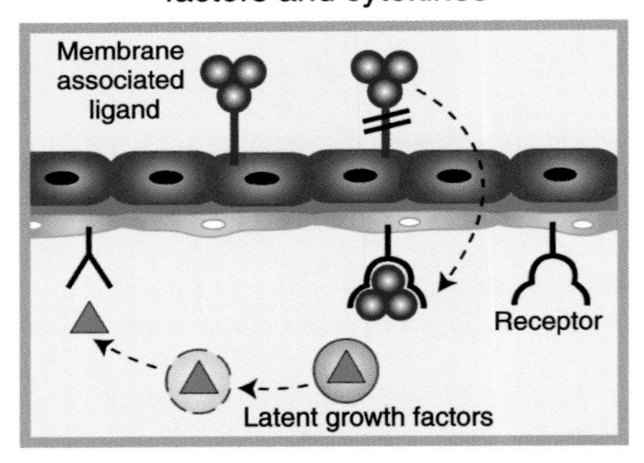

Figure 10.2 Multiple functions of MMPs in cancer progression. (Counterclockwise) MMPs degrade components of ECM, facilitating angiogenesis, tumor cell invasion and metastasis. MMPs modulate the interactions between tumor cells by cleaving E-cadherin, and between tumor cells and ECM by processing integrins, which also enhances the invasiveness of tumor cells. MMPs also process and activate signaling molecules, including growth factors and cytokines, making these factors more accessible to target cells by either liberating them from the ECM (e.g., VEGF and bFGF) and inhibitory complexes (e.g., TGF-β), or by shedding them from cell surface (e.g., HB-EGF) [1]. (Reprinted with permission ©(2009) American Society of Clinical Oncology. All rights reserved).

The first MMPI to reach clinical trials was batimastat, a broad-spectrum, competitive peptidomimetic administered intraperitoneally or intrapleurally to cancer patients [8]. Because the natural role for MMPs is to remodel the basement membrane and extracellular environment, in pathological processes such as tumor cell migration, angiogenesis, and invasion, inhibiting these processes was believed to be critical for combating metastasis [13]. Batimastat, for which phase I clinical trial results were made available in 1994, is a hydroxamate-containing substituted analog of the residues on the domain adjacent to a principle cleavage site in type I collagen [14]. Although batimastat was highly potent against MMPs-1, MMP-2, MMP-3, MMP-7, MMP-9, and MMP-14 and was efficacious during clinical trials targeting malignant ascites and malignant effusion, development was ultimately halted as it is highly insoluble and exhibits very low bioavailability when administered orally; therefore, the drug had very limited potential for use in most clinical settings [15]. Chemical modification of batimastat led to the development of its analog, marimastat, which retains the potency of its predecessor but demonstrates excellent oral bioavailability [16]. While marimastat gave some evidence of efficacy in delaying disease progression in Phase II clinical trials, significance could not be established in randomized Phase III clinical trials. This was compounded by dose-limiting toxicity related to frequent musculoskeletal pain and inflammation. [17].

In hindsight, it is now understood that the pre-clinical animal tests for these peptidomimetic drugs were initiated at early stages of tumor growth whereas the human studies included patients at all stages of disease progression. During these preclinical animal studies, batimastat reduced angiogenic islands during a prevention study by 49% and successfully reduced tumor burden by 83% during an intervention trial. However, it had no effect on regression of large tumors or invasive carcinomas [18]. Taken together, this suggests that the clinical trials may not have been appropriately focused. Further, the fact that only three MMPs had been identified (and were poorly characterized) when the first MMPI entered clinical trials also contributed to the disappointing human trials, as physiological knowledge of MMP functions and expression patterns (in terms of substrate preference and spatial/temporal expression) was severely limited [19]. For MMPIs to be efficacious in targeting cancer while avoiding severe off- target effects, future clinical studies must evaluate the patient's stage of cancer prior to drug administration, especially if they are to be used as part of the first line of defense [20].

5.3 Structure-based MMPI drug design

Structure-based drug design follows a practical paradigm: the target binding site is first identified and characterized, a library of compounds is "docked" into this binding site, and the predicted ability of the compounds binding is "scored". The top scoring "hits" are ranked and then additional molecular simulations and biochemical assays are performed to further narrow down the list. The highest performing hits are then optimized to attempt to improve potency, drug delivery, stability, etc. [21].

As crystal structures of MMPs became available, laboratories turned to structure-based drug design and began diversifying the groups that chelate the catalytic

Zn^{2+} ion. These groups now include sulphydryls, carboxylates, phosphonates, and heterocyclic groups; however, they all are reported to coordinate the Zn^{2+} with a 50- to 2000-fold decreased potency compared to hydroxamate-containing compounds [5, 10]. The current accepted model of binding small molecule drugs to any enzyme is the idea of an "induced fit". In this model, the ligand interacts with the active site to induce a structural modification of the enzyme, thus allowing the ligand to fit snugly into the binding site [22]. Generally, structure-based drug design yields small-molecule compounds which are more selective than peptidomimetics. For example, the drug may inhibit a set of MMPs that share substrate preference, such as inhibiting all gelatinases without targeting stromelysins or matrilysins. Additionally, structure-based drugs are often designed in such a manner that they exhibit improved bioavailability with more acceptable pharmacokinetics [23].

One of the first examples of a structure-based inhibitor was the non-peptidic compound CGS 27023A (Novartis, also referred to as MMI 270), which selectively targets MMP-2, MMP-8, and MMP-9. CGS 27023A was designed with an isopropyl substituent that is α to the Zn^{2+}-chelating hydroxamic acid moiety and likely also binds in the S1′ subsite. The function of this moiety is intended to slow down the metabolism of the zinc-binding function. CGS 27023A also has a bulkier moiety, which is believed to bind within a shallow, solvent-exposed pocket on the enzyme's surface (termed S2′), which helps anchor the compound and increase specificity. An arylsulfonyl group also occupies (although does not fill) the specificity S1′ pocket [24]. *In vitro* studies in tumor cell lines did not indicate any antiproliferative activity for CGS 27023A, though it was observed to significantly reduce tumor burden in rat tumor models of breast and endometrial cancer compared with controls, while also demonstrating antimetastatic and antiangiogenic effects [25]. However, similar to its peptidomimetic predecessors, CGS 27023A was also terminated following clinical trials for patients with non-small cell lung carcinoma due to poorly tolerated joint and muscle pain [12].

Several other hydroxamate-based small-molecule compounds made their way into clinical trials, including prinomastat (Pfizer) and tanomastat (Bayer). Prinomastat is an optimized version of CGS 27023A developed for use as an antiangiogenic which retains the arylsulfonyl and hydroxamate [26]. Ultimately, Pfizer halted its phase III clinical trials due to the drug's lack of effectiveness in patients with late-stage cancer. Reportedly however, the musculoskeletal symptoms experienced by the other MMPIs developed at the time were less of an issue [27]. Tanomastat uses a carboxylate group to chelate the catalytic Zn^{2+} ion and also incorporates a biphenyl segment which binds into deep S1′ pockets. Because it is more selective than the prior experimental compounds (it is a better inhibitor of MMP-2, MMP-3, and MMP-9 than it is of MMP-1 or MMP-13), the drug was investigated in clinical trials for treatment of solid tumors, rheumatoid arthritis, and for prevention of organ transplant rejection [26]. Although the drug was well-tolerated and performed well during *in vitro* and *in vivo* preclinical studies, none of the clinical trials provided improved progression-free or overall survival [28].

As the molecular mechanisms and roles of MMPs in various pathologies are elucidated, opportunities are being identified for development of MMPIs that may be highly selective for the MMP at the root of the disease, without causing undesirable side effects. For example, MMP-13 is highly implicated in the degradation of collagen that characterizes the tissue destruction in osteoarthritis (OA) [29]. Current therapies for OA include pain management, glucosamine and chondroitin sulfate therapies, corticosteroid injections, or surgical intervention [30]; however, these treatment options are short-term or ineffective and manage only pain and inflammation. All of these options fail to address the pathophysiological and biochemical mechanisms involved with cartilage degeneration causing the induction of pain in arthritic joints. Recently, a unique pocket has been identified in MMP-13 that protrudes off the deep S1′ specificity pocket (referred to as the S1′* pocket), the conformational space of which is not accessible to other MMPs. Exploitation of this pocket may allow for the design of efficacious and highly selective inhibitors [31]. Small molecule inhibitors capable of diminishing or preventing destruction of tissue in joints that can be delivered intra-articularly via injection would accomplish two major goals: targeting the molecular pathways responsible for the major pathology of the disease while also reducing or eliminating system toxicity typically associated with most oral compounds for OA tested to date (due to extremely poor diffusion into synovium) [23]. A practical limitation to this approach is the need for repeated intra-articular drug injections.

Fragment-based drug design (FBDD) is based on the idea that small chemical compounds (fragments) may bind to a drug target through high quality interactions, although they are rarely potent. Once such a fragment has been identified and its binding mode understood, the fragment will be elaborated to develop a more potent lead compound [32]. In some cases, multiple fragments binding at different sites can be connected by a linker to obtain a single inhibitor exhibiting greater potency than any of the individual fragments. Such was the case when a zinc binding fragment (acetohydroxamic acid) with high potency was designed. A lipophilic S1′-interacting fragment (paramethoxybenzenesulfonylamide) was identified separately. These two groups were then tethered together by a single covalent bond and the resulting compound has been shown to selectively and potently inhibit MMP-12 in *in vitro* studies [33]. This drug is especially attractive for pulmonary pathologies that commonly present with overexpression of MMP-12, including emphysema and COPD [34]. Similar fragment-based drug design processes have targeted MMP-3, which is implicated in OA, rheumatoid arthritis, aortic and abdominal aneurysms, injured respiratory epithelial cells, gastrointestinal lesions (Crohn's disease, peptic ulcers, ulcerative colitis), colorectal cancer, head and neck carcinoma, basal cell carcinoma, bronchial and lung squamous cell carcinoma, and esophageal squamous cell carcinoma [34]. The rational design was based on a library of MMP-3-binding compounds and involved the linkage of biphenol and hydroxamic acid fragments to form a compound that is computationally predicted to bind very tightly [35–37]. While these compounds are certainly attractive, significant work remains to determine the efficacy and degree of selectivity of these compounds.

5.4 Mechanism-based MMPI design

Mechanism-based inhibition, also known as suicide inhibition, is a form of irreversible enzyme inhibition that occurs when an enzyme's active site covalently binds with an inhibitor during "normal" catalysis. The resulting adduct is rendered catalytically inactive and forms a stable and irreversible enzyme-inhibitor complex [38]. Gelatinases (MMP-2 and -9), which have been implicated in tumor metastasis and angiogenesis, were the first targets for developing mechanism-based inhibitors following the discovery of the gelatinase-selective compound SB-3ct. SB-3ct was designed to coordinate the active-site Zn^{2+} with a thiirane, activating it for subsequent modification. The activation of the Zn^{2+} allows a biphenyl moiety to move to fit in the deep hydrophobic P1′ subsite. This slow-binding mode then brings the sulfur of the thiirane into the coordination sphere of the Zn^{2+} that allows it to act as a Lewis acid. After covalently binding the sulfur, a global conformational change occurs, resulting in restructuring the enzyme back to the proenzyme state [39]. Despite low aqueous solubility and significant metabolism, *in vivo* animal experiments indicate the compound is a potent inhibitor of liver, prostate, breast, and T-cell lymphoma metastases and is also capable of increasing survival when tested in an aggressive mouse model of T-cell lymphoma [40, 41]. Though initially studies were aimed at cancer, recent studies have expanded the focus to include the use of the compound for CNS-related disorders such as traumatic injury, ischemic stroke, dementia-associated inflammation, neurogenesis, and penetration of the blood-brain barrier [42]. Although the research around this compound has been extensive, with hopes of improving the pharmacodynamics and pharmacokinetics, development and optimization are still ongoing in pre-clinical studies and are yet to reach clinical trials. To date, mechanism-based inhibitors have focused primarily on the gelatinases and the compounds tested have been SB-3ct derivatives.

5.5 Allosteric MMPI design

A growing trend in the field of drug design emphasizes the surface residues distal from the catalytic site as well as accessory domains as targets for allosteric inhibition. Such a trend is true for MMPIs as well. The rational design of such allosteric inhibitors relies on how the global structure, not just the catalytic domain, of each MMP interacts with endogenous substrates, ligands, and effectors. Most enzymes, MMPs included, are characterized by exosites which can be unique to a specific isoform. Exosites are auxiliary motifs which bind proteins in domains other than the active site and which mediate binding-partner interactions and/or facilitate substrate localization. Such actions determine the affinity, efficiency, and specificity for substrate cleavage [43]. As our basic understanding of the structural and biophysical aspects governing MMP molecular mechanisms develops, this knowledge can be leveraged for the design of selective MMPIs to block functional interactions such as those with co-factors and binding partners unique to a specific MMP [44]. For example, the hemopexin (PEX) domain of membrane-type 1 matrix metalloproteinase (MT1-MMP or MMP-14) is a four-bladed β-propeller which confers the ability to interact with other extracellular matrix components and cell adhesion molecules [45]. It is the region responsible for its association with CD44, an interaction which induces the intracellular cytoskeletal rearrangements responsible for cell migration and assembly of invasion machinery [46]. The association is also required for proteolysis of proMMP-2 by MMP-14; over-expression of MMP-14 and increased activation of MMP-2 are hallmarks of aggressive tumors [17, 47]. Compounds which bind to the PEX domain and prevent dimerization have been shown to significantly decrease tumor size and reduce MMP-14 and MMP-2-mediated cell scattering/invasion, angiogenesis, and tumor metastasis both *in vitro* and in animal cancer experiments [48, 49].

There is very low sequence similarity among the prodomains of MMPs, and thus targeting these domains is another attractive mechanism for inhibition. The latent pro-MMP-9 is rendered inactive through a cysteine-switch mechanism, in which a Cys in the propeptide complexes with the catalytic Zn^{2+} to occlude the active site [50]. Prospective research has been ongoing into the development of small-molecule "processing inhibitors" which can bind the latent, inactive proMMP-9, stabilize it, and inhibit its proteolytic activation [51]. Although specificity has not been established, it is believed to be much higher due to the individuality of MMP prodomains.

MMPs -2 and MMP-9 are the only two MMPs that contain a collagen binding domain (CBD). Other collagen-cleaving MMPs, MMP-1, MMP-8, MMP-13, MMP-14, and MMP-18, lack this domain. A synthetic peptide compound with high homology to a short segment of the α1(I) collagen chain binds to this exosite domain and blocks substrate access to the neighboring catalytic site. Effects on MMP-2 *in vitro* indicated inhibition of the enzyme which was not mirrored when tested on MMP-8 [52]. The CBD is a unique domain for targeting of MMP-2 and likely MMP-9, though additional work is warranted to verify the selectivity of such drugs.

High throughput screens have now been developed to screen for inhibitors binding in exosites that can confer allosteric inhibition. As an example, fluorescence resonance energy transfer (FRET) experiments using triple-helical peptides (THP) as substrates (FRET-THP) have been developed to measure collagenolytic MMP activities. Because THPs have distinct conformational features that interact with certain MMP exosites, these substrates can be used to identify non-active site-binding inhibitors. Screening of 65,000 small molecules led to the identification of four compounds that bound to MMP-13 only, with one of these four inhibitors exhibiting greater inhibition against MMP-13 triple-helical peptidase activity compared with single-stranded peptidase activity. As the hydrolysis of triple-helical collagen is implicated in a variety of pathologies, including tumor metastasis, arthritis, kidney disease, periodontal disease, and tissue ulcerations, high throughput screens which can lead to the identity of selective exosite inhibitors are extremely valuable [53].

5.6 Macromolecular MMP inhibitors

After the failure of MMPIs in clinical trials, inhibitor development expanded to include macromolecules with a focus on optimizing natural TIMPs. However, despite the high potency of TIMPs, they lack selectivity and possess other significant biological functions, which would likely lead to unwanted side effects. Additionally, production of these compounds is riddled with technical difficulties [10]. Attention was also given to development of monoclonal antibody derivatives, which are promising due to a potential for high specificity. Antibodies are attractive because they can be targeted to any exposed domain on an MMP, including binding in the active site to prevent substrates from binding or to the diverse ancillary domains that are allosterically linked to the catalytic site [19, 44]. REGA-3G12 is a murine monoclonal antibody generated via hybridoma technology against human MMP-9 produced by neutrophils [54]. The antibody binds with high affinity and inhibits MMP-9 but has no effect on the related gelatinase MMP-2, making it the most selective MMP-9 inhibitor to date. REGA-3G12 recognizes the amino terminal part of the catalytic domain surface but likely does not interact with the catalytic zinc [55]. As inappropriate expression or activity of MMP-9 has been observed in many pathological processes, including asthma, COPD, inflammatory diseases, diabetes, organ-specific autoimmune inflammation, atherosclerosis, aneurisms and ischemia, thrombosis, and many types of cancer, it is a highly desirable target for inhibition [56]. Antibodies binding with the hemopexin domain of MMP-14 have been shown to bind with nanomolar affinity and were not found to associate with the other collagen-binding proteins used as controls [57]. However, additional research will be necessary for this antibody to determine the degree of selectivity as well as *in vivo* efficacy.

The potential clinical use of the epitope-binding domain of antibodies (fragment antigen-binding compounds [Fabs]) has also attracted interest. Fabs are advantageous because, as they are smaller, they have greater tissue distribution and better penetration than full-size antibodies. However, Fabs lack the stabilizing domains of antibodies and therefore are rapidly degraded *in vivo*, leading to shorter circulating half-lives [58]. Despite this potential shortcoming, using a Fab-displaying phage library, a potent and highly selective inhibitor of MMP-14, DX-2400, was identified. DX-2400 is a Fab which is highly selective for the catalytic domain of MMP-14. The compound was shown to efficiently block activation of proMMP-2 and reduce invasion activity of cancer cell lines *in vitro*. In cancer cell lines positive for MMP-14 expression, this Fab decreased tumor growth and tumor vascularization when tested *in vivo*. Notably, these effects were not mirrored in cancer cell lines negative for MMP-14, suggesting the high selectivity of DX-2400 [59]. Future work focusing on optimization of Fabs such as DX-2400 to increase systemic availability may allow for success in clinical trials.

As biotechnology and computational biology programs develop and become accessible, the technology is being applied for the development of innovative drugs that have the potential to address specificity issues while achieving high potency and efficacy. A fusion protein that contains a ten amino acid residue sequence of a MMP-2 selective inhibitory peptide (APP-IP) linked to the N-terminus of TIMP-2 was designed by incorporating a multitude of techniques discussed in this chapter thus far. The APP-IP and TIMP-2 regions of the fusion protein are designed to interact with the active site and the hemopexin-like domain of MMP-2, respectively. In this manner, selectivity for MMP-2 is greatly increased compared to either subunit individually and is capable of inhibiting catalytic activity while preventing binding of endogenous partners. APP-IP-TIMP-2 exerts its strong inhibitory effect against MMP-2 but with significantly weaker affects on MMP-1, MMP-3, MMP-7, MMP-8, MMP-9, or MMP-14. In HT-1080 cells, after concanavalin A stimulation, the fusion protein inhibited the activation of pro-MMP-2, degradation of type IV collagen, and cell migration [60]. As MMP-2 expression or activity is known to be dysregulated in glioma cell invasion, tumor growth, angiogenesis and metastasis, atherosclerosis, and endometriosis [34], targeting the enzyme with such high selectivity is quite enticing.

5.7 Chemically-Modified tetracyclines

Clinical MMPI development is currently focused heavily on chemically-modified tetracyclines. Decades after the discovery of tetracycline as an antibiotic, it was observed that some tetracycline analogs, notably doxycycline, inhibit the activity of collagenases (MMP-1, MMP-2, MMP-8, MMP-9, and MMP-13) via a mechanism unrelated to its antimicrobial activities [61]. While the precise inhibitory mechanism remains incompletely defined, it is believed that the enzymes are directly inhibited with a coinciding down-regulation in their expression. Models indicate the drug likely does not interact with the catalytic Zn^{2+} ion but more likely with the structural zinc and/or calcium atoms within the protein, destabilizing the entire tertiary structure [62]. An especially tempting reason for using modified tetracyclines as a starting point for compound development is that their safety, toxicity, and efficacy profiles are well known and members of this class have been used safely for years as antibiotics [63].

Doxycyline is the only clinically approved MMPI and is prescribed to treat periodontal disease. Approved in 2001, the drug is an adjunctive treatment in tablet form that is taken orally twice daily. Doxycycline is a broad-spectrum MMPI systemically available with only minor adverse reactions including headache and flu-like symptoms; however, the drug is contraindicated for pregnant women. MMPs are produced by both infiltrating and resident cells of the periodontium, playing a role in appropriate physiological events (such as tooth eruption) as well as in some pathological events (such as periodontitis). Periodontal disease is the result of tissue inflammation that disrupts the balance between MMPs and TIMPs, as well as the types of MMPs present. This alteration is ultimately responsible for the degradation of the connective tissue at periodontal ligaments, allowing for migration and extension of the inflamed periodontal pocket [64, 65]. Specifically, MMP-3, MMP-8, MMP-9 and MMP-13 are known to become pathologically altered [65]. An additional function of doxycycline is to inhibit TNF-α and IL-8 production, thereby targeting the MMP expression pathway upstream to reduce overall levels [66]. Co-administration of doxycycline with an NSAID synergistically reduced MMPs in the gingiva of patients, while enhancing efficacy with an overall improvement on quality of life [67].

As the safety and efficacy profiles were well evaluated for the use of doxycyline as an MMPI in periodontal disease, studies of the drug began for many varied diseases. Open label phase IV clinical studies for the use of doxycycline to treat multiple sclerosis (MS) have recently been completed, during which a significant decrease was observed in the number of contrast-enhancing lesions on brain MRIs. This indicates that co-administration of doxycycline with interferon β-1a (standard therapy option for MS) reduces the inflammatory cascade of MS and thus stabilizes the blood-brain barrier. The co-therapy was safe and well tolerated [62]. Both cerebrospinal fluid (CSF) and intrathecal levels of active MMP-9 have been shown to be more prominent in patients with MS than in patients with non-inflammatory neurological disorders. Similarly, CSF and serum active MMP-9/TIMP-1 ratios are increased in untreated patients [68, 69].

Polycystic ovary syndrome (PCOS) is the most common gynecological disorder in women of reproductive age and is characterized by elevated MMP-9 serum levels. Although serum measurements of MMP-9 are an artifact of *in vitro* leukocyte enzyme release, the test appears to provide clinical relevance. MMP-9 mediates the cyclical remodeling of the extracellular matrix necessary for the menstrual cycle. Deregulation of MMP-9 expression coincides with the chronic anovulation, hyperandrogenism, insulin resistance, and increased risk for miscarriage associated with the syndrome [70]. Phase III clinical trials are currently recruiting to evaluate the use of doxycycline to treat PCOS.

Mouse studies indicate that doxycycline, because of its broad spectrum inhibition capabilities, inhibits activity of tissue MMPs and can therefore attenuate the decrease in the collagen content in aortas of animals haploinsufficient for collagen III. Inhibiting MMP-mediated degradation of collagen, which is important for vascular resistance to deformation, may prevent the development of stress-induced pathologies in blood vessels. It has thus been proposed that doxycycline may be beneficial for treatment of Vascular Ehlers-Danlos syndrome, a heritable and life-threatening condition resulting in a defect in connective tissue due to mutations in the type III pro-collagen gene [71]. Because it is also known that MMPs are involved in remodeling of these resistance vessels in chronic hypertension, the drug was also studied in relation to ischemia. After transient cerebral ischemia, treatment with doxycyline in spontaneously hypersensitive rats attenuates hypertensive remodeling and is associated with increased pial blood flow to the infarcted hemisphere, reducing overall damage due to ischemia. Increases in the outer and lumen diameters and cross-sectional areas were observed, indicating an attenuation in vascular remodeling [72]. Two early phase clinical trials have shown minocycline, a more lipophilic tetracycline-analog than doxycycline, to be safe and have a potentially protective role in prevention from acute ischemic stroke. Larger and more comprehensive human clinical trials are yet to be conducted to confirm the safety and efficacy of the drug for this disease [73].

5.8 Alternative approaches

Small interfering RNAs (siRNAs), which are capable of sequence-specific post-transcriptional gene silencing, have potential to be utilized as therapeutics. Development is still in early stages and studies of potency/efficacy are still quite immature. To date, numerous cell culture studies have indicated that administration of siRNAs targeting MMPs decreased expression of multiple MMPs and, when targeted to cancer cells, reduced cell migration and decreased invasion and angiogenic signaling. These have been tested with promising results in chondrosarcoma (siMMP-1), glioblastoma (siMMP-2), and lung cancer (siMMP-2) cell lines. It is believed that siRNA technologies will allow for more rapid drug development, reducing the time from siRNA design to clinical trials [74–76]. Though promising, there is a risk of off-target effects which may compromise the specificity of siRNA, if sufficient sequence identity with non-targeted mRNA transcripts causes their knockdown as well. Fortunately however, computational programs have been developed to assess such risk with simply the click of a button. While siRNAs hold potential for use, notably for topical application (e.g., inhibiting MMP-1 or -3 in skin to reduce photo aging or the progression of skin cancer [77]), special drug delivery systems would have to be developed, as siRNAs will not readily cross the capillary endothelium and are not hydrophobic enough to pass directly through membranes. Further, siRNAs are at elevated risk compared to most of the previously mentioned drug classes for filtration by the kidneys, phagocytosis, aggregating with serum proteins, and nuclease degradation [78].

Off-target inhibitors may also be utilized to indirectly decrease MMP expression/activity; indeed, this is how chemically-modified tetracyclines were identified as potential therapeutics. The drug letrozole is a reversible non-steroidal inhibitor of P450 aromatase, which indirectly represses MMP-2 and -9 expression. This drug is FDA approved for post-surgery breast cancer patients to reduce metastasis [79]. Tofacitinib is another approved drug which is used to treat rheumatoid arthritis. The drug works by inhibiting Jak3; inhibition of this pathway decreases IL-6 levels, which corresponds to decreased MMP-3 levels. Patients on this drug report a significant improvement in disease progression, ultimately resulting in suppression of cartilage destruction [80]. A challenge to the development of off-target inhibitors is that the precise mechanisms and signaling pathways that induce MMP expression remain poorly defined. Additionally, targeting identified MMP transcription factors is extraordinarily difficult due to the intranuclear localization of these factors and the fact that, in general, transcription factors lack sites in which small molecules can bind [19].

Alternative, non-medicinal approaches to disease management are also gaining popularity in disease prevention as well as for use as an adjuvant to standard treatment options. Compounds in certain foods, notably foods typical to Mediterranean or Asian diets, are known to decrease MMP expression and activity while promoting cardiovascular, immune, and gastrointestinal health. *In vitro* studies have shown that olive oil and red wine polyphenols reduce inflammatory angiogenesis through inhibition of MMP-9 and Cox-2, a pro-inflammatory enzyme, both known to be upregulated in certain cancers and inflammatory diseases [81]. Curcumin, the principal polyphenol component of turmeric, is a powerful antioxidant that has been studied in multiple clinical trials targeting a multitude of diseases and has been shown to significantly reduce MMP-9 levels with some reduction of MMP-2 and MMP-14 as well [82, 83]. Catechins, the polyphenols common in green tea, are also antioxidants capable of reducing MMP-2 and MMP-9 levels [84]. Although the impact on MMP activity and expression is only described in this text, all of these compounds exhibit additional protective effects on multiple signaling and metabolic pathways. These molecular events may synergistically contribute to reduced tumor growth and cancer cell apoptosis as well as inhibition of invasion, angiogenesis, and metastasis. These have been demonstrated in various animal models, but preventive activity for any of the above compounds have not been consistently observed in human studies.

5.9 MMPs as anti-targets

Because of the critical role that MMPs play in a wide range of diseases, the failure of clinical trials with early, promising MMPIs was an unexpected disappointment. It has become clear that one reason for the failure of these trials is related to the complex effects of MMP activity. The family of MMPs collectively acts on a wide range of substrates including cytokines, cryptic growth factors, and extracellular matrix components and they are expressed under certain normal conditions in healthy individuals. Currently, the diversity of MMP substrates is unclear as each MMP family member has a particular degradome that can overlap with other MMP degradomes and it is therefore difficult to discern one MMP's activity from another in an *in vivo* context [85]. The cleavage activity of these enzymes can negatively influence disease progression, such as promoting metastasis in the case of cancer, but can alternatively have an anti-tumoral effect; in some cases MMP activity can serve to mediate inhibitory pathways in disease. For this reason, in addition to the importance of MMPs in normal biological activities, some MMPs may be considered 'antitargets' [20]. They can have favorable effects on disease progression, depending upon the context in which the specific MMP is present [20]. The strategy of broad spectrum inhibition of these proteases was a clinical failure due to inhibition of antitargets in addition to the proper targets, resulting in no significant benefit to patients, exerting unwanted side effects, and potentially speeding the course of disease progression by effectively decreasing any inhibitory effects that the antitargets were responsible for [20]. Hence, it is prudent to develop MMP-specific inhibitors to avoid the risk of impeding the activity of antitargets, while still targeting the MMP that is actively contributing to the pathology.

To date, ten MMP family members have been validated as antitargets that naturally impede disease progression [20]. MMPs that have been assigned antitarget function include MMP-2, MMP-3, MMP-7, MMP-8, MMP-9, MMP-11, MMP-12, MMP-20, MMP-26, and MMP-28. Additional MMPs have the potential to be antitargets but are yet to be validated, including MMP-14 and -24 [20]. For those that have been validated, the antitarget effects have been documented in inflammatory diseases, wound healing, angiogenesis, and cancer. For example, MMP-2 and MMP-9 have been found to play a crucial role in lung diseases such as asthma, as they are responsible for mediating the clearance of immune cells from the lung tissue ("airway remodeling") [86]. This resolves inflammation in the case of allergic asthma and without MMP-2 and MMP-9 function, lung tissues retain the immune cell effectors aberrantly and this results in destruction of tissue [86, 87]. MMP-8 is a crucial protease involved in dampening the immune response in arthritic joints, terminating the activity of infiltrating neutrophils after an initial influx to the joints, thereby reducing the severity of arthritic symptoms [88]. Inhibition of MMP-12 can result in an increase in angiogenesis yet conversely, overexpression of MMP-12 in certain cancer cell lines has been found to reduce tumor volume by inhibition of angiogenesis [89, 90]. MMP-26 is capable of contributing to invasion in the case of colon cancer, but its expression in early stages of breast cancer correlates to longer overall patient survival [91, 92]. Invasion and metastasis of gastric cancer cells can be mediated by MMP-28, but in the context of myocardial infarction (MI), MMP-28 serves to mediate the necessary immune response to clear debris and promote scar formation and therefore healing [93, 94]. Furthermore, the temporal parameters of MMP expression influence disease status. For example, following an MI, infiltration of neutrophils followed by macrophages into the damaged tissue leads to early release of MMP-9 from neutrophils and a subsequent release of MMP-2 from macrophages [95]. This has been proposed to be beneficial to healing, as the presence of these enzymes can clear ECM and necrotic myocytes, but high levels that accumulate over a short time period following the MI may cause cardiac rupture and death soon after the initial insult [95]. Interestingly, reduction of MMP-9 expression in the long-term following a cardiac ischemic event disrupts proper healing and leads to cardiac failure [96]. In this case, MMP-2 and MMP-9 can act as both targets (in the short-term) and anti-targets (in the long-term), emphasizing the importance of understanding not only the biological activity of MMPs in disease, but also the timing of MMP expression and activity.

Because of the contributions of MMP activity to either disease control or progression, it is of utmost importance to design specific MMPIs. Ideally, future MMPIs should be able to inhibit the target MMP and avoid interference with other MMPs at a rate of three log units difference in K_i compared to the target MMP [97, 98]. One possibility to solve the problem of selectivity in MMP inhibition is to use a fragment-based approach to inhibitor design. As previously mentioned, a potent inhibitor of MMP-2 was designed utilizing this approach. The kinetic profile of the inhibitor was such that the inhibitor demonstrated effectiveness against MMP-2 and -13 but, importantly, it did not inhibit the antitargets MMP-1, MMP-7, or MMP-8 [99]. This work demonstrates that it is possible to design an effective, selective inhibitor of MMPs. Pursuing the development of specific inhibitors of MMPs is a clinically relevant endeavor, as it is clear that MMP effects on disease progression is dependent on timing and disease context. While it is important to design MMP-specific inhibitors, judicious application of these inhibitors will be another key to clinical effectiveness. The timing of MMPI use in relationship to disease progression will be vital, as exemplified by differential MMP expression in late and early stages of diseases such as cancer and ischemic injury [92, 95, 96].

5.10 Conclusions

MMPIs are promising drugs for a diverse set of pathologies. As biotechnology and drug screening and design methods advance, potent and efficacious therapeutics will develop in stride. As evidenced by the disappointing results of the initial clinical trials that focused on small molecules with broad-spectrum inhibition, successful MMPIs will likely be highly selective for the MMP of interest. In order to achieve such selectivity, these drugs must be designed to target unique domains on a protease. These may include the S1′ specificity site, exosites which participate in necessary actions for enzymatic activity, or allosteric domains. Further, clinical trials must be designed with great care to ensure patients are provided with MMPIs at the right *time*, for example, during early stages of cancer, in order to contain tumors, as opposed to late-stage disease. In doing so, we can provide patients with therapeutics that target only the overactive protease at the appropriate stage in their disease while reducing off-target adverse effects. The collective result of the proper design and use of MMPIs will be a significant improvement in both patient quality of life and overall survival of some of the most challenging diseases facing humanity.

References

1. Rodríguez, D., Morrison, C.J. & Overall, C.M. (2010) Matrix metalloproteinases: What do they not do? New substrates and biological roles identified by murine models and proteomics. *Biochimica et Biophysica Acta (BBA) - Molecular Cell Research*, 1803 (1), 39–54.
2. Vihinen, P. & Kähäri, V.M. (2002) Matrix metalloproteinases in cancer: prognostic markers and therapeutic targets. *International Journal of Cancer*, 99 (2), 157–166.
3. Kevorkian, L., Young, D.A., Darrah, C. *et al.* (2004) Expression profiling of metalloproteinases and their inhibitors in cartilage. *Arthritis & Rheumatism*, 50 (1), 131–141.
4. Molet, S., Belleguic, C., Lena, H. *et al.* (2005) Increase in macrophage elastase (MMP-12) in lungs from patients with chronic obstructive pulmonary disease. *Inflammation Research*, 54 (1), 31–36.
5. Dormán, G. *et al.* (2010) Matrix metalloprotease inhibitors: a critical appraisal of design principles and proposed therapeutic utility. *Drugs*, 70 (8), 949–964.
6. Jacobsen, J.A., Major Jourden, J.L., Miller, M.T. *et al.* (2010) To bind zinc or not to bind zinc: An examination of innovative approaches to improved metalloproteinase inhibition. *Biochimica et Biophysica Acta (BBA) - Molecular Cell Research*, 1803 (1), 72–94.
7. Fragai, M., Nativi, C., Richichi, B. *et al.* (2005) Design *in silico*, synthesis and binding evaluation of a carbohydrate-based scaffold for structurally novel inhibitors of matrix metalloproteinases. *ChemBioChem*, 6 (8), 1345–1349.
8. Hidalgo, M. & Eckhardt, S.G. (2001) Development of matrix metalloproteinase inhibitors in cancer therapy. *Journal of the National Cancer Institute*, 93 (3), 178–193.
9. Botos, I., Scapozza, L., Zhang, D. *et al.* (1996) Batimastat, a potent matrix mealloproteinase inhibitor, exhibits an unexpected mode of binding. *Proceedings of the National Academy of Sciences*, 93 (7), 2749–2754.
10. Hu, J., Van den Steen, P.E., Sang, Q.X.A. *et al.* (2007) Matrix metalloproteinase inhibitors as therapy for inflammatory and vascular diseases. *Nature Reviews. Drug Discovery*, 6 (6), 480–498.
11. Beckett, R.P., Davidson, A.H., Drummond, A.H. *et al.* (1996) Recent advances in matrix metalloproteinase inhibitor research. *Drug Discovery Today*, 1 (1), 16–26.
12. Rao, B.G. (2005) Recent developments in the design of specific matrix metalloproteinase inhibitors aided by structural and computational studies. *Current Pharmaceutical Design*, 11 (3), 295–322.
13. Gialeli, C., Theocharis, A.D. & Karamanos, N.K. (2011) Roles of matrix metalloproteinases in cancer progression and their pharmacological targeting. *FEBS Journal*, 278 (1), 16–27.
14. Macaulay, V.M., O'Byrne, K.J., Saunders, M.P. *et al.* (1999) Phase I study of intrapleural batimastat (BB-94), a matrix metalloproteinase inhibitor, in the treatment of malignant pleural effusions. *Clinical Cancer Research*, 5 (3), 513–520.
15. Rasmussen, H.S. & McCann, P.P. (1997) Matrix metalloproteinase inhibition as a novel anticancer strategy: a review with special focus on batimastat and marimastat. *Pharmacology & Therapeutics*, 75 (1), 69–75.
16. Steward, W.P. (1999) Marimastat (BB2516): current status of development. *Cancer Chemotherapy and Pharmacology*, 43 (1), S56–S60.
17. Coussens, L.M., Fingleton, B. & Matrisian, L.M. (2002) Matrix metalloproteinase inhibitors and cancer—trials and tribulations. *Science*, 295 (5564), 2387–2392.
18. McCawley, L.J. & Matrisian, L.M. (2000) Matrix metalloproteinases: multifunctional contributors to tumor progression. *Molecular Medicine Today*, 6 (4), 149–156.
19. Overall, C.M. & Lopez-Otin, C. (2002) Strategies for MMP inhibition in cancer: innovations for the post-trial era. *Nature Reviews. Cancer*, 2 (9), 657–672.
20. Dufour, A. & Overall, C.M. (2013) Missing the target: matrix metalloproteinase antitargets in inflammation and cancer. *Trends in Pharmacological Sciences*, 34 (4), 233–242.
21. Kalyaanamoorthy, S. & Chen, Y.-P.P. (2011) Structure-based drug design to augment hit discovery. *Drug Discovery Today*, 16 (17–18), 831–839.
22. Bursavich, M.G. & Rich, D.H. (2002) Designing non-peptide peptidomimetics in the 21st century: inhibitors targeting conformational ensembles. *Journal of Medicinal Chemistry*, 45 (3), 541–558.
23. Gege, C., Bao, B., Bluhm, H. *et al.* (2012) Discovery and evaluation of a non-Zn chelating, selective matrix metalloproteinase 13 (MMP-13) inhibitor for potential intra-articular treatment of osteoarthritis. *Journal of Medicinal Chemistry*, 55 (2), 709–16.
24. Supuran, C.T., Casini, A. & Scozzafava, A. (2003) Protease inhibitors of the sulfonamide type: Anticancer, antiinflammatory, and antiviral agents. *Medicinal Research Reviews*, 23 (5), 535–558.
25. Wood, J., Schnell, C., Cozens, R. *et al.* (1998) CGS 27023A, a potent and orally active matrix metalloprotease inhibitor with antitumor activity. *Proceedings of the American Association for Cancer Research*, 39 (83a).

26. Fisher, J.F. & Mobashery, S. (2006) Recent advances in MMP inhibitor design. *Cancer Metastasis Reviews*, 25 (1), 115–136.
27. Phase III Clinical Trial of Prinomastat Halted. 2000.
28. Hirte, H., Vergote, I.B., Jeffrey, J.R. *et al.* (2006) A phase III randomized trial of BAY 12–9566 (tanomastat) as maintenance therapy in patients with advanced ovarian cancer responsive to primary surgery and paclitaxel/platinum containing chemotherapy: A national cancer institute of canada clinical trials group study. *Gynecologic Oncology*, 102 (2), 300–308.
29. Goldring, M., Otero, M., Plumb, D.A. *et al.* (2011) Roles of inflammatory and anabolic cytokines in cartilage metabolism: signals and multiple effectors converge upon MMP-13 regulation in osteoarthritis. *European Cells And Materials Journal*, 21, 202–220.
30. Sovani, S. & Grogan, S.P. (2013) Osteoarthritis: detection, pathophysiology, and current/future treatment strategies. *Orthopaedic Nursing*, 32 (1), 25–36.
31. Engel, C.K., Pirard, B., Schimanski, S. *et al.* (2005) Structural basis for the highly selective inhibition of MMP-13. *Chemistry & Biology*, 12 (2), 181–189.
32. Murray, C.W., Verdonk, M.L. & Rees, D.C. (2012) Experiences in fragment-based drug discovery. *Trends in Pharmacological Sciences*, 33 (5), 224–232.
33. Borsi, V., Calderone, V., Fragai, M. *et al.* (2010) Entropic contribution to the linking coefficient in fragment based drug design: a case study. *Journal of Medicinal Chemistry*, 53 (10), 4285–4289.
34. Amălinei, C., CăruntuI.D., Giuşcă, S.E. *et al.* (2010) Matrix metalloproteinases involvement in pathologic conditions. *Romanian Journal of Morphology and Embryology*, 51 (2), 215–228.
35. Wilfong, E.M., Du, Y. & Toone, E.J. (2012) An enthalpic basis of additivity in biphenyl hydroxamic acid ligands for stromelysin-1. *Bioorganic & Medicinal Chemistry Letters*, 22 (20), 6521–6524.
36. Olejniczak, E.T., Hajduk, P.J., Marcotte, P.A. *et al.* (1997) Stromelysin inhibitors designed from weakly bound fragments: effects of linking and cooperativity. *Journal of the American Chemical Society*, 119 (25), 5828–5832.
37. Hajduk, P.J., Sheppard, G., Nettesheim, D.G. *et al.* (1997) Discovery of potent nonpeptide inhibitors of stromelysin using SAR by NMR. *Journal of the American Chemical Society*, 119 (25), 5818–5827.
38. Abeles, R.H. & Maycock, A.L. (1976) Suicide enzyme inactivators. *Accounts of Chemical Research*, 9 (9), 313–319.
39. Brown, S., Bernardo, M.M., Li, Z.H. *et al.* (2000) Potent and selective mechanism-based inhibition of gelatinases. *Journal of the American Chemical Society*, 122 (28), 6799–6800.
40. Krüger, A., Arlt, M.J.E., Gerg, M. *et al.* (2005) Antimetastatic activity of a novel mechanism-based gelatinase inhibitor. *Cancer Research*, 65 (9), 3523–3526.
41. Gooyit, M., Lee, M., Shroeder, V.A. *et al.* (2011) Selective water-soluble gelatinase inhibitor prodrugs. *Journal of Medicinal Chemistry*, 54 (19), 6676–6690.
42. Verslegers, M., Lemmens, K., Van Hove, I. *et al.* (2013) Matrix metalloproteinase-2 and -9 as promising benefactors in development, plasticity and repair of the nervous system. *Progress in Neurobiology*, 105 (0), 60–78.
43. Morrison, C.J., Butler, G.S., Rodriguez, D. *et al.* (2009) Matrix metalloproteinase proteomics: substrates, targets, and therapy. *Current Opinion in Cell Biology*, 21 (5), 645–653.
44. Sela-Passwell, N., Rosenblum, G., Shoham, T. *et al.* (2010) Structural and functional bases for allosteric control of MMP activities: can it pave the path for selective inhibition? *Biochimica et Biophysica Acta*, 1803 (1), 29–38.
45. Lee, M.-H. & Murphy, G. (2004) Matrix metalloproteinases at a glance. *Journal of Cell Science*, 117 (18), 4015–4016.
46. Mori, H., Tomari, T., Koshikawa, N. *et al.* (2002) CD44 directs membrane-type 1 matrix metalloproteinase to lamellipodia by associating with its hemopexin-like domain. *The EMBO Journal*, 21 (15), 3949–3959.
47. Lichte, A., Kolkenbrock, H. & Tschesche, H. (1996) The recombinant catalytic domain of membrane-type matrix metalloproteinase-1 (MT1-MMP) induces activation of progelatinase A and progelatinase A complexed with TIMP-2. *FEBS Letters*, 397 (2), 277–282.
48. Zarrabi, K., Dufour, A., Li, J. *et al.* (2011) Inhibition of matrix metalloproteinase 14 (MMP-14)-mediated cancer cell migration. *Journal of Biological Chemistry*, 286 (38), 33167–33177.
49. Remacle, A.G., Golubkov, V.S., Shiryaev, S.A. *et al.* (2012) Novel MT1-MMP small-molecule inhibitors based on insights into hemopexin domain function in tumor growth. *Cancer Research*, 72 (9), 2339–2349.
50. Morgunova, E., Tuuttila, A., Bergmann, U. *et al.* (1999) Structure of human pro-matrix metalloproteinase-2: activation mechanism revealed. *Science*, 284 (5420), 1667–1670.
51. Leonard, K.A., Alexander, R.S., Barbay, J.K. et al., Crystal Structure of the Pro Form of a Matrix Metalloproteinase and an Allosteric Processing Inhibitor. 2013, US Patent 20,130,040,360.

52. Xu, X., Chen, Z., Wang, Y. *et al.* (2007) Inhibition of MMP-2 gelatinolysis by targeting exodomain-substrate interactions. *The Biochemical Journal*, 406, 147–155.
53. Lauer-Fields, J.L., Minond, D., Chase, P.S. *et al.* (2009) High throughput screening of potentially selective MMP-13 exosite inhibitors utilizing a triple-helical FRET substrate. *Bioorganic & Medicinal Chemistry*, 17 (3), 990–1005.
54. Zhou, N., Paemen, L., Opdenakker, G. *et al.* (1997) Cloning and expression in Escherichia coli of a human gelatinase B-inhibitory single-chain immunoglobulin variable fragment (scFv). *FEBS Letters*, 414 (3), 562–566.
55. Martens, E., Leyssen, A., Van Aelst, I. *et al.* (2007) A monoclonal antibody inhibits gelatinase B/MMP-9 by selective binding to part of the catalytic domain and not to the fibronectin or zinc binding domains. *Biochimica et Biophysica Acta (BBA) - General Subjects*, 1770 (2), 178–186.
56. Vandooren, J., Van den Steen, P.E. & Opdenakker, G. (2013) Biochemistry and molecular biology of gelatinase B or matrix metalloproteinase-9 (MMP-9): The next decade. *Critical Reviews in Biochemistry and Molecular Biology*, 48 (3), 222–272.
57. Basu, B., Correa de Sampaio, P., Mohammed, H. *et al.* (2012) Inhibition of MT1-MMP activity using functional antibody fragments selected against its hemopexin domain. *The International Journal of Biochemistry & Cell Biology*, 44 (2), 393–403.
58. Nelson, A.L. (2010) Antibody fragments: hope and hype. *MAbs. Landes Bioscience*, 2 (1), 77–83.
59. Devy, L., Huang, L., Naa, L. *et al.* (2009) Selective inhibition of matrix metalloproteinase-14 blocks tumor growth, invasion, and angiogenesis. *Cancer Research*, 69 (4), 1517–1526.
60. Higashi, S., Hirose, T., Takeuchi, T. *et al.* (2013) Molecular design of a highly selective and strong protein inhibitor against matrix metalloproteinase-2 (MMP-2). *Journal of Biological Chemistry*, 288 (13), 9066–9076.
61. Hanemaaijer, R., Visser, H., Koolwijk, P. *et al.* (1998) Inhibition of MMP synthesis by doxycycline and chemically modified tetracyclines (CMTs) in human endothelial cells. *Advances in Dental Research*, 12 (1), 114–118.
62. Stoilova, T., Colombo, L., Forloni, G. *et al.* (2013) A new face for old antibiotics: tetracyclines in treatment of amyloidoses. *Journal of Medicinal Chemistry*, 56 (15), 5987–6006.
63. Jones, C.H. & Petersen, P.J. (2005) Tigecycline: a review of preclinical and clinical studies of the first-in-class glycylcycline antibiotic. *Drugs Today*, 41 (10), 637.
64. Ryan, M.E. & Golub, L.M. (2000) Modulation of matrix metalloproteinase activities in periodontitis as a treatment strategy. *Journal of Periodontology*, 24 (1), 226–238.
65. Preshaw, P.M., Hefti, A.F., Jepsen, S. *et al.* (2004) Subantimicrobial dose doxycycline as adjunctive treatment for periodontitis. *Journal of Clinical Periodontology*, 31 (9), 697–707.
66. Perdigão, J., Reis, A. & Loguercio, A.D. (2013) Dentin adhesion and MMPs: a comprehensive review. *Journal of Esthetic and Restorative Dentistry*, 25 (4), 219–241.
67. Lee, H.-M., Ciancio, S.G., Tüter, G. *et al.* (2004) Subantimicrobial dose doxycycline efficacy as a matrix metalloproteinase inhibitor in chronic periodontitis patients is enhanced when combined with a non-steroidal anti-inflammatory drug. *Journal of Periodontology*, 75 (3), 453–463.
68. Fainardi, E., Castellazzi, M., Bellini, T. *et al.* (2006) Cerebrospinal fluid and serum levels and intrathecal production of active matrix metalloproteinase-9 (MMP-9) as markers of disease activity in patients with multiple sclerosis. *Multiple Sclerosis (Houndmills, Basingstoke, England)*, 12 (3), 294–301.
69. Minagar, A., Alexander, J.S., Schwendimann, R.N. *et al.* (2008) Combination therapy with interferon beta-1a and doxycycline in multiple sclerosis: An open-label trial. *Archives of Neurology*, 65 (2), 199–204.
70. Liu, B., Cai, L.Y., Lv, H.M. *et al.* (2008) Raised serum levels of matrix metalloproteinase-9 in women with polycystic ovary syndrome and its association with insulin-like growth factor binding protein-1. *Gynecological Endocrinology*, 24 (5), 285–288.
71. Briest, W., Cooper, T.K., Tae, H.J. *et al.* (2011) Doxycycline ameliorates the susceptibility to aortic lesions in a mouse model for the vascular type of ehlers-danlos syndrome. *Journal of Pharmacology and Experimental Therapeutics*, 337 (3), 621–627.
72. Pires, P.W., Rogers, C.T., McClain, J.L. *et al.* (2011) Doxycycline, a matrix metalloprotease inhibitor, reduces vascular remodeling and damage after cerebral ischemia in stroke-prone spontaneously hypertensive rats. *American Journal of Physiology - Heart and Circulatory Physiology*, 301 (1), H87–H97.
73. Fagan, S.C., Cronic, L.E. & Hess, D.C. (2011) Minocycline development for acute ischemic stroke. *Translational stroke research*, 2 (2), 202–208.
74. Yuan, J., Dutton, C.M. & Scully, S.P. (2005) RNAi mediated MMP-1 silencing inhibits human chondrosarcoma invasion. *Journal of Orthopaedic Research*, 23 (6), 1467–1474.
75. Badiga, A.V., Chetty, C, Kesanakurti, D. *et al.* (2011) MMP-2 siRNA inhibits radiation-enhanced invasiveness in glioma cells. *PLoS ONE*, 6 (6), e20614.

76. Chetty, C., Bhoopathi, P., Joseph, P. *et al.* (2006) Adenovirus-mediated small interfering RNA against matrix metalloproteinase-2 suppresses tumor growth and lung metastasis in mice. *Molecular Cancer Therapeutics*, 5 (9), 2289–2299.
77. Polte, T. & Tyrrell, R.M. (2004) Involvement of lipid peroxidation and organic peroxides in UVA-induced matrix metalloproteinase-1 expression. *Free Radical Biology and Medicine*, 36 (12), 1566–1574.
78. Whitehead, K.A., Langer, R. & Anderson, D.G. (2009) Knocking down barriers: advances in siRNA delivery. *Nature Reviews. Drug Discovery*, 8 (2), 129–138.
79. Gialeli, C., Theocharis, A.D. & Karamanos, N.K. (2011) Roles of matrix metalloproteinases in cancer progression and their pharmacological targeting. *The FEBS Journal*, 278 (1), 16–27.
80. Scott, L.J. (2013) Tofacitinib: a review of its use in adult patients with rheumatoid arthritis. *Drugs*, 73 (8), 857–874.
81. Scoditti, E., Calabriso, N., Massaro, M. *et al.* (2012) Mediterranean diet polyphenols reduce inflammatory angiogenesis through MMP-9 and COX-2 inhibition in human vascular endothelial cells: a potentially protective mechanism in atherosclerotic vascular disease and cancer. *Archives of Biochemistry and Biophysics*, 527 (2), 81–9.
82. Basnet, P. & Skalko-Basnet, N. (2011) Curcumin: an anti-inflammatory molecule from a curry spice on the path to cancer treatment. *Molecules*, 16 (6), 4567–4598.
83. Mannello, F. (2006) Natural bio-drugs as matrix metalloproteinase inhibitors: new perspectives on the horizon? *Recent Patents on Anti-Cancer Drug Discovery*, 1 (1), 91–103.
84. Yang, C.S. & Wang, X. (2010) Green tea and cancer prevention. *Nutrition and Cancer*, 62 (7), 931–937.
85. Rodriguez, D., Morrison, C.J. & Overall, C.M. (2010) Matrix metalloproteinases: what do they not do? New substrates and biological roles identified by murine models and proteomics. *Biochimica et Biophysica Acta*, 1803 (1), 39–54.
86. Corry, D.B., Rishi, K., Kanellis, J. *et al.* (2002) Decreased allergic lung inflammatory cell egression and increased susceptibility to asphyxiation in MMP2-deficiency. *Nature Immunology*, 3 (4), 347–353.
87. Corry, D.B., Kiss, A., Song, L.Z. *et al.* (2004) Overlapping and independent contributions of MMP2 and MMP9 to lung allergic inflammatory cell egression through decreased CC chemokines. *The FASEB Journal*, 18 (9), 995–997.
88. Cox, J.H., Starr, A.E., Kappelhoff, R. *et al.* (2010) Matrix metalloproteinase 8 deficiency in mice exacerbates inflammatory arthritis through delayed neutrophil apoptosis and reduced caspase 11 expression. *Arthritis and Rheumatism*, 62 (12), 3645–3655.
89. Margheri, F., Serratí, S., Lapucci, A. *et al.* (2010) Modulation of the angiogenic phenotype of normal and systemic sclerosis endothelial cells by gain–loss of function of pentraxin 3 and matrix metalloproteinase 12. *Arthritis & Rheumatism*, 62 (8), 2488–2498.
90. Gorrin-Rivas, M.J., Arii, S., Furutani, M. *et al.* (2000) Mouse macrophage metalloelastase gene transfer into a murine melanoma suppresses primary tumor growth by halting angiogenesis. *Clinical Cancer Research*, 6 (5), 1647–1654.
91. Gutschalk, C.M., Yanamandra, A.K., Linde, N. *et al.* (2013) GM-CSF enhances tumor invasion by elevated MMP-2, −9, and −26 expression. *Cancer Medicine*, 2 (2), 117–129.
92. Savinov, A.Y., Remacle, A.G., Golubkov, V.S. *et al.* (2006) Matrix metalloproteinase 26 proteolysis of the NH2-terminal domain of the estrogen receptor beta correlates with the survival of breast cancer patients. *Cancer Research*, 66 (5), 2716–2724.
93. Jian, P., Yanfang, T., Zhuan, Z. *et al.* (2011) MMP28 (epilysin) as a novel promoter of invasion and metastasis in gastric cancer. *BMC Cancer*, 11, 200.
94. Ma, Y., Halade, G.V., Zhang, J. *et al.* (2013) Matrix metalloproteinase-28 deletion exacerbates cardiac dysfunction and rupture after myocardial infarction in mice by inhibiting M2 macrophage activation. *Circulation Research*, 112 (4), 675–688.
95. Tao, Z.Y., Cavasin, M.A., Yang, F. *et al.* (2004) Temporal changes in matrix metalloproteinase expression and inflammatory response associated with cardiac rupture after myocardial infarction in mice. *Life Sciences*, 74 (12), 1561–1572.
96. Heymans, S., Luttun, A., Nuyens, D. *et al.* (1999) Inhibition of plasminogen activators or matrix metalloproteinases prevents cardiac rupture but impairs therapeutic angiogenesis and causes cardiac failure. *Nature Medicine*, 5 (10), 1135–1142.
97. Rubino, M.T., Agamennone, M., Campestre, C. *et al.* (2009) Synthesis, SAR, and biological evaluation of α-sulfonylphosphonic acids as selective matrix metalloproteinase inhibitors. *ChemMedChem*, 4 (3), 352–362.
98. Overall, C.M. & Kleifeld, O. (2006) Towards third generation matrix metalloproteinase inhibitors for cancer therapy. *British Journal of Cancer*, 94 (7), 941–946.
99. Zapico, J.M., Serra, P., García-Sanmartín, J. *et al.* (2011) Potent "clicked" MMP2 inhibitors: synthesis, molecular modeling and biological exploration. *Organic and Biomolecular Chemistry*, 9 (12), 4587–4599.

6 Matrix Metalloproteinase Modification of Extracellular Matrix-Mediated Signaling

Howard C. Crawford[1] and M. Sharon Stack[2]

[1] Department of Cancer Biology, Mayo Clinic, Jacksonville, FL, USA
[2] Harper Cancer Research Institute, University of Notre Dame, Notre Dame, IN, USA

6.1 Introduction

Our understanding of matrix metalloproteinases (MMPs) and their functions has gradually evolved over the past several decades. The components of the extracellular matrix (ECM), a complex group of proteins critical for maintaining tissue architecture, were initially considered the primary MMP substrates. The ability of the MMP family to collectively cleave the entirety of the ECM supported the common assumption that their primary function was bulk ECM degradation. Indeed, in the remodeling processes associated with development and wound healing and in invasion of cancer cells into the surrounding stroma, MMP expression is robustly induced in both the epithelia and stromal tissue and MMP inhibition affects these processes dramatically [1–3].

Even so, by researching the molecular details of tissue remodeling and tumor invasion, new models of MMP function have emerged to enhance the earlier dogma. We now know that MMPs function as more than enzymatic battering rams, blasting through collagen-rich barriers for invading cancer cells, and as more than housemaids, cleaning up useless proteinaceous detritus within an involuting tissue. Most MMPs are associated with the plasma membrane either through interaction with transmembrane proteins or through an encoded transmembrane domain [3]. On the one hand, this is consistent with the original models of MMP function, as it allows cells to localize proteolysis to the immediate path of invading filopodia and it ensures that these potentially destructive activities would remain under tight cellular control. But it also put the MMPs at a hotspot of cellular signaling, whereby they could elicit the shedding of membrane-bound proto-growth factors, cleave membrane receptors leading to signal activation or suppression, or interrupt cell–cell adhesion as observed in advanced cancers. Indeed, several of the “soluble”, secreted MMPs and their transmembrane cousins effect the shedding of several such factors, including EGFR ligands and ErbB receptors, TNF family members and cadherins, among many others [2, 3].

Matrix Metalloproteinase Biology, First Edition. Edited by Irit Sagi and Jean P. Gaffney.

Genetic knockouts of the MMPs supported this more subtle method of affecting cellular behavior and led us into a renaissance in MMP biology, introducing the principle of proteolysis being the ultimate post-translational protein modification within a complex signaling network.

Coevolving with these newly discovered roles of MMPs was our understanding of the signaling functions of the ECM itself. The ECM acts as a reservoir for otherwise soluble growth factors and cytokines, waiting to be liberated by proteolysis. Furthermore, the complex array of proteins that make up the ECM have their own signaling receptors on the cell surface, usually comprised of integrin family heterodimers. Interaction with these receptors can be disrupted by MMP cleavage of either the ECM ligand or its receptor, but can alternatively be transformed by the creation of neoepitopes liberated by proteolysis that create a distinct ligand/receptor interaction. Finally, as a part of its most basic function of establishing tissue architecture, the density and composition of the ECM places different degrees of torsional stress upon the cell which conveys critical signaling cues. It follows that modification of the ECM by MMP cleavage alters these stress signals in ways that we have only begun to appreciate. In this chapter, we focus on these many ECM-mediated signaling functions of the MMP family.

6.2 The extracellular matrix as a source for signaling ligands

Interdependent tissue compartments in mammalian organisms are separated by the ECM which serves as a structural scaffold and physically delineates tissue boundaries to maintain tissue architecture. These matrices are variably comprised of glycoproteins and proteoglycans, including collagens, laminins, fibronectin, and vitronectin. Basement membrane type ECMs underlie epithelial and endothelial cells and are comprised predominantly of laminin isoforms, type IV collagen, heparin sulfate proteoglycans, and other relatively low abundance matrix components such as nidogen [4]. In contrast, fibrillar collagens types I and III are more abundant in interstitial matrices, together with fibronectin, tenascin, and proteoglycans [5]. Tissue specificity is dictated, in part, by defined differences in matrix protein composition and post-translational modifications (such as cross-linking and glycosylation).

Several traditionally soluble growth factors have a natural affinity for specific ECM components. Vascular endothelial growth factor (VEGF) is often embedded in the basement membrane and ECM by interacting with heparin sulfate proteoglycans (HSPGs) [6]. The "latency associated peptide" (LAP) is complexed with transforming growth factor beta (TGF-β) which together covalently associate with latent TGF-β-binding protein (LTBP), itself a member of the fibrillin family. This trimeric complex binds tightly to several ECM components, including fibrillin, fibronectin, and decorin [7]. HSPGs also bind to fibroblast growth factors (FGFs) [8], hepatocyte growth factor (HGF) [9] and heparin binding EGF-like growth factor (HBEGF) [10] where they can act as co-factors for receptor activation [9–11]. Thus, specific ECM makeup has the potential to create distinct signaling microenvironments in addition to providing structure, stiffness, and topology to maintain tissue architecture and regulate the behavior of matrix-adherent cells [12].

The enzymatic release of embedded growth factors from the ECM is an elegant method of orchestrating a rapid response within actively remodeling and neoplastic tissues. Bergers and colleagues showed that neoplastic cells could use a variety of proteases to release VEGF from the surrounding matrix [13], greatly enhancing vascularization of the tumor tissue, thereby providing both nutrients to and escape routes from primary tumors. MMPs that have been shown to release VEGF include MMP-1, MMP-3, MMP-7, MMP-9, MMP-16, and MMP-19 [13, 14]. The broad substrate specificities of these MMPs suggest that VEGF liberation is likely due to general basement membrane degradation. In contrast, the LAP peptide of TGFβ can be clipped by MMP-2, MMP-9, MMP-13, and MMP-14 [15–17], leading to the release of the mature soluble TGFβ ligand. Other MMPs, including MMP-2, MMP-3, and MMP-7, can directly cleave the basement membrane protein decorin to liberate TGFβ stores from the ECM [18], though presumably the LAP still must be cleaved. Whichever method is employed in a given context, TGFβ1 activation can both slow epithelial cell proliferation and promote collagen-rich fibrosis [19]. These two processes are important in the later stages of wound healing in many tissues, essentially establishing a negative feedback loop that would use ECM degradation as signal that the wound pathology has progressed sufficiently. However, in the context of tumor progression, TGFβ liberation can also enhance epithelial tumor aggressiveness by inducing an epithelial-to-mesenchymal transition [19] while still promoting fibrosis, which can enhance tumor cell invasion and associated inflammation, as discussed later. These pro-angiogenesis and pro-fibrosis activities released by MMPs combined with MMP cleavage of HSPGs, which can release pro-survival/proliferation factors such as FGF, HGF, and HBEGF, ECM degradation, have the potential to greatly influence a variety of tissue pathologies.

Apart from the growth factors they sequester, ECM components are signaling molecules in their own right. Cells interact with ECM components predominantly through the integrin family of cell adhesion molecules. Integrins comprise a family of 24 distinct transmembrane proteins that assemble as heterodimers (comprised of an α and a β partner) to recognize specific matrix components [20, 21]. Additionally, the cytoplasmic tails of integrins interact through adhesion plaque components to couple to cytoskeletal proteins, thereby providing a conduit for information transfer from the ECM into the cell. Integrin regulation of Rho family GTPases controls actin dynamics, while activation of focal adhesion kinase (FAK) and Src kinases influences gene expression [22, 23]. Distinct hierarchies of cellular responses are transduced through integrins depending on the nature of specific ligand–integrin binding events that result in integrin occupancy, integrin aggregation (lateral clustering in the plane of the cell membrane) or both occupancy and aggregation [24, 25]. Even in a quiescent state, basement membrane ligation of integrin receptors is critical for cell survival. Loss of this contact through a variety of mechanisms, including MMP-mediated matrix degradation, can induce a form of apoptosis known as "anoikis", providing a classic example of how ECM proteolysis coordinates signal cessation [26].

Accompanying the destruction of traditional ECM receptor ligands is the creation of "neo-epitopes" that may ligate distinct integrin heterodimers or transmembrane receptors for other traditionally soluble ligands. An example of this is laminin 5, a major component of the epithelial cell basement membrane that can be processed by MMP14 and MMP2 to release an EGF-domain containing fragment, domain III (DIII). This DIII fragment can bind and activate the EGF receptor, promoting cellular proliferation and migration in adjacent cells, including invading tumor cells [27]. EGF-domains can also be released from thrombospondin by MMP processing [28].

Another powerful mode of ECM-derived, MMP-dependent signaling is the liberation of soluble peptides that can act distally. The most famous example of this would be endostatin, a peptide with anti-angiogenic activity that was found in the circulation of tumor-bearing mice [29]. Surprisingly, once purified, it was identified as a fragment of the α1 chain of Collagen XVIII, a basement membrane component [29], which could be released via cleavage by MMP-3, MMP-9, MMP-7, MMP-12, MMP-13, and MMP-14 [30]. Analogously, tumstatin is an anti-angiogenic peptide generated by MMP-9 cleavage of the α3 chain of Collagen IV [31]. Though both endostatin and tumstatin block angiogenesis, endostatin blocks endothelial cell migration by binding to integrin α5β1 whereas tumstatin induces endothelial cell apoptosis by binding to integrin αvβ3 [32]. Osteopontin (OPN) is a secreted phosphoprotein that can incorporate into the ECM and interact with integrin receptors [33]. Its expression has been associated with bone maturation and remodeling [34], inflammatory disease [35], aggressive cancers [33], and maintenance of cancer stem cells [36]. OPN can enhance tumor cell metastasis *in vivo* and can act as a macrophage, T-cell, and fibroblast chemoattractant [37, 38]. While OPN is itself an adhesive ligand, proteolysis by MMP-3 and MMP-7 enhances both its adhesive and pro-migratory properties [39]. Additionally, the macrophage chemoattractant activity of OPN is released by MMP-9 or thrombin cleavage [40].

Epithelial and stromal cells both contribute profoundly to inflammation in remodeling and neoplastic tissue by direct synthesis and secretion of pro-inflammatory chemokines, some of which can also be embedded in the ECM [41]. However, ECM fragmentation, possibly acting as an early harbinger of tissue damage, can also act as a proinflammatory signal. For instance, MMP-12 synthesized by macrophages can cleave elastin so as to create peptides that act as monocyte and fibroblast chemoattractants [42], the latter possibly enhancing fibrosis. The major ECM component in fibrosis, collagen I, can be cleaved by MMP-9 to generate glutamine-leucine-arginine (ELR) containing peptides. These peptides can act as neutrophil chemoattractants by mimicking the ELR domain in CXC chemokines [43]. Neutrophils are inevitably followed to inflammatory sites by macrophages, providing additional MMP-12. Thus, together with the other signals, MMP cleavage of ECM components may contribute to some of the pathologies of a persistent, self-sustaining fibro-inflammatory stroma.

6.3 ECM and mechanosensory signal transduction

Rather than functioning simply as either an inert scaffold or a reservoir of neo-ligands, the ECM also provides a physical source of epigenetically modified guidance cues to direct cell behavior during development and disease progression. In addition to the biochemical stimuli discussed above, cells can also be guided by gradients in ECM stiffness [44]. This phenomenon, termed "durotaxis", was first observed in fibroblasts [45] migrating preferentially toward stiffer matrices. Indeed, interesting new data show that cells can follow stiffness gradients even when the substratum constitutive properties are constant [46–48]. Thus, the response of cells to ECM ligand presentation is dependent on matrix material properties that are coupled to matrix stiffness, as well as matrix composition. This provides a mechanism whereby matrix stiffness, although not genetically encoded, may nevertheless initiate and drive disease. Excessive tissue remodeling, such as occurs in many disease states including cancer, alters matrix stiffness. Changes in matrix organization, such as the orientation of collagen fibrils, have also been shown to promote malignant transformation [49, 50]. To this end, it is interesting to note that most tumors represent an unregulated growth of epithelial cells, and early progression of such carcinomas is characterized by a disruption of normal boundaries between epithelial structures bounded by basement membranes and connective tissues with distinct material properties [51].

Cells in tissues are subjected to an array of mechanical forces generated in part by cell-matrix and cell–cell interactions that alter cytoskeletal dynamics. This information integrates with "traditional" biochemical cues (such as interaction of a soluble ligand with a cellular receptor) to modulate cell behavior [52–54]. Cells grown in two-dimensional culture, even in the presence of matrix proteins, often fail to exhibit properties of differentiation and tissue assembly, highlighting the critical role for matrix architecture in cell growth and phenotypic differentiation [12, 55]. Indeed, traditional tissue culture surfaces (plastic, glass) are significantly stiffer than biologic materials, including bone [54, 55], and likely signal aberrant gene expression pathways. This is demonstrated by studies comparing mammary epithelial cells grown in traditional two- dimensional culture relative to more physiologically relevant compliant basement membrane three-dimensional cultures, wherein cells assumed a more normal acinar phenotype [55]. Furthermore, by tuning the stiffness of the matrix, it was demonstrated that these cells responded to altered matrix stiffness by regulating the expression of greater than 1000 genes [54, 56]. Enhancing matrix stiffness to values similar to that measured in human breast cancer stroma was sufficient to induce pre-malignant changes in these cells. The concept that physical changes in the matrix microenvironment may drive aberrant differentiation was further supported by studies showing that changes in ECM cross-linking can alter breast cancer growth *in vivo* [57, 58]. Recent computer simulations provide additional evidence that alterations in ECM structure or mechanical properties can actively drive cancer progression by destabilizing tissue structure, accelerating neoplastic transformation, and uncontrolled growth [59].

Besides interacting with monomeric ECM ligands, cells are capable of sensing the spatial presentation of ECM ligands on the nanoscale and responding via alterations in integrin organization, clustering, and integrin-dependent activation of sub-cellular signaling pathways [60, 61]. In this manner, cells can probe the extracellular environment to generate distinct cellular responses dependent on the tissue-specific physical properties of the matrix, including rigidity, porosity, and topography [62].

Ample evidence suggests that integrin-mediated ECM adhesion is inherently a mechanosensory process wherein integrins can function as mechanotransducers, propagating physical signals from the ECM. In addition to ligand binding, force can also induce integrin activation and clustering, leading to phosphorylation of focal adhesion kinase to perpetuate downstream signaling [63–65]. This force-dependent activation enables rapid cellular response to changing force environments. Cells are also induced to alter the synthesis and secretion of ECM proteins as well as the expression of ECM-degrading proteinases in response to mechanical loading [54]. In normal tissues, this dynamic interplay regulates differentiation, whereas aberrant responses to mechanical stress can lead to pathologic changes in matrix deposition and degradation.

6.4 Matrix remodeling and modification of mechano-sensory signaling

MMPs have historically been implicated in pathologic processes such as tumor invasion and metastasis by catalyzing removal of physical barriers to tissue penetration. More recently this MMP-catalyzed tissue clearing has been shown to promote tumor cell proliferation by removing constraints on cytoskeletal changes necessary to drive proliferative responses [66]. Emerging data also suggest that localized (nanoscale to microscale) MMP-catalyzed remodeling of ECM surfaces is sufficient to change the physical properties of the ECM, altering mechanical cues perceived by the cell and subsequent activation of sub-cellular signal transduction pathways that control gene expression and ultimately impact cell behavior [12].

Early reports on the functional interplay between matrix structure, MMP expression and matrix dynamics showed that altering cell shape or engaging cellular integrins with anti-integrin antibodies induced MMP expression [67–69]. Subsequent studies using intact matrix proteins or matrix-derived peptides showed differential effects on MMP expression, suggesting that tissue-invasive cells can receive distinct signals from intact versus degraded matrix that may function to dictate subsequent processing of the ECM [68–72]. Modeling hierarchical interaction of cells with ECM using linear peptides, complex triple helical peptides, or bead-immobilized anti-integrin antibodies has provided additional insight into the physical nature of the integrin signal relative to the proteinase response [73, 74] and represents a tractable model system with which to probe cellular interaction with intact three-dimensional matrices versus peptides resulting from MMP-dependent matrix processing. This is exemplified by studies of interaction of a variety of tumor cells with two-dimensional or three-dimensional collagen, denatured collagen (gelatin), soluble integrin subunit-specific antibodies (that ligate integrins), and bead-immobilized integrin subunit-specific antibodies (that ligate and aggregate integrins) [75]. Induced expression of the membrane-anchored collagenase membrane type 1 MMP (MT1-MMP, MMP-14) was shown to require multi-valent ligation and aggregation of collagen binding integrins either by three-dimensional collagen gels or bead-immobilized anti-integrin antibodies, whereas a cellular MMP response was not elicited by denatured or proteolyzed matrix or soluble anti-integrin antibodies [76–79].

These *in vitro* experimental models can be used to understand and interpret more complex *in vivo* systems wherein it has been shown that solid tumors often exhibit heterogeneity in ECM microarchitecture with areas of both loose and dense connective tissue surrounding the tumor [80, 81]. Enhanced collagen deposition and crosslinking during tumor progression increase the overall density and stiffness of the tumor stroma and cancer cells can further reorganize the ECM through collagen bundling [81]. A desmoplastic response often accompanies tumor progression, characterized by significant changes in matrix deposition, stiffness, and cross-linking [54, 82]. These changes in tumor-associated matrix microarchitecture and the resultant signals transduced through cellular integrins may contribute to localized changes in MMP expression and subsequent MMP-catalyzed matrix degradation to form microtracks that provide a pathway for directed cell migration and may ultimately influence metastasis [81, 83, 84].

6.5 Conclusions and future directions

Many important questions remain regarding how tissue material properties are altered in disease and the impact of MMP-dependent changes in ECM biomechanics on cell-matrix cell behavior. Add to this many additional layers of complexity contributed by the simultaneous MMP-dependent induction of neo-ligand/receptor interactions, coupled with the shedding of membrane bound signals, and it is clear that the ultimate biological response is dictated by complex cellular integration of the totality of the signals present as a cell probes its immediate environment. While the dominant signal may dictate a relatively linear response, crosstalk between these pathways will undoubtedly modify the cellular response in unanticipated ways. For instance, forces applied on cells by the ECM can be transduced to the nucleus through integrin-cytoskeleton interactions to alter chromatin conformation and subsequent gene expression [52–54, 85], providing a sensitive epigenetic mechanism whereby localized ECM proteolysis can contribute to disease pathogenesis. Clearly, a more nuanced and mechanistic understanding of how localized changes in matrix deposition versus matrix proteolytic remodeling drive disease progression is warranted. Ultimately, integrative multi-disciplinary studies will address these questions by combining the tools and approaches from engineering, cell and animal biology, and computational modeling.

References

1. Jodele, S., Blavier, L., Yoon, J.M. & DeClerck, Y.A. (2006) Modifying the soil to affect the seed: role of stromal-derived matrix metalloproteinases in cancer progression. *Cancer and Metastasis Reviews*, 25, 35–43.
2. Page-McCaw, A., Ewald, A.J. & Werb, Z. (2007) Matrix metalloproteinases and the regulation of tissue remodelling. *Nature Reviews. Molecular Cell Biology*, 8, 221–33.
3. Parks, W.C., Wilson, C.L. & Lopez-Boado, Y.S. (2004) Matrix metalloproteinases as modulators of inflammation and innate immunity. *Nature Reviews. Immunology*, 4, 617–29.
4. Yurchenco, P.D. & Schittny, J.C. (1990) Molecular architecture of basement membranes. *The FASEB Journal*, 4, 1577–90.
5. Bosman, F.T. & Stamenkovic, I. (2003) Functional structure and composition of the extracellular matrix. *The Journal of Pathology*, 200, 423–8.
6. Robinson, C.J. & Stringer, S.E. (2001) The splice variants of vascular endothelial growth factor (VEGF) and their receptors. *Journal of Cell Science*, 114, 853–65.
7. Saharinen, J., Hyytiainen, M., Taipale, J. & Keski-Oja, J. (1999) Latent transforming growth factor-beta binding proteins (LTBPs)–structural extracellular matrix proteins for targeting TGF-beta action. *Cytokine and Growth Factor Reviews*, 10, 99–117.

8. Presta, M., Dell'Era, P., Mitola, S., Moroni, E., Ronca, R. & Rusnati, M. (2005) Fibroblast growth factor/fibroblast growth factor receptor system in angiogenesis. *Cytokine and Growth Factor Reviews*, 16, 159–78.
9. Cecchi, F., Pajalunga, D., Fowler, C.A. *et al.* (2012) Targeted disruption of heparan sulfate interaction with hepatocyte and vascular endothelial growth factors blocks normal and oncogenic signaling. *Cancer Cell*, 22, 250–62.
10. Iwamoto, R., Mine, N., Kawaguchi, T., Minami, S., Saeki, K. & Mekada, E. (2010) HB-EGF function in cardiac valve development requires interaction with heparan sulfate proteoglycans. *Development*, 137, 2205–14.
11. Jakobsson, L., Kreuger, J., Holmborn, K. *et al.* (2006) Heparan sulfate in trans potentiates VEGFR-mediated angiogenesis. *Developmental Cell*, 10, 625–34.
12. Karsdal, M.A., Nielsen, M.J., Sand, J.M. *et al.* (2013) Extracellular matrix remodeling: the common denominator in connective tissue diseases. Possibilities for evaluation and current understanding of the matrix as more than a passive architecture, but a key player in tissue failure. *Assay and Drug Development Technologies*, 11, 70–92.
13. Bergers, G., Brekken, R., McMahon, G. *et al.* (2000) Matrix metalloproteinase-9 triggers the angiogenic switch during carcinogenesis. *Nature Cell Biology*, 2, 737–44.
14. Lee, S., Jilani, S.M., Nikolova, G.V., Carpizo, D. & Iruela-Arispe, M.L. (2005) Processing of VEGF-A by matrix metalloproteinases regulates bioavailability and vascular patterning in tumors. *The Journal of Cell Biology*, 169, 681–91.
15. Mu, D., Cambier, S., Fjellbirkeland, L. *et al.* (2002) The integrin alpha(v)beta8 mediates epithelial homeostasis through MT1-MMP-dependent activation of TGF-beta1. *The Journal of Cell Biology*, 157, 493–507.
16. Dangelo, M., Sarment, D.P., Billings, P.C. & Pacifici, M. (2001) Activation of transforming growth factor beta in chondrocytes undergoing endochondral ossification. *Journal Of Bone and Mineral Research*, 16, 2339–47.
17. Yu, Q. & Stamenkovic, I. (2000) Cell surface-localized matrix metalloproteinase-9 proteolytically activates TGF-beta and promotes tumor invasion and angiogenesis. *Genes and Development*, 14, 163–76.
18. Imai, K., Hiramatsu, A., Fukushima, D., Pierschbacher, M.D. & Okada, Y. (1997) Degradation of decorin by matrix metalloproteinases: identification of the cleavage sites, kinetic analyses and transforming growth factor-beta1 release. *The Biochemical Journal*, 322(Pt 3), 809–14.
19. Bierie, B. & Moses, H.L. (2006) Tumour microenvironment: TGFbeta: the molecular Jekyll and Hyde of cancer. *Nature Reviews. Cancer*, 6, 506–20.
20. Hynes, R.O. (2002) Integrins: bidirectional, allosteric signaling machines. *Cell*, 110, 673–87.
21. Hynes, R.O. (2009) The extracellular matrix: not just pretty fibrils. *Science*, 326, 1216–9.
22. Yamada, K.M. & Miyamoto, S. (1995) Integrin transmembrane signaling and cytoskeletal control. *Current Opinion in Cell Biology*, 7, 681–9.
23. Schwartz, M.A. & Ingber, D.E. (1994) Integrating with integrins. *Molecular Biology of the Cell*, 5, 389–93.
24. Miyamoto, S., Akiyama, S.K. & Yamada, K.M. (1995) Synergistic roles for receptor occupancy and aggregation in integrin transmembrane function. *Science*, 267, 883–5.
25. Miyamoto, S., Teramoto, H., Coso, O.A. *et al.* (1995) Integrin function: molecular hierarchies of cytoskeletal and signaling molecules. *The Journal of Cell Biology*, 131, 791–805.
26. Egeblad, M. & Werb, Z. (2002) New functions for the matrix metalloproteinases in cancer progression. *Nature Reviews. Cancer*, 2, 161–74.
27. Schenk, S., Hintermann, E., Bilban, M. *et al.* (2003) Binding to EGF receptor of a laminin-5 EGF-like fragment liberated during MMP-dependent mammary gland involution. *The Journal of Cell Biology*, 161, 197–209.
28. Liu, A., Garg, P., Yang, S. *et al.* (2009) Epidermal growth factor-like repeats of thrombospondins activate phospholipase Cgamma and increase epithelial cell migration through indirect epidermal growth factor receptor activation. *The Journal of Biological Chemistry*, 284, 6389–402.
29. O'Reilly, M.S., Boehm, T., Shing, Y. *et al.* (1997) Endostatin: an endogenous inhibitor of angiogenesis and tumor growth. *Cell*, 88, 277–85.
30. Ferreras, M., Felbor, U., Lenhard, T., Olsen, B.R. & Delaisse, J. (2000) Generation and degradation of human endostatin proteins by various proteinases. *FEBS Letters*, 486, 247–51.
31. Hamano, Y., Zeisberg, M., Sugimoto, H. *et al.* (2003) Physiological levels of tumstatin, a fragment of collagen IV alpha3 chain, are generated by MMP-9 proteolysis and suppress angiogenesis via alphaV beta3 integrin. *Cancer Cell*, 3, 589–601.

32. Sudhakar, A., Sugimoto, H., Yang, C., Lively, J., Zeisberg, M. & Kalluri, R. (2003) Human tumstatin and human endostatin exhibit distinct antiangiogenic activities mediated by alpha v beta 3 and alpha 5 beta 1 integrins. *Proceedings of the National Academy of Sciences of the United States of America*, 100, 4766–71.
33. Anborgh, P.H., Mutrie, J.C., Tuck, A.B. & Chambers, A.F. (2010) Role of the metastasis-promoting protein osteopontin in the tumour microenvironment. *Journal of Cellular and Molecular Medicine*, 14, 2037–44.
34. Staines, K.A., MacRae, V.E. & Farquharson, C. (2012) The importance of the SIBLING family of proteins on skeletal mineralisation and bone remodelling. *The Journal of Endocrinology*, 214, 241–55.
35. Morimoto, J., Kon, S., Matsui, Y. & Uede, T. (2010) Osteopontin; as a target molecule for the treatment of inflammatory diseases. *Current Drug Targets*, 11, 494–505.
36. Pietras, A., Katz, A.M., Ekstrom, E.J. *et al.* (2014) Osteopontin-CD44 Signaling in the Glioma Perivascular Niche Enhances Cancer Stem Cell Phenotypes and Promotes Aggressive Tumor Growth. *Cell Stem Cell*, 14, 357–69.
37. O'Regan, A.W., Chupp, G.L., Lowry, J.A., Goetschkes, M., Mulligan, N. & Berman, J.S. (1999) Osteopontin is associated with T cells in sarcoid granulomas and has T cell adhesive and cytokine-like properties in vitro. *The Journal of Immunology*, 162, 1024–31.
38. Crawford, H.C., Matrisian, L.M. & Liaw, L. (1998) Distinct roles of osteopontin in host defense activity and tumor survival during squamous cell carcinoma progression in vivo. *Cancer Research*, 58, 5206–15.
39. Agnihotri, R., Crawford, H.C., Haro, H., Matrisian, L.M., Havrda, M.C. & Liaw, L. (2001) Osteopontin, a novel substrate for matrix metalloproteinase-3 (stromelysin-1) and matrix metalloproteinase-7 (matrilysin). *The Journal of Biological Chemistry*, 276, 28261–7.
40. Tan, T.K., Zheng, G., Hsu, T.T. *et al.* (2013) Matrix metalloproteinase-9 of tubular and macrophage origin contributes to the pathogenesis of renal fibrosis via macrophage recruitment through osteopontin cleavage. *Laboratory Investigation*, 93, 434–49.
41. Rodriguez, D., Morrison, C.J. & Overall, C.M. (2010) Matrix metalloproteinases: what do they not do? New substrates and biological roles identified by murine models and proteomics. *Biochimica et Biophysica Acta*, 1803, 39–54.
42. Nenan, S., Planquois, J.M., Berna, P. *et al.* (2005) Analysis of the inflammatory response induced by rhMMP-12 catalytic domain instilled in mouse airways. *International Immunopharmacology*, 5, 511–24.
43. Weathington, N.M., van Houwelingen, A.H., Noerager, B.D. *et al.* (2006) A novel peptide CXCR ligand derived from extracellular matrix degradation during airway inflammation. *Nature Medicine*, 12, 317–23.
44. Roca-Cusachs, P., Sunyer, R. & Trepat, X. (2013) Mechanical guidance of cell migration: lessons from chemotaxis. *Current Opinion in Cell Biology*, 25, 543–9.
45. Lo, C.M., Wang, H.B., Dembo, M. & Wang, Y.L. (2000) Cell movement is guided by the rigidity of the substrate. *Biophysical Journal*, 79, 144–52.
46. Maloney, J.M., Walton, E.B., Bruce, C.M. & Van Vliet, K.J. (2008) Influence of finite thickness and stiffness on cellular adhesion-induced deformation of compliant substrata. *Physical Review. E, Statistical, Nonlinear, and Soft Matter Physics*, 78, 041923.
47. Kuo, C.H., Xian, J., Brenton, J.D., Franze, K. & Sivaniah, E. (2012) Complex stiffness gradient substrates for studying mechanotactic cell migration. *Advanced Materials*, 24, 6059–64.
48. Choi, Y.S., Vincent, L.G., Lee, A.R. *et al.* (2012) The alignment and fusion assembly of adipose-derived stem cells on mechanically patterned matrices. *Biomaterials*, 33, 6943–51.
49. Clarijs, R., Ruiter, D.J. & De Waal, R.M. (2003) Pathophysiological implications of stroma pattern formation in uveal melanoma. *Journal of Cellular Physiology*, 194, 267–71.
50. Ruiter, D., Bogenrieder, T., Elder, D. & Herlyn, M. (2002) Melanoma-stroma interactions: structural and functional aspects. *The Lancet Oncology*, 3, 35–43.
51. Ingber, D.E. (2008) Can cancer be reversed by engineering the tumor microenvironment? *Seminars in Cancer Biology*, 18, 356–64.
52. Ingber, D.E. (2003) Tensegrity I. Cell structure and hierarchical systems biology. *The Journal of Cell Science*, 116, 1157–73.
53. Ingber, D.E. (2003) Tensegrity II. How structural networks influence cellular information processing networks. *The Journal of Cell Science*, 116, 1397–408.
54. Butcher, D.T., Alliston, T. & Weaver, V.M. (2009) A tense situation: forcing tumour progression. *Nature Reviews. Cancer*, 9, 108–22.
55. Paszek, M.J., Zahir, N., Johnson, K.R. *et al.* (2005) Tensional homeostasis and the malignant phenotype. *Cancer Cell*, 8, 241–54.
56. Rizki, A., Weaver, V.M., Lee, S.Y. *et al.* (2008) A human breast cell model of preinvasive to invasive transition. *Cancer Research*, 68, 1378–87.
57. Levental, K.R., Yu, H., Kass, L. *et al.* (2009) Matrix crosslinking forces tumor progression by enhancing integrin signaling. *Cell*, 139, 891–906.

58. Sternlicht, M.D., Bissell, M.J. & Werb, Z. (2000) The matrix metalloproteinase stromelysin-1 acts as a natural mammary tumor promoter. *Oncogene*, 19, 1102–13.
59. Werfel, J., Krause, S., Bischof, A.G. *et al.* (2013) How changes in extracellular matrix mechanics and gene expression variability might combine to drive cancer progression. *PLoS ONE*, 8, e76122.
60. Geblinger, D., Addadi, L. & Geiger, B. (2010) Nano-topography sensing by osteoclasts. *Journal of Cell Science*, 123, 1503–10.
61. Dalby, M.J., Riehle, M.O., Johnstone, H., Affrossman, S. & Curtis, A.S. (2002) In vitro reaction of endothelial cells to polymer demixed nanotopography. *Biomaterials*, 23, 2945–54.
62. Lu, P., Weaver, V.M. & Werb, Z. (2012) The extracellular matrix: a dynamic niche in cancer progression. *The Journal of Cell Biology*, 196, 395–406.
63. Galbraith, C.G., Yamada, K.M. & Sheetz, M.P. (2002) The relationship between force and focal complex development. *The Journal of Cell Biology*, 159, 695–705.
64. Riveline, D., Zamir, E., Balaban, N.Q. *et al.* (2001) Focal contacts as mechanosensors: externally applied local mechanical force induces growth of focal contacts by an mDia1-dependent and ROCK-independent mechanism. *The Journal of Cell Biology*, 153, 1175–86.
65. Clark, E.A., King, W.G., Brugge, J.S., Symons, M. & Hynes, R.O. (1998) Integrin-mediated signals regulated by members of the rho family of GTPases. *The Journal of Cell Biology*, 142, 573–86.
66. Hotary, K.B., Allen, E.D., Brooks, P.C., Datta, N.S., Long, M.W. & Weiss, S.J. (2003) Membrane type I matrix metalloproteinase usurps tumor growth control imposed by the three-dimensional extracellular matrix. *Cell*, 114, 33–45.
67. Aggeler, J., Frisch, S.M. & Werb, Z. (1984) Changes in cell shape correlate with collagenase gene expression in rabbit synovial fibroblasts. *The Journal of Cell Biology*, 98, 1662–71.
68. Werb, Z., Tremble, P.M., Behrendtsen, O., Crowley, E. & Damsky, C.H. (1989) Signal transduction through the fibronectin receptor induces collagenase and stromelysin gene expression. *The Journal of Cell Biology*, 109, 877–89.
69. Seftor, R.E., Seftor, E.A., Gehlsen, K.R. *et al.* (1992) Role of the alpha v beta 3 integrin in human melanoma cell invasion. *Proceedings of the National Academy of Sciences of the United States of America*, 89, 1557–61.
70. Terranova, V.P., Williams, J.E., Liotta, L.A. & Martin, G.R. (1984) Modulation of the metastatic activity of melanoma cells by laminin and fibronectin. *Science*, 226, 982–5.
71. Kanemoto, T., Reich, R., Royce, L. *et al.* (1990) Identification of an amino acid sequence from the laminin A chain that stimulates metastasis and collagenase IV production. *Proceedings of the National Academy of Sciences of the United States of America*, 87, 2279–83.
72. Stack, S., Gray, R.D. & Pizzo, S.V. (1991) Modulation of plasminogen activation and type IV collagenase activity by a synthetic peptide derived from the laminin A chain. *Biochemistry*, 30, 2073–7.
73. Lauer, J.L., Gendron, C.M. & Fields, G.B. (1998) Effect of ligand conformation on melanoma cell alpha3beta1 integrin-mediated signal transduction events: implications for a collagen structural modulation mechanism of tumor cell invasion. *Biochemistry*, 37, 5279–87.
74. Baronas-Lowell, D., Lauer-Fields, J.L., Borgia, J.A. *et al.* (2004) Differential modulation of human melanoma cell metalloproteinase expression by alpha2beta1 integrin and CD44 triple-helical ligands derived from type IV collagen. *The Journal of Biological Chemistry*, 279, 43503–13.
75. Munshi, H.G. & Stack, M.S. (2006) Reciprocal interactions between adhesion receptor signaling and MMP regulation. *Cancer and Metastasis Reviews*, 25, 45–56.
76. Azzam, H.S. & Thompson, E.W. (1992) Collagen-induced activation of the M(r) 72,000 type IV collagenase in normal and malignant human fibroblastoid cells. *Cancer Research*, 52, 4540–4.
77. Gilles, C., Polette, M., Seiki, M., Birembaut, P. & Thompson, E.W. (1997) Implication of collagen type I-induced membrane-type 1-matrix metalloproteinase expression and matrix metalloproteinase-2 activation in the metastatic progression of breast carcinoma. *Laboratory Investigation*, 76, 651–60.
78. Ellerbroek, S.M., Fishman, D.A., Kearns, A.S., Bafetti, L.M. & Stack, M.S. (1999) Ovarian carcinoma regulation of matrix metalloproteinase-2 and membrane type 1 matrix metalloproteinase through beta1 integrin. *Cancer Research*, 59, 1635–41.
79. Ellerbroek, S.M., Wu, Y.I., Overall, C.M. & Stack, M.S. (2001) Functional interplay between type I collagen and cell surface matrix metalloproteinase activity. *The Journal of Biological Chemistry*, 276, 24833–42.
80. Bordeleau, F., Tang, L.N. & Reinhart-King, C.A. (2013) Topographical guidance of 3D tumor cell migration at an interface of collagen densities. *Physical Biology*, 10, 065004.
81. Bordeleau, F., Alcoser, T.A. & Reinhart-King, C.A. (2014) Physical biology in cancer. 5. The rocky road of metastasis: the role of cytoskeletal mechanics in cell migratory response to 3D matrix topography. *American Journal of Physiology. Cell Physiology*, 306, C110–20.

82. Shields, M.A., Dangi-Garimella, S., Krantz, S.B., Bentrem, D.J. & Munshi, H.G. (2011) Pancreatic cancer cells respond to type I collagen by inducing snail expression to promote membrane type 1 matrix metalloproteinase-dependent collagen invasion. *The Journal of Biological Chemistry*, 286, 10495–504.
83. Friedl, P. & Wolf, K. (2003) Tumour-cell invasion and migration: diversity and escape mechanisms. *Nature Reviews. Cancer*, 3, 362–74.
84. Kraning-Rush, C.M., Carey, S.P., Lampi, M.C. & Reinhart-King, C.A. (2013) Microfabricated collagen tracks facilitate single cell metastatic invasion in 3D. *Integrative Biology*, 5, 606–16.
85. Tang, Y., Rowe, R.G., Botvinick, E.L. *et al.* (2013) MT1-MMP-dependent control of skeletal stem cell commitment via a beta1-integrin/YAP/TAZ signaling axis. *Developmental Cell*, 25, 402–16.

7 Meprin and ADAM Metalloproteases: Two Sides of the Same Coin?

Christoph Becker-Pauly[1] and Stefan Rose-John[1]

[1] *Biochemisches Institut, Medizinische Fakultät, Christian-Albrechts-Universität zu Kiel, Kiel, Germany*

7.1 Introduction

Proteolysis is an irreversible posttranslational modification that regulates tissue homeostasis and cell signaling events in health and disease. Recent studies have demonstrated that many proteases do not act as single players but are rather concerted in a complex network of other activators, inhibitors, and regulatory molecules, in the so-called protease web [1, 2]. Understanding pathological conditions requires functional characterization of each of the 578 human proteolytic enzymes, for further elucidation of potential therapeutic targets. Figuratively speaking, proteases are single components of complex machinery, with different impact on functionality. A car would not break down when the front bumper is removed, but obviously the safety of vehicle occupants under certain conditions would be reduced. Along the same line, genetic modifications in mice resulting in the loss of single proteases mostly reveal mild phenotypes, but might lead to severe defects when challenging the animals. One of the few candidates that show dramatic consequences for viability and subsequent embryonic/early lethality when deleted are the metalloproteases ADAM (a disintegrin and metalloprotease) 10 and ADAM17. On the one hand this allows for the identification of obvious biological functions of the enzymes, but on the other hand, in terms of therapeutic strategies, application of specific ADAM10/17 inhibitors under certain pathological conditions has to be considered very carefully. In this regard, it became obvious that the identification of regulatory molecules influencing the activity state of proteolytic enzymes might unravel molecular tools for clinical applications. Recently it was demonstrated that iRhom1/2, which are inactive homologues of transmembrane rhomboid proteases, guide ADAM17 from the ER to the Golgi, where subsequent activation occurs by furin [3–5]. Other important proteins for ADAM regulation are the tetraspanins [6]. ADAM10 binds to tetraspanin-15, which promotes the release of the enzyme from the ER [7].

Matrix Metalloproteinase Biology, First Edition. Edited by Irit Sagi and Jean P. Gaffney.

Taken together, knowing the molecular compounds involved in the protease web, including enzymes, activators, inhibitors, and other regulatory proteins, will help to identify the best targets for therapeutic treatment under certain pathological conditions.

7.2 Meprin metalloproteases

Meprin metalloproteases were first discovered when patients showed remaining proteolytic activity after pancreas surgery. In the clinic, the PABA-peptide was applied as a marker for chymotryptic activity, but was surprisingly cleaved by meprins, too [8]. These findings were published in 1980 and therefore meprins were initially named PABA-peptide hydrolases (PPH). At the same time, strong proteolytic activity of unknown origin was described for the mouse kidney, which could be assigned to the same enzymes, which explains the name meprin: metalloprotease from renal tissue [9].

Meprins, which are only found in vertebrates [10, 11], are members of the astacin family of zinc-endopeptidases and the metzincin superfamily. They are characterized by the conserved motif HExxHxxGxxHxxxRxDR [12]. Three major groups of metzincins exist: the matrix metalloproteinases (MMPs), the adamalysins, including the ADAMs (a disintegrin and metalloproteases domain), and the astacins. The meprins belong to the astacins [13].

Seven different astacin proteases are found in humans: meprin α and meprin β, BMP-1 (bone morphogenetic protein-1), mTld (mammalian tolloid), Tll-1 and Tll-2 (tolloid like-1 and -2), and ovastacin. The astacins have homologues catalytic domains, but differ in their exosite regions. BMP-1 and the tolloids were found to be important in dorsal/ventral patterning during embryogenesis [14] and ovastacin prevents polyspermy during fertilization [15, 16]. Meprin α and meprin β are important for the proteolytic maturation of fibrillar collagens [17, 18] and additionally involved in neurodegeneration, inflammation, and cancer [19].

7.3 Structure of meprin α and meprin β

Among astacins, the meprins exhibit unique structural features. Meprin α is the largest secreted protease known, forming complex homooligomers of molecular weights up to 6 MDa [20]. Meprin β is the only membrane-bound member of the astacin family. However, the protein can be shed from the cell surface by ADAM10 and ADAM17 [21, 22]. Meprins have only been identified in vertebrates, indicating specific functions within this subphylum. The genes of human meprin α and meprin β are located on chromosomes 6 and 8, respectively [23], encoding for an amino-terminal signal peptide important for secretion, followed by an propeptide that has to be cleaved for activation, the astacin-like catalytic domain, a MAM (meprin A5 protein tyrosine phosphatase μ) and a TRAF domain (tumour necrosis factor receptor associated factor), which are known to mediate protein-protein interactions, followed by an EGF-like (epidermal growth factor-like) domain, a C-terminal transmembrane domain and a cytosolic part, which can be phosphorylated in meprin β [19].

Meprin α, but not meprin β, contains an additional inserted domain (I-domain), which is proteolytically cleaved by furin within the Golgi [24]. The released ectodomain further oligomerizes, while the EGF-like, the transmembrane, and cytosolic domain remain membrane-bound (Fig. 7.1). Interestingly, isolated soluble forms of the EGF-like domain of meprin α were identified in human hemofiltrates [25]. The function, however, has not yet been elucidated.

Recently, the crystal structures of inactive and mature human meprin β have been solved [26]. It turned out that the protease forms dimers which are stabilized by an intermolecular disulfide bond between the MAM domains. Interestingly, the

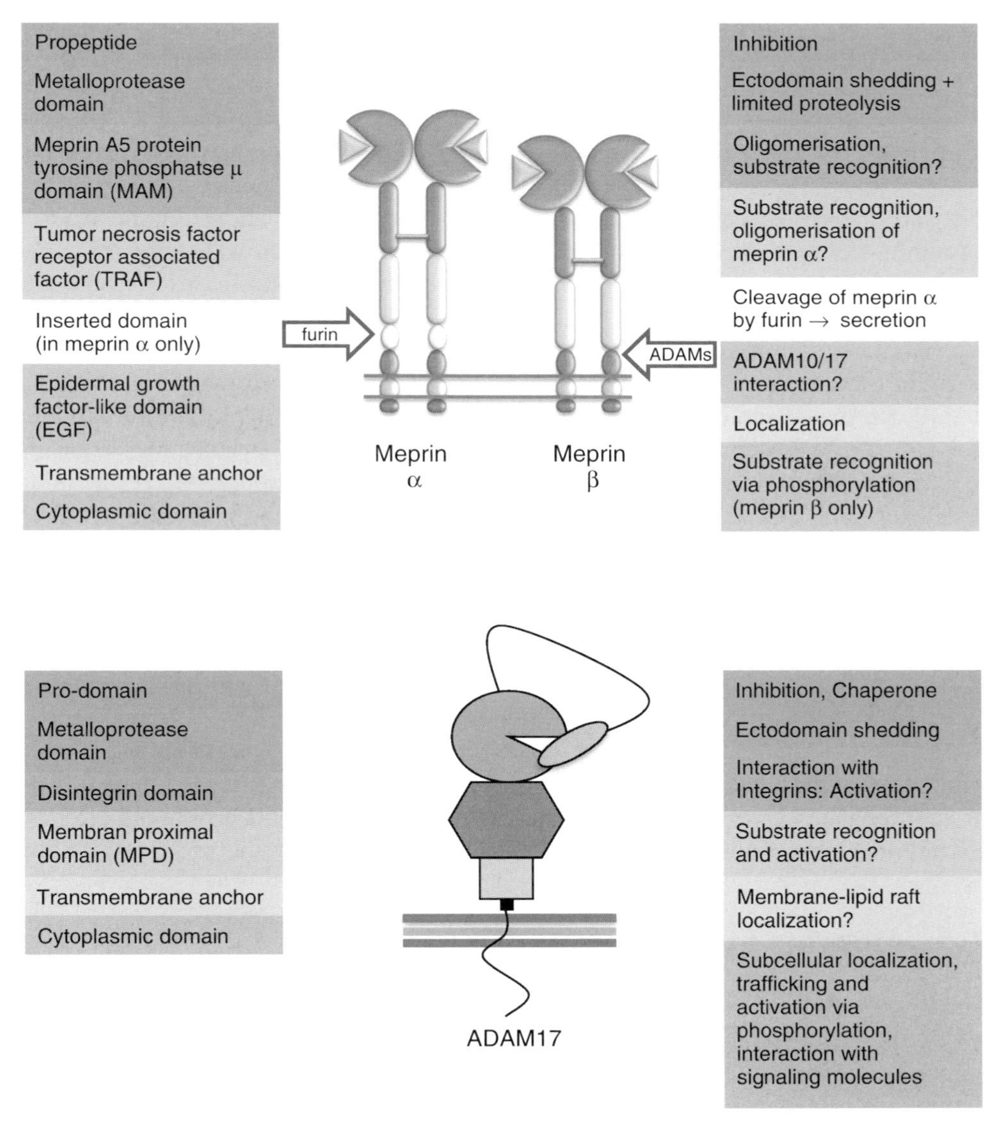

Figure 7.1 Domain structure and function of meprin α, meprin β and ADAM17. The functions of the domains are indicated in the figure. (*See insert for color representation of this figure.*)

orientation of the catalytic domains and the active site cleft demonstrates how meprin β acts at the cell surface as a sheddase, cleaving transmembrane proteins, such as the amyloid precursor protein (APP) [27]. Meprins are highly glycosylated proteins, forming sugar channels in an elongated active site cleft probably influencing substrate recognition and binding [26]. To gain full activity, meprins have to be activated proteolytically, because they have an N-terminal propeptide that blocks the active site. Meprin α and meprin β contain a conserved sequence motif, which gives rise to enzymatic activation by tryptic serine proteases. Several studies have revealed that pancreatic trypsin performs this activation in the gut [17, 28]. Outside the intestinal tract, meprin α, but not meprin β, can be efficiently activated by plasmin. In skin, the human tissue kallikrein-related peptidase 4 (KLK4), KLK5, and KLK8, were shown to cleave off the propeptide of meprin β, while maturation of meprin α is done by KLK5 only [29].

7.4 Proteomics for the identification of meprin substrates

The latest mass spectrometry based proteomics techniques allow for the determination of cleavage specificity and the identification of specific substrates of certain proteases [30]. By employing proteomic identification of protease cleavage sites (PICS), we were able to reveal a unique preference of meprin α and meprin β for negatively charged amino acid residues around the cleavage site, particularly at the P1′ position (nomenclature by Schechter & Berger) [31]. This information is especially helpful for the design of specific activity-based probes and potent inhibitors to monitor and regulate meprin activity, respectively. Additionally, it helped to determine meprin-mediated cleavage events in another proteomics approach. A technique called terminal amine isotopic labeling of substrates (TAILS) has allowed for the identification of more than 100 potential substrates for meprin α and meprin β, such as procollagens and the amyloid precursor protein (APP) [32]. These are described in more detail later in this chapter. 70% of the corresponding cleavage sites contained aspartate or glutamate, nicely reflecting the unique specificity of meprins. This is structurally based on positively charged arginine residues within the active site clefts of meprins, nicely seen in the crystal structure of active meprin β and the corresponding model of meprin α [26]. Meprin β indeed is capable of cleaving peptides consisting exclusively of negatively charged amino acids [17], which is due to additional positively charged residues within the prolonged active site cleft that are not present in meprin α.

7.5 Meprins in health and disease

Many substrates described for meprin α and meprin β were investigated *in vitro*, simply by incubating isolated proteins with the proteases. This led to the conclusion that meprins cleave compounds of the extracellular matrix such as laminin-V, collagen IV, fibronectin or nidogen 1, but also growth factors, cytokines and peptide hormones, including bradykinin, angiotensins, and gastrin [33]. These rather crude *in vitro* proteolysis assays might be misleading. Giving proteases the chance to hydrolyze peptide bonds under artificial conditions will result in non-physiological cleavage events that would be otherwise prevented in complex cellular protein networks. Hence, identification of substrates in cellular systems, as done by TAILS, and the analysis of appropriate animal models, for example meprin knock-out mice, is important and provides better knowledge about the biological functions of meprin α and meprin β in health and disease.

7.6 Proteolytic back-and-forth of meprins and ADAMs

Based on the TAILS data, we found an interesting proteolytic network of meprin β and ADAMs with consequences for the processing of the APP [17]. A cleavage product of APP, the amyloid beta peptide (Aβ), is found at high levels in Alzheimer's disease brains, where it promotes neurotoxicity [34]. Aβ is released during the amyloidogenic pathway, through cleavage of APP by a β-secretase, predominantly BACE1, and γ-secretase. We found bivalent cleavage of APP by meprin β at the β-secretase site and in the ectodomain, releasing N-terminal fragments of 11 and 22 kDa (APP11 and APP22) [27, 35]. Indeed, the 11 kDa peptide was previously observed in human brain lysates [36], although its function remains ambiguous. ADAM10, the constitutive α-secretase, cleaves APP within the Aβ sequence, thereby releasing the soluble ectodomain sAPPα, which prevents the formation of neurotoxic Aβ plaques [37, 38]. We found that meprin β can activate the α-secretase ADAM10 through cleavage of the propeptide [22]. ADAM10 is initially cleaved by proprotein convertases during the secretory pathway, but the prodomain still binds to the catalytic domain of ADAM10 and acts as an inhibitor [39]. Cleavage of ADAM10 by meprin β releases the propeptide, resulting in increased proteolytic activity. The potential protective role that the activation of ADAM10 plays in the progression of AD has to be further investigated, especially with regard to the counteracting β-secretase activity of meprin β. However, ADAM10/17 are sheddases of meprin β, releasing the ectodomain from the cell surface, which is then no longer capable of releasing Aβ [27]. Thus, the meprin β-ADAM-axis might contribute to the constitutive processing of APP in the non-amyloidogenic pathway. Interestingly, a recent study provides evidence that not only ADAM10, but also ADAM17 might play a protective role in the progression of Alzheimer's and prion disease [40]. Hence, the lack of ADAM10/17 activity would lead to decreased α-cleavage of APP and enhanced meprin β activity at the cell surface, thereby increasing Aβ levels.

7.7 Collagen fibril formation

The most striking biological function of meprins derived from proteomics data and knock-out mice phenotypes is the C- and N-procollagen proteinase activity. Meprin α and meprin β cleave off the globular propeptides of collagen I+III, thereby inducing its assembly and fibril formation *in vivo* [17, 18]. Interestingly, the C-terminal cleavage sites in the proα1(I) chain generated by both meprins were identified as Ala1218/Asp1219, identical to the BMP-1 cleavage site. We further determined meprin-mediated cleavage at Tyr1108/Asp1109 in the proα2(I) chain, some residues

N-terminal to the known BMP-1 site [40]. BMP-1 and tolloid proteases were thought to be the only essential C-procollagen proteinases. However, knock-out mice lacking these enzymes still have collagen fibrils, indicating that other proteases must exist to compensate the loss of activity. The efficiency of meprin procollagen proteinase activity was further demonstrated by de novo fibril formation experiments visualized by electron microscopy [17]. Only full maturation of procollagen I by meprins, and not the partial C-terminal processing by BMP-1, resulted in spontaneous collagen fibril assembly. The collagen fibrils in BMP-1 knock-out embryos exhibit smaller diameters and an abnormally organized arrangement compared with WT animals [14, 41]. This particular phenotype is seen in meprin α and meprin β knock-out mice, too. Decreased procollagen I conversion in skin and primary fibroblasts isolated from meprin- deficient mice provide further evidence that both enzymes are important for collagen maturation *in vivo*.

Identification of the meprin cleavage sites at the N-propeptides revealed position Tyr81/Asp82 in the α2 chain, an amino acid region known to cause Ehlers–Danlos syndrome VIIB when deleted. This results in reduced N-terminal processing of procollagen I and subsequent disordered collagen assembly, resulting in fibrils that are more loosely and randomly organized and exhibiting smaller diameters. Again, this phenotype is similar to the phenotype observed in meprin α and meprin β knock-out mice [17].

It was previously shown that meprins are overexpressed in fibrotic skin, characterized by massive deposition of fibrillar collagen [18]. Such conditions are further promoted by reduced activity of collagen degrading enzymes such as MMPs. In our TAILS approach, we identified MMP1 as a substrate of meprin α and meprin β. The cleavage site is in close proximity to the active site cleft and pre-incubation with meprins indeed revealed inactivation of MMP1 [17]. This makes sense with regard to fibrosis and the observed phenotype in meprin knock-out mice, where increased MMP-1 activity might contribute to decreased collagen deposition in skin.

Overall, the data demonstrate that meprins are important for the assembly of collagen fibrils and the integrity of the connective tissue. This is supported by a morpholino-induced meprin knock-down in zebrafish embryos, which revealed severe defects in organogenesis and tail morphology [42]. Moreover, meprins are likely associated with inherited collagen disorders and keloids/hypertrophic scars, which make them promising candidates for therapeutic applications to limit fibrosis or related diseases.

7.8 Angiogenesis and cancer

In contrast to the knock-down of meprin β in zebrafish embryos, meprin α morphants did not show such general deformations and organ failure but revealed dramatic impairments of the vascular system [42]. In the morpholino-treated fishes we observed erythrocyte accumulation due to largely diminished vascular system. Comparable phenotypes were described for VEGF-A morphants [43], revealing a possible role of meprin α in angiogenesis. Interestingly, in the TAILS approach, VEGF-A was found to be a substrate of meprin α and biochemical analysis validated this proteolytic event [22]. Angiogenesis is an important biological process, but additionally has an impact on tumor progression. Meprins were found to be up-regulated in breast and colon carcinoma, correlating with increased tumor growth [44–46]. Indeed, treatment of cultured human breast carcinoma cells with actinonin, a potent meprin inhibitor, resulted in decreased invasiveness [44].

7.9 Inflammation

Meprins are strongly expressed in human and rodent intestine and in mice, additionally in the kidney, where it makes up to 5% of the total cellular protein [33]. Therefore, meprin research with regard to pathophysiological conditions was focused on these organs. Several mouse studies revealed a strong correlation of altered meprin expression with acute kidney injury (AKI) and nephritis [47, 48]. For instance, down-regulation of meprins was found in chronic pathologies, such as experimental diabetes, adriamycin-induced nephropathy, hydronephrosis, in collagen IVA3 knock-out mice that develop Alport's syndrome, and in passive Heymann nephritis [17]. Increased levels or mislocation of meprin α and meprin β are related to acute renal failure. It has been demonstrated in mouse models of ischemia, reperfusion injury, and acute renal failure that meprins appear untypically at the basolateral side of tubular epithelial cells, leading to cytotoxicity and tissue damage [49–51]. This is supported by the effect of the meprin inhibitor actinonin, which prevented renal pathology in these mouse models [50], but also reduced the late organ-damaging effects of sepsis [52]. Additionally, the lack of meprin activity in corresponding knock-out mice significantly protected the animals against renal ischemia-reperfusion injury and bladder inflammation [53].

In chronic inflammation of the intestine, such as inflammatory bowel disease (IBD), meprins were identified as susceptibility genes. This is based on SNPs (single nucleotide polymorphisms) and a decreased mRNA expression in the intestinal epithelium of patients suffering from IBD [54]. Meprin α knock-out mice treated with DSS (dextran sodium sulphate) showed increased inflammation and intestinal injury compared with wild-type animals, characterized by more severe changes in clinical symptoms associated with IBD.

These particular observations together with our recent findings that meprins are capable of activating ADAMs provide evidence of a proteolytic cascade under certain inflammatory conditions (Fig.7.2).

7.10 ADAM Proteases

In 1997 researchers from two biotechnological companies, Immunex and Glaxo, reported in *Nature* the cloning of a human cDNA coding for a TNFα converting enzyme, TACE. TNFα is a type II membrane protein, which is cleaved by a metalloprotease, released into the bloodstream to act as a systemic inflammatory cytokine. The cloned cDNA was predicted to code for an 824 amino acid protein belonging to the disintegrin family of metalloproteases (ADAM) [55, 56]. The founding member of this membrane protein family was fertilin, which was shown to be responsible for

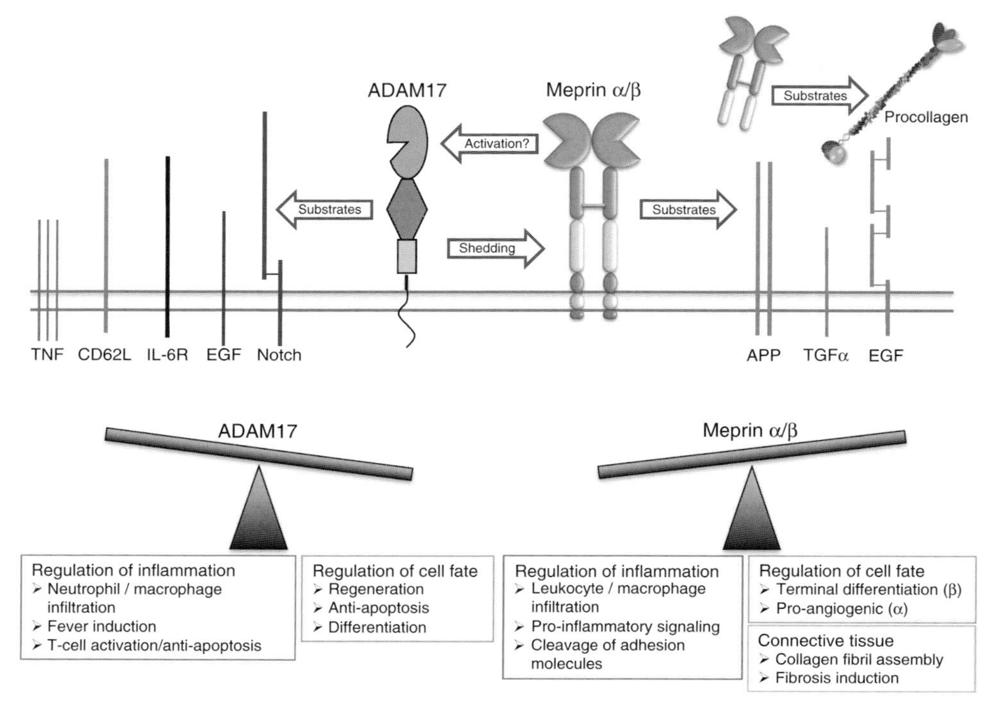

Figure 7.2 Physiological functions of ADAM17, meprin α, and meprin β. Both proteases orchestrate different processes in development and during the activation of the immune system. (*See insert for color representation of this figure.*)

the fusion of membranes of sperm and egg [57]. The protein TACE was later named ADAM17 [58].

ADAM17 consists of a signal peptide, a pro-domain, a metalloprotease domain, a disintegrin domain, a membrane proximal domain, a transmembrane domain, and a short cytoplasmic domain (Fig. 7.1).

Using cells lacking a functional ADAM17 gene, it was shown that ADAM17 was responsible for most of the TNFα release from T-cells and myeloid cells, although other proteases with TNFα converting activity could not be completely excluded [55].

Surprisingly, when ADAM17 knock-out mice were generated, it turned out that ADAM17$^{-/-}$ mice were not viable. Most of the mice died between embryonic day 17.5 and the first day after birth and only less than 1% of the mice survived and showed significantly reduced body weight [59]. The striking similarities of the ADAM17$^{-/-}$ animals with mice lacking TGFα, one of several ligands for the EGF-R [60, 61], led to the conclusion that ADAM17 was also responsible for the cleavage of this transmembrane growth factor [59]. Now we know that ADAM17 can cleave at least 76 different substrates ranging from growth factors and cytokines to receptors and cell adhesion proteins [62].

Several years earlier it had already been shown that the cellular receptor for the pro-inflammatory cytokine Interleukin-6 (IL-6R) was shed from the cellular membrane to give rise to a soluble IL-6R [63, 64]. Interestingly, this soluble IL-6R could still bind its ligand and the IL-6/soluble IL-6R was shown to render cells, which by themselves were unresponsive to IL-6 since they did not express the IL-6R, responsive to the cytokine IL-6. This process was named IL-6 'trans-signaling' [65]. It turned out that the protease responsible for most of the cleavage of the IL-6R was ADAM17, although other proteases such as ADAM10 were also capable of cleaving the IL-6R protein and releasing the soluble IL-6R [66–68]. Shedding of the IL-6R by ADAM17 was strongly induced by the phorbol ester PMA [63, 64, 66]. Cleavage was shown to occur close to the transmembrane domain of the human IL-6R at the sequence SLAVQ_{357}/D_{358}SSSV [69].

7.11 The ADAM family of proteases

As already mentioned, ADAM17 belongs to a family of metalloproteases with high homology to soluble zinc-dependent proteases found in the venom of snakes [58, 70, 71]. The three dimensional structure of the catalytic domain of ADAM17 has been solved and turned out to be most similar to the catalytic domain of the snake venom protease adamalysin II [72]. The family of ADAM proteases consists of 30 members, all characterized by a protease and disintegrin domain, although about half of the ADAM family members have lost their protease activity in evolution [67]. The ADAM family member most related to ADAM17 is the protease ADAM10. This protease is also essential for life as ADAM10$^{-/-}$ mice die early during embryogenesis [73]. One of the main substrates of ADAM10 is the transmembrane protein Notch, which is important for many developmental processes [73]. Besides Notch, ADAM10 has been shown to cleave various substrates, including the Alzheimer protein APP, CD44, E-Cadherin, N-Cadherin and L1 [74]. Other ADAM family members with proven shedding activity include ADAM8, ADAM12, and ADAM15 [75].

7.12 Orchestration of different pathways by ADAM17

Therefore, the *in vivo* role of ADAM17 seems to feed into at least two different pathways; ADAM17 cleaves membrane bound TNFα and membrane expressed IL-6R. Both soluble TNFα [76] and soluble IL-6R [62, 77, 78] act in a pro-inflammatory way. On the other hand, ligands of the EGF-R [79] and Notch [80] play a role in differentiation, proliferation and wound-healing and are also important mediators of tumor growth. Therefore, ADAM17 orchestrates very different cellular responses, highlighting the importance of this proteolytic pathway (Fig.7.2) [81].

7.13 Regulation of ADAM17 activity

How is the activity of the ADAM17 protease regulated? ADAM17 is transcribed in nearly all cells of the body at similar levels. In most cells, however, ADAM17 is not expressed at the cell surface but is found in the endoplasmic reticulum (ER). Phosphorylation of the cytoplasmic tail of ADAM17 has been described to lead to transport of the protein to the cell surface [82]. Interestingly, during inflammatory states and in tumors, ADAM17 is mostly expressed at the plasma membrane [83, 84]. Recently, it has been found that the inactive protease iRhom2 is needed for trafficking of ADAM17 to the cell surface [3, 4, 85]. It turned out that iRhom2 is mainly expressed in myeloid cells. In other cells, ADAM17 trafficking to the plasma membrane depends on the related protein iRhom1 [86].

Regulated trafficking to the cell surface is, however, not the only mechanism by which the activity of ADAM17 is regulated. It was noted early on that the activity of ADAM17 was strongly stimulated by treatment of cells with the phorbol ester PMA [64]. It is known that PMA acts by binding to protein kinase C (PKC) and inducing its translocation to the cell membrane, which leads the activation of the kinase activity [87]. Consequently, the PMA-mediated stimulation of ADAM17 activity could be enhanced by overexpression of PKCα [63]. Subsequently, it was found that treatment of myeloid cells with limited concentrations of bacterial pore forming toxins, which insert into the membrane, induced rapid shedding of the proteins IL-6R and CD14 [88]. Depletion of cell membranes of cholesterol also induced rapid shedding of the IL-6R [68]. These experiments led to the hypothesis that changes of membrane homeostasis, either by translocation of PKC to the membrane, by membrane insertion of bacterial toxins, or by cholesterol depletion lead to activation of ADAM17. Consequently, we could demonstrate that induction of apoptosis, which is known to cause the exposure of phosphatidyl serine to the outer leaflet of the plasma membrane, resulted in activation of ADAM17 and subsequent shedding of the IL-6R [89]. Interestingly, this activation of ADAM17 was caspase-dependent and occurred when apoptosis of cells was induced by the DNA damaging agent doxorubicin, by stimulation of the Fas receptor and by growth factor deprivation [89].

Rapid activation of ADAM17 by PMA is independent of the cytoplasmatic tail of the protein [67]. These results were confirmed with additional stimuli such as IL-1β, and the activator of MAP-kinases, anisomycin [90]. The authors concluded that the rapid activation mechanism of ADAM17 did not rely on its cytoplasmic domain [90].

These experiments are in contrast to the report that phosphorylation of the cytoplasmatic tail of ADAM17 was necessary for the transport of the protein to the cell surface [82]. Likewise, it was reported by Xu et al. that the activity of ADAM17 was induced by stimulation of MAP-kinases via a shift from a dimeric to a monomeric conformation of the ADAM17 protein and consequent dissociation from the tissue inhibitor of metalloproteinase-3 (TIMP3) protein leading to the activation of the protease [4]. Although these discrepancies could not be finally resolved, it is possible that rapid stimulation of ADAM17 is mechanistically different from physiologic regulation of ADAM17 trafficking and activation. This notion is supported by a recent report, which links phosphorylation of the cytoplasmatic portion of ADAM17 by 3-phosphoinositide–dependent kinase-1 with the induction of internalization of the protein [40].

The activity of ADAM17 is also regulated by the cellular redox environment. Willems et al. showed that the activity of thiol isomerases, namely protein disulphide isomerase (PDI), maintained ADAM17 in an inactivated state [91]. The group of Grötzinger could recently explain the molecular mechanism of this inactivation. They solved the three dimensional structure of the membrane proximal domain of ADAM17 by NMR spectroscopy. They further demonstrated the direct interaction of PDI with the membrane proximal domain (MPD) of ADAM17 by tandem-mass spectrometry. The MPD of ADAM17 had earlier been implicated with dimerization and substrate recognition [83, 92]. PDI was shown to catalyze an isomerization of disulfide bridges within the MPD, resulting in a structural change from an open active state to an inactive closed conformation. The authors speculate that PDI thereby executes a molecular switch governing substrate accessibility and thereby shedding activity [93].

7.14 Role of ADAM17 *in vivo*

As already mentioned, ADAM17$^{-/-}$ mice have defects in the mammary epithelium, lung, eye, hair, heart, skin, and the vascular system. Most of the animals die between E17.5 and a few days after birth and could thus not be used to study the physiological role of this enzyme [59]. The group of Blobel generated floxed ADAM17 mice, which allowed the ablation of the ADAM17 gene in various tissues [94]. Mice which were ADAM17 deficient in myeloid cells were protected from LPS induced endotoxemia [94]. ADAM17 deficiency in leukocytes upon bacterial challenge led to increased survival of mice due to faster neutrophil recruitment and bacterial clearance [95]. When the ADAM17 gene was ablated in osteochondroprogenitor cells, the animals lived for a shorter time and had increased numbers of osteoclasts leading to symptoms of osteoporosis [96]. Inactivation of the ADAM17 gene in hepatocytes showed that ADAM17 plays a role in the cellular protection of apoptosis [97]. Our group had chosen a different strategy to study the physiological role of ADAM17. We generated hypomorphic ADAM17 mice, which exhibited about 5% of normal ADAM17 expression levels in all tissues [79]. Using these mice we could show that ADAM17 plays a protective role in inflammatory bowel disease. The ligand of the EGF receptor (EGF-R) TGFα is necessary for regeneration of the intestinal epithelium during inflammation. Failure of ADAM17 to cleave TGFα led to a breakdown of the intestinal barrier with subsequent infiltration of immune cells and severe inflammation [79].

Our results were confirmed by a study in which mice were subjected to a random mutagenesis screen for exaggerated susceptibility to inflammatory bowel disease. In this screen, a mouse with a mutation within the metalloprotease domain of the ADAM17 gene was identified [98].

7.15 Role of ADAM17 in humans

Interestingly, a patient with symptoms of inflammatory skin and bowel disease was shown to carry a homozygous loss-of-function mutation in the ADAM17 gene [99]. Blood cells of this patient could not be stimulated to cleave membrane bound TNFα. The fact that in the patient, complete lack of ADAM17 results in a much milder phenotype than in mice [59] might point to a compensation mechanism in humans, which is not present in mice. This might be good news for companies thinking about therapeutically targeting ADAM17. A blockade of ADAM17 might seem to be a good strategy for the treatment of inflammatory diseases and cancer. From the severe phenotype seen in mice it was concluded that complete absence of ADAM17 was not compatible with life. The relatively mild symptoms seen in the identified patient might lead to a reconsideration of the feasibility of therapeutic strategies aimed at the inactivation of ADAM17 [99].

In this regard, it will be interesting to investigate the proteolytic interaction of ADAM17 and meprin β in terms of activation and shedding. These proteases were independently found to be associated with chronic intestinal inflammation and the generation of appropriate tools for specific regulation of their activity might improve current clinical strategies.

References

1. Dufour, A. & Overall, C.M. (2013) Missing the target: matrix metalloproteinase antitargets in inflammation and cancer. *Trends in Pharmacological Sciences*, 34, 233–242.
2. Overall, C.M. & Blobel, C.P. (2007) In search of partners: linking extracellular proteases to substrates. *Nature Reviews. Molecular Cell Biology*, 8, 245–257.
3. Adrain, C., Zettl, M., Christova, Y., Taylor, N. & Freeman, M. (2012) Tumor necrosis factor signaling requires iRhom2 to promote trafficking and activation of TACE. *Science*, 335, 225–228.
4. McIlwain, D.R., Lang, P.A., Maretzky, T. *et al.* (2012) iRhom2 regulation of TACE controls TNF-mediated protection against Listeria and responses to LPS. *Science*, 335, 229–232.
5. Maretzky, T., McIlwain, D.R., Issuree, P.D. *et al.* (2013) iRhom2 controls the substrate selectivity of stimulated ADAM17-dependent ectodomain shedding. *Proceedings of the National Academy of Sciences of the United States of America*, 110, 11433–11438.
6. Yanez-Mo, M., Gutierrez-Lopez, M.D. & Cabanas, C. (2011) Functional interplay between tetraspanins and proteases. *Cellular and Molecular Life Sciences*, 68, 3323–3335.
7. Prox, J., Willenbrock, M., Weber, S. *et al.* (2012) Tetraspanin15 regulates cellular trafficking and activity of the ectodomain sheddase ADAM10. *Cellular and Molecular Life Sciences*, 69, 2919–2932.
8. Sterchi, E., Green, J. & Lentze, M.J. (1982) Non-pancreatic hydrolysis of N-benzoyl-L-tyrosyl-p-aminobenzoic acid (PABA-peptide) in the human small intestine. *Clinical Science*, 62, 557–560.
9. Beynon, R., Shannon, J. & Bond, J. (1981) Purification and characterization of a metallo-endoproteinase from mouse kidney. *Biochemical Journal*, 199, 591–598.
10. Becker-Pauly, C., Bruns, B.C., Damm, O. *et al.* (2009) News from an ancient world: two novel astacin metalloproteases from the horseshoe crab. *Journal of Molecular Biology*, 385, 236–248.
11. Schutte, A., Lottaz, D., Sterchi, E.E., Stocker, W. & Becker-Pauly, C. (2007) Two alpha subunits and one beta subunit of meprin zinc-endopeptidases are differentially expressed in the zebrafish Danio rerio. *Biological Chemistry*, 388, 523–531.
12. Gomis-Ruth, F.X., Trillo-Muyo, S. & Stocker, W. (2012) Functional and structural insights into astacin metallopeptidases. *Biological Chemistry*, 393, 1027–1041.
13. Gomis-Rüth, F.X. (2003) Structural aspects of the metzincin clan of metalloendopeptidases. *Molecular Biotechnology*, 24, 157–202.
14. Hopkins, D.R., Keles, S. & Greenspan, D.S. (2007) The bone morphogenetic protein 1/Tolloid-like metalloproteinases. *Matrix Biology*, 26, 508–523.
15. Avella, M.A., Xiong, B. & Dean, J. (2013) The molecular basis of gamete recognition in mice and humans. *Molecular Human Reproduction*, 19, 279–289.
16. Burkart, A.D., Xiong, B., Baibakov, B., Jimenez-Movilla, M. & Dean, J. (2012) Ovastacin, a cortical granule protease, cleaves ZP2 in the zona pellucida to prevent polyspermy. *The Journal of Cell Biology*, 197, 37–44.
17. Broder, C., Arnold, P., Vadon-Le Goff, S. *et al.* (2013) Metalloproteases meprin alpha and meprin beta are C- and N-procollagen proteinases important for collagen assembly and tensile strength. *Proceedings of the National Academy of Sciences of the United States of America*, 110, 14219–14224.
18. Kronenberg, D., Bruns, B.C., Moali, C. *et al.* (2010) Processing of Procollagen III by Meprins: New Players in Extracellular Matrix Assembly? *Journal of Investigative Dermatology*.
19. Broder, C. & Becker-Pauly, C. (2013) The metalloproteases meprin alpha and meprin beta: unique enzymes in inflammation, neurodegeneration, cancer and fibrosis. *The Biochemical Journal*, 450, 253–264.

20. Bertenshaw, G.P., Norcum, M.T. & Bond, J.S. (2003) Structure of homo- and hetero-oligomeric meprin metalloproteases. Dimers, tetramers, and high molecular mass multimers. *The Journal of Biological Chemistry*, 278, 2522–2532.
21. Hahn, D., Pischitzis, A., Roesmann, S. *et al.* (2003) Phorbol 12-myristate 13-acetate-induced ectodomain shedding and phosphorylation of the human meprinbeta metalloprotease. *The Journal of Biological Chemistry*, 278, 42829–42839.
22. Jefferson, T., Auf dem Keller, U., Bellac, C. *et al.* (2013) The substrate degradome of meprin metalloproteases reveals an unexpected proteolytic link between meprin beta and ADAM10. *Cellular and Molecular Life Sciences*, 70, 309–333.
23. Hahn, D., Illisson, R., Metspalu, A. & Sterchi, E.E. (2000) Human N-benzoyl-L-tyrosyl-p-aminobenzoic acid hydrolase (human meprin): genomic structure of the alpha and beta subunits. *The Biochemical Journal*, 346 (Pt 1), 83–91.
24. Hahn, D., Lottaz, D. & Sterchi, E.E. (1997) C-cytosolic and transmembrane domains of the N-benzoyl-L-tyrosyl-p-aminobenzoic acid hydrolase alpha subunit (human meprin alpha) are essential for its retention in the endoplasmic reticulum and C-terminal processing. *European Journal of Biochemistry*, 247, 933–941.
25. Richter, R., Schulz-Knappe, P., Schrader, M. *et al.* (1999) Composition of the peptide fraction in human blood plasma: database of circulating human peptides. *Journal of Chromatography B: Biomedical Sciences and Applications*, 726, 25–35.
26. Arolas, J.L., Broder, C., Jefferson, T. *et al.* (2012) Structural basis for the sheddase function of human meprin beta metalloproteinase at the plasma membrane. *Proceedings of the National Academy of Sciences of the United States of America*, 109, 16131–16136.
27. Bien, J., Jefferson, T., Causevic, M. *et al.* (2012) The metalloprotease meprin beta generates amino terminal-truncated amyloid beta peptide species. *The Journal of Biological Chemistry*, 287, 33304–33313.
28. Becker, C., Kruse, M.N., Slotty, K.A. *et al.* (2003) Differences in the activation mechanism between the alpha and beta subunits of human meprin. *Biological Chemistry*, 384, 825–831.
29. Ohler, A., Debela, M., Wagner, S., Magdolen, V. & Becker-Pauly, C. (2010) Analyzing the protease web in skin: meprin metalloproteases are activated specifically by KLK4, 5 and 8 vice versa leading to processing of proKLK7 thereby triggering its activation. *Biological Chemistry*, 391, 455–460.
30. Auf dem Keller, U. & Schilling, O. (2010) Proteomic techniques and activity-based probes for the system-wide study of Proteolysis. *Biochimie*, 92, 1705–1714.
31. Schechter, I. & Berger, A. (1967) On the size of the active site in proteases. I. Papain. *Biochemical and Biophysical Research Communications*, 27, 157–162.
32. Becker-Pauly, C., Barre, O., Schilling, O. *et al.* (2011) Proteomic analysis reveal an acidic prime side specificity for the astacin metalloprotease family reflected by physiological substrates. *Molecular and Cellular Proteomics*, 10 (M111), 009233.
33. Sterchi, E.E., Stocker, W. & Bond, J.S. (2008) Meprins, membrane-bound and secreted astacin metalloproteinases. *Molecular Aspects of Medicine*, 29, 309–328.
34. Selkoe, D.J. (2001) Alzheimer's disease results from the cerebral accumulation and cytotoxicity of amyloid beta-protein. *Journal of Alzheimer's Disease*, 3, 75–80.
35. Jefferson, T., Causevic, M., auf dem Keller, U. *et al.* (2011) Metalloprotease meprin beta generates nontoxic N-terminal amyloid precursor protein fragments in vivo. *The Journal of Biological Chemistry*, 286, 27741–27750.
36. Portelius, E., Brinkmalm, G., Tran, A. *et al.* (2010) Identification of novel N-terminal fragments of amyloid precursor protein in cerebrospinal fluid. *Experimental Neurology*, 223, 351–358.
37. Prox, J., Bernreuther, C., Altmeppen, H. *et al.* (2013) Postnatal disruption of the disintegrin/metalloproteinase ADAM10 in brain causes epileptic seizures, learning deficits, altered spine morphology, and defective synaptic functions. *The Journal of Neuroscience*, 33, 12915–12928.
38. Kuhn, P.H., Wang, H., Dislich, B. *et al.* (2010) ADAM10 is the physiologically relevant, constitutive alpha-secretase of the amyloid precursor protein in primary neurons. *The EMBO Journal*, 29, 3020–3032.
39. Moss, M.L., Bomar, M., Liu, Q. *et al.* (2007) The ADAM10 prodomain is a specific inhibitor of ADAM10 proteolytic activity and inhibits cellular shedding events. *The Journal of Biological Chemistry*, 282, 35712–35721.
40. Pietri, M., Dakowski, C., Hannaoui, S. *et al.* (2013) PDK1 decreases TACE-mediated alpha-secretase activity and promotes disease progression in prion and Alzheimer's diseases. *Nature Medicine*, 19, 1124–1131.
41. Pappano, W.N., Steiglitz, B.M., Scott, I.C., Keene, D.R. & Greenspan, D.S. (2003) Use of Bmp1/Tll1 doubly homozygous null mice and proteomics to identify and validate in vivo substrates of bone morphogenetic protein 1/tolloid-like metalloproteinases. *Molecular and Cellular Biology*, 23, 4428–4438.

42. Schutte, A., Hedrich, J., Stocker, W. & Becker-Pauly, C. (2010) Let it flow: Morpholino knockdown in zebrafish embryos reveals a pro-angiogenic effect of the metalloprotease meprin alpha2. *PLoS ONE*, 5, e8835.
43. Nasevicius, A., Larson, J. & Ekker, S.C. (2000) Distinct requirements for zebrafish angiogenesis revealed by a VEGF-A morphant. *Yeast (Chichester, England)*, 17, 294–301.
44. Matters, G.L., Manni, A. & Bond, J.S. (2005) Inhibitors of polyamine biosynthesis decrease the expression of the metalloproteases meprin alpha and MMP-7 in hormone-independent human breast cancer cells. *Clinical and Experimental Metastasis*, 22, 331–339.
45. Lottaz, D., Maurer, C.A., Hahn, D., Buchler, M.W. & Sterchi, E.E. (1999) Nonpolarized secretion of human meprin alpha in colorectal cancer generates an increased proteolytic potential in the stroma. *Cancer Research*, 59, 1127–1133.
46. Lottaz, D., Maurer, C.A., Noel, A. *et al.* (2011) Enhanced activity of meprin-alpha, a pro-migratory and pro-angiogenic protease, in colorectal cancer. *PLoS ONE*, 6, e26450.
47. Carmago, S., Shah, S.V. & Walker, P.D. (2002) Meprin, a brush-border enzyme, plays an important role in hypoxic/ischemic acute renal tubular injury in rats. *Kidney International*, 61, 959–966.
48. Herzog, C., Seth, R., Shah, S.V. & Kaushal, G.P. (2007) Role of meprin A in renal tubular epithelial cell injury. *Kidney International*, 71, 1009–1018.
49. Holly, M.K., Dear, J.W., Hu, X. *et al.* (2006) Biomarker and drug-target discovery using proteomics in a new rat model of sepsis-induced acute renal failure. *Kidney International*, 70, 496–506.
50. Takayama, J., Takaoka, M., Yamamoto, S., Nohara, A., Ohkita, M. & Matsumura, Y. (2008) Actinonin, a meprin inhibitor, protects ischemic acute kidney injury in male but not in female rats. *European Journal of Pharmacology*, 581, 157–163.
51. Trachtman, H., Valderrama, E., Dietrich, J.M. & Bond, J.S. (1995) The role of meprin A in the pathogenesis of acute renal failure. *Biochemical and Biophysical Research Communications*, 208, 498–505.
52. Wang, Z., Herzog, C., Kaushal, G.P., Gokden, N. & Mayeux, P.R. (2010) Actinonin, a meprin a inhibitor, protects the renal microcirculation during sepsis. *Shock*.
53. Bylander, J., Li, Q., Ramesh, G., Zhang, B., Reeves, W.B. & Bond, J.S. (2008) Targeted disruption of the meprin metalloproteinase beta gene protects against renal ischemia-reperfusion injury in mice. *American Journal of Physiology. Renal Physiology*, 294, F480–490.
54. Banerjee, S., Oneda, B., Yap, L.M. *et al.* (2009) MEP1A allele for meprin A metalloprotease is a susceptibility gene for inflammatory bowel disease. *Mucosal Immunology*, 2, 220–231.
55. Black, R.A., Rauch, C.T., Kozlosky, C.J. *et al.* (1997) A metalloproteinase disintegrin that releases tumour-necrosis factor-alpha from cells. *Nature*, 385, 729–733.
56. Moss, M.L., Jin, S.L., Becherer, J.D. *et al.* (1997) Structural features and biochemical properties of TNF-alpha converting enzyme (TACE). *Journal of Neuroimmunology*, 72, 127–129.
57. Blobel, C.P., Wolfsberg, T.G., Turck, C.W., Myles, D.G., Primakoff, P. & White, J.M. (1992) A potential fusion peptide and an integrin ligand domain in a protein active in sperm-egg fusion. *Nature*, 356, 248–252.
58. Black, R.A. & White, J.M. (1998) ADAMs: focus on the protease domain. *Current Opinion in Cell Biology*, 10, 654–659.
59. Peschon, J.J., Slack, J.L., Reddy, P. *et al.* (1998) An essential role for ectodomain shedding in mammalian development. *Science*, 282, 1281–1284.
60. Luetteke, N.C., Qiu, T.H., Peiffer, R.L., Oliver, P., Smithies, O. & Lee, D.C. (1993) TGF alpha deficiency results in hair follicle and eye abnormalities in targeted and waved-1 mice. *Cell*, 73, 263–278.
61. Mann, G.B., Fowler, K.J., Gabriel, A., Nice, E.C., Williams, R.L. & Dunn, A.R. (1993) Mice with a null mutation of the TGF alpha gene have abnormal skin architecture, wavy hair, and curly whiskers and often develop corneal inflammation. *Cell*, 73, 249–261.
62. Jones, S.A., Scheller, J. & Rose-John, S. (2011) Therapeutic strategies for the clinical blockade of IL-6/gp130 signaling. *The Journal of Clinical Investigation*, 121, 3375–3383.
63. Müllberg, J., Schooltink, H., Stoyan, T., Heinrich, P.C. & Rose-John, S. (1992) Protein kinase C activity is rate limiting for shedding of the interleukin-6 receptor. *Biochemical and Biophysical Research Communications*, 189, 794–800.
64. Müllberg, J., Schooltink, H., Stoyan, T. *et al.* (1993) The soluble interleukin-6 receptor is generated by shedding. *European Journal of Immunology*, 23, 473–480.
65. Rose-John, S. & Heinrich, P.C. (1994) Soluble receptors for cytokines and growth factors: generation and biological function. *The Biochemical Journal*, 300, 281–290.
66. Müllberg, J., Durie, F.H., Otten Evans, C. *et al.* (1995) A metalloprotease inhibitor blocks shedding of the IL-6 receptor and the p60 TNF receptor. *The Journal of Immunology*, 155, 5198–5205.
67. Althoff, K., Reddy, P., Peschon, J., Voltz, N., Rose-John, S. & Müllberg, J. (2000) Shedding of interleukin-6 receptor and tumor necrosis factor alpha. Contribution of the stalk sequence to the cleavage pattern of transmembrane proteins. *European Journal of Biochemistry*, 267, 2624–2631.

68. Matthews, V., Schuster, B., Schütze, S. *et al.* (2003) Cellular cholesterol depletion triggers shedding of the human interleukin-6 receptor by ADAM10 and ADAM17 (TACE). *The Journal of Biological Chemistry*, 278, 38829–38839.
69. Müllberg, J., Oberthur, W., Lottspeich, F. *et al.* (1994) The soluble human IL-6 receptor. Mutational characterization of the proteolytic cleavage site. *The Journal of Immunology*, 152, 4958–4968.
70. Stöcker, W., Grams, F., Baumann, U. *et al.* (1995) The metzincins--topological and sequential relations between the astacins, adamalysins, serralysins, and matrixins (collagenases) define a superfamily of zinc-peptidases. *Protein Sciences*, 4, 823–840.
71. Wolfsberg, T.G., Straight, P.D., Gerena, R.L. *et al.* (1995) ADAM, a widely distributed and developmentally regulated gene family encoding membrane proteins with a disintegrin and metalloprotease domain. *Developments in Biologicals*, 169, 278–283.
72. Maskos, K., Fernandez Catalan, C., Huber, R. *et al.* (1998) Crystal structure of the catalytic domain of human tumor necrosis factor-alpha-converting enzyme. *Proceedings of the National Academy of Sciences of the United States of America*, 95, 3408–3412.
73. Hartmann, D., De Strooper, B., Serneels, L. *et al.* (2002) The disintegrin/metalloprotease ADAM 10 is es-sential for Notch signalling but not for alpha-secretase activity in fibroblasts. *Human Molecular Genetics*, 11, 2615–2624.
74. Weber, S. & Saftig, P. (2012) Ectodomain shedding and ADAMs in development. *Development*, 139, 3693–3709.
75. Blobel, C.P. (2005) ADAMs: key components in EGFR signalling and development. *Nature Reviews. Molecular Cell Biology*, 6, 32–43.
76. Van Hauwermeiren, F., Vandenbroucke, R.E. & Libert, C. (2011) Treatment of TNF mediated diseases by selective inhibition of soluble TNF or TNFR1. *Cytokine and Growth Factor Reviews*, 22, 311–319.
77. Scheller, J., Chalaris, A., Schmidt-Arras, D. & Rose-John, S. (2011) The pro- and anti-inflammatory properties of the cytokine interleukin-6. *Biochimica et Biophysica Acta*, 1813, 878–888.
78. Waetzig, G.H. & Rose-John, S. (2012) Hitting a complex target: an update on interleukin-6 trans-signalling. *Expert Opinion on Therapeutic Targets*, 16, 225–236.
79. Chalaris, A., Adam, N., Sina, C. *et al.* (2010) Critical role of the disintegrin metalloprotease ADAM17 for intestinal inflammation and regeneration in mice. *The Journal of Experimental Medicine*, 207, 1617–1624.
80. Burghardt, S., Erhardt, A., Claass, B. *et al.* (2013) Hepatocytes contribute to immune regulation in the liver by activation of the Notch signaling pathway in T cells. *The Journal Immunology*, 191 (11), 5574–82.
81. Scheller, J., Chalaris, A., Garbers, C. & Rose-John, S. (2011) ADAM17: a molecular switch to control inflammation and tissue regeneration. *Trends in Immunology*, 32, 380–387.
82. Soond, S.M., Everson, B., Riches, D.W. & Murphy, G. (2005) ERK-mediated phosphorylation of Thr735 in TNFalpha-converting enzyme and its potential role in TACE protein trafficking. *Journal of Cell Science*, 118, 2371–2380.
83. Lorenzen, I., Lokau, J., Dusterhoft, S. *et al.* (2012) The membrane-proximal domain of A Disintegrin and Metalloprotease 17 (ADAM17) is responsible for recognition of the interleukin-6 receptor and interleukin-1 receptor II. *FEBS Letters*, 586, 1093–1100.
84. Trad, A., Hansen, H.P., Shomali, M. *et al.* (2012) ADAM17-overexpressing breast cancer cells selectively targeted by antibody-toxin conjugates. *Cancer Immunology, Immunotherapy*.
85. Siggs, O.M., Xiao, N., Wang, Y. *et al.* (2012) iRhom2 is required for the secretion of mouse TNFalpha. *Blood*, 119, 5769–5771.
86. Issuree, P.D., Maretzky, T., McIlwain, D.R. *et al.* (2013) iRHOM2 is a critical pathogenic mediator of inflammatory arthritis. *The Journal of Clinical Investigation*, 123, 928–932.
87. Rose-John, S., Dietrich, A. & Marks, F. (1988) Molecular cloning of mouse protein kinase C (PKC) cDNA from Swiss 3T3 fibroblasts. *Gene*, 74, 465–471.
88. Walev, I., Vollmer, P., Palmer, M., Bhakdi, S. & Rose-John, S. (1996) Pore-forming toxins trigger shedding of receptors for interleukin 6 and lipopolysaccharide. *Proceedings of the National Academy of Sciences of the United States of America*, 93, 7882–7887.
89. Chalaris, A., Rabe, B., Paliga, K. *et al.* (2007) Apoptosis is a natural stimulus of IL6R shedding and contributes to the pro-inflammatory trans-signaling function of neutrophils. *Blood*, 110, 1748–1755.
90. Hall, K.C. & Blobel, C.P. (2012) Interleukin-1 stimulates ADAM17 through a mechanism independent of its cytoplasmic domain or phosphorylation at threonine 735. *PLoS ONE*, 7, e31600.
91. Willems, S.H., Tape, C.J., Stanley, P.L. *et al.* (2010) Thiol isomerases negatively regulate the cellular shedding activity of ADAM17. *The Biochemical Journal*, 428, 439–450.
92. Lorenzen, I., Trad, A. & Grotzinger, J. (2011) Multimerisation of A disintegrin and metalloprotease protein-17 (ADAM17) is mediated by its EGF-like domain. *Biochemical and Biophysical Research Communications*, 415, 330–336.

93. Düsterhöft, S., Jung, S., Hung, C.-W. *et al.* (2013) The membrane-proximal domain of ADAM17 represents the putative molecular switch of its shedding activity operated by protein-disulfide isomerase. *Journal of the American Chemical Society*, 135, 5776–5781.
94. Horiuchi, K., Kimura, T., Miyamoto, T. *et al.* (2007) Cutting edge: TNF-α-converting enzyme (TACE/ADAM17) inactivation in mouse myeloid cells prevents lethality from endotoxin shock. *The Journal of Immunology*, 179, 2686–2689.
95. Long, C., Hosseinkhani, M.R., Wang, Y., Sriramarao, P. & Walcheck, B. (2012) ADAM17 activation in circulating neutrophils following bacterial challenge impairs their recruitment. *Journal of Leukocyte Biology*, 92, 667–672.
96. Weskamp, G., Mendelson, K., Swendeman, S. *et al.* (2010) Pathological neovascularization is reduced by inactivation of ADAM17 in endothelial cells but not in pericytes. *Circulation Research*, 106, 932–940.
97. Murthy, A., Defamie, V., Smookler, D.S. *et al.* (2010) Ectodomain shedding of EGFR ligands and TNFR1 dictates hepatocyte apoptosis during fulminant hepatitis in mice. *The Journal of Clinical Investigation*, 120, 2731–2744.
98. Brandl, K., Sun, L., Neppl, C. *et al.* (2010) MyD88 signaling in nonhematopoietic cells protects mice against induced colitis by regulating specific EGF receptor ligands. *Proceedings of the National Academy of Sciences of the United States of America*, 107, 19967–19972.
99. Blaydon, D.C., Biancheri, P., Di, W.L. *et al.* (2011) Inflammatory skin and bowel disease linked to ADAM17 deletion. *The New England Journal of Medicine*, 365, 1502–1508.

8 Subtracting Matrix out of the Equation: New Key Roles of Matrix Metalloproteinases in Innate Immunity and Disease

Antoine Dufour and Christopher M. Overall

Department of Oral Biological & Medical Sciences and Department of Biochemistry and Molecular Biology, Centre for Blood Research, University of British Columbia, Vancouver, British Columbia, Canada

8.1 The tale of a Frog's tail

Extracellular matrices (ECM) and connective tissues are constantly being synthesized and then dynamically remodeled to maintain proper tissue homeostasis. These biological events are accompanied by or accomplished by proteolytic activity, respectively, and are tightly regulated at many levels. From the viewpoint of ECM degradation, two basic pathways perform this biological process: an intracellular pathway where matrix is internalized and degraded by lysosomal proteases in the phagolysosome and an extracellular pathway where secreted proteases can remodel the ECM [1]. In healthy tissues, the intracellular pathway will often predominate (except in the case of bone matrix), where it is under tight cellular control, whereas in pathological processes and inflammation, aberrant tissue destruction or degradation causes a shift to the extracellular pathway [1]. Several enzymes have been described to orchestrate these events, for example, the cathepsins located in the lysosomes that operate optimally at low pH and complete the intracellular pathway. Another class of enzymes included in ECM degradation that execute the extracellular pathway are the matrix metalloproteinases (MMPs).

Understanding these biological processes has proven to be more complex than initially described and several groups have now demonstrated extracellular roles for cathepsins and intracellular roles for MMPs, which is counter-intuitive [2–5]. The MMP story began 50 years ago in 1962, when Gross and Lapiere first identified a collagenase (later known as matrix metalloproteinase-1; MMP1) that degraded the collagen amidst apoptosing cells of tadpoles' tails that together lead to tail resporption [6]. In the 1980s, MMPs were identified as prime drug targets for cancer therapies based on the concept that cancer cells overexpress MMPs and this was supposed to increase their ECM degradation capabilities, thus clearing a path in the surrounding matrix, facilitating cancer intravasation into the bloodstream and subsequent metastasis. This was a reasonable hypothesis then, and drug companies cannot be blamed

Matrix Metalloproteinase Biology, First Edition. Edited by Irit Sagi and Jean P. Gaffney.

for embarking on a high profile drug development program. However, the failure in phase III clinical trials [7] has damped enthusiasm on all fronts for these proteases and even proteases in general as drug targets, but this is a naive view. In the 1980s, only three human MMPs (out of 24) were known and their roles were only associated with matrix degradation and angiogenesis [7]. As their name suggests, MMPs were discovered as ECM protein degraders but we now know that MMPs are involved in a plethora of biological processes other than ECM remodeling, and many more interesting processes where they exert higher order control of cellular responses in the extracellular and more recently transciptionally in the intracellular milieus, respectively. In this chapter, the impact of the non-proteolytic roles of MMPs on various diseases is discussed. The non-ECM MMP substrates are described and the methods by which novel MMP substrates are studied and identified are presented. Thus, MMPs have evolved to be even more fascinating and essential regulators of cell function than before in the drug development heyday. After more than 50 years of investigation, it appears that just like tadpole metamorphosis to frogs, and the more dramatic metamorphosis of a frog to a handsome prince upon a princess's kiss, MMPs too have morphed from dowdy matrix remodellers to princes of the cell signaling realm.

8.2 The MMP family

Over the past decade, new evidence on the roles of MMPs in disease has forced a rewriting of the story. In humans, there are 23 MMPs, including a gene duplication of MMP23 (present as MMP23A and MMP23B), whereas in mice, there are 24 due to a duplication of MMP1 (MMP1A and MMP1B) but a lack of MMP23B and MMP26. All MMPs employ a Zn^{2+} ion in their active site to catalyze proteolytic activity [7]. Secreted MMPs typically contain five peptides or domains: (i) a signal peptide which directs the enzyme to the secretory pathway; (ii) an approximately 80–90 amino acid prodomain that folds over the active site to confer latency upon the enzyme (also termed proMMP); (iii) a zinc-containing catalytic domain of approximately 160 residues where mutation of the glutamic acid in the active site ablates proteolytic activity; (iv) a linker or hinge region of approximately 15–65 residues; and (v) an approximately 200 amino acid hemopexin domain that mediates protein–protein interactions, including exosites that confer specificity for substrates in concert with the active site [8–10]. In addition to these five archetypal peptides and domains, six MMPs are anchored to the cell membrane either by a glycosylphosphatidylinositol (GPI) anchor or a hydrophobic transmembrane sequence followed by a short cytoplasmic domain [9, 10]. In order to maintain control, MMPs are tightly regulated by the tissue inhibitors of metalloproteinases (TIMP1, TIMP2, TIMP3, and TIMP4) which bind and inhibit all MMPs, to different degrees, with 1:1 stoichiometry [10]. Several other inhibitors of MMPs include β-amyloid precursor protein, α2-macroglobulin, endostatin, procollagen C-terminal proteinase enhancer, the non-collagenous NC1 domain of type IV collagen, tissue factor pathway inhibitor-2 and the reversion-inducing cysteine-rich protein with kazal motifs (RECK) that inhibit MMPs in different tissue locations or in plasma [7, 10].

8.3 Making the cut as immune regulators

Proteases are often described by the substrates they can cleave: for example, the collagenases (MMP1, MMP8, and MMP13) cleave native collagen, the gelatinases (MMP2 and MMP9) cleave gelatin (denatured collagen), and macrophage elastase (MMP12) cleaves elastin. But what happens when a collagenase can also cleave dozens of substrates other than collagen or the metalloelastase (MMP12) cleaves IFNα among over 300 new substrates? [11] At the turn of the last century, several groups reported that MMPs were not only matrix-degrading enzymes, but were critical in regulating immune responses and inflammatory processes. In landmark papers, in 1999, Wilson et al. [12] demonstrated that *Mmp7*$^{-/-}$ mice had decreased antimicrobial activity against *Escherichia coli* and *Salmonella typhimurium* due to the absence of MMP7 cleavage and shedding of antibacterial defensins in the intestinal epithelium. Soon after, in 2000, McQuibban et al. [13] demonstrated by the first use of the yeast two-hybrid system for protease substrates that MMP2-bound CCL7/monocyte chemoattractant protein-3 (MCP3) on the hemopexin domain presents the bound substrate to the catalytic domain for more efficient cleavage. Biologically, the result was that MMP2 dampens the inflammatory response *in vivo*, controlling chemotaxis and ablating calcium fluxes by cleaving CCL7 to generate a potent chemokine receptor antagonist. With time, MMPs were demonstrated to precisely cleave most (and likely all) chemokines. In humans, chemokines are a family of 54 proteins involved in leukocyte chemotaxis, inflammation, and immune effector cell recruitment. For example, to date, MMPs have been demonstrated to efficiently cleave all 14 monocyte/macrophage chemoattractant chemokines (CC-motif) ligand (CCL) at a startling 165 different sites (Table 8.1) [14, 15]. The expression of MMPs and their substrate repertoires greatly differ, based on the tissue location, context of the micro-environment (healthy versus inflamed tissue) and the cellular provenance (epithelial, fibroblasts, neutrophils, macrophages, etc.). Different cells produce differing arrays of MMPs and chemokines in order to control various scenarios during immune and inflammatory responses. Another important factor is the temporal presence of MMPs and chemokines: for example, the roles of MMPs during the first hours following triggering of a wound healing response differ from those 2–3 days post-injury. This provides a partial explanation as to why so many MMPs can cleave most if not all the chemokines, turning these chemoattractant cytokines into either agonists or antagonists in response to tissue challenge and guiding the tissue under stress or injury to adapt in all stages of the immune or inflammatory response. Thus, different MMPs orchestrate overlapping but independent contributions to immune chemotactic cross talk.

The picture that has emerged for the key roles of MMPs in regulating chemokine activity and consequently regulating innate immunity is as follows. The neutrophil-specific MMPs MMP25 [15] and MMP8 [16] activate the neutrophil chemoattractant CXCL5, whereas MMP8, but not MMP25, cleaves and activates CXCL8 directly [16] or indirectly by cleaving and inactivating alpha 1 protease inhibitor (serpin A1) enabling neutrophil elastase to efficiently perform the activation cleavage of CXCL5 and CXCL8 *in vivo* [17]. Thus the neutrophil utilizes MMP25 and MMP8 to form a feed forward mechanism to chemoattract neutrophils.

Table 8.1 MMP-truncation products of CC chemokines.

	MMP1	MMP2	MMP3	MMP7	MMP8	MMP9	MMP12	MMP13	MMP14	MMP25
CCL1 (1–73)	no cleavage	no cleavage	(8–73)	(7–73) (7–70)	no cleavage	no cleavage	(7–73)	(6–73)	no cleavage	no cleavage
CCL2 (1–76)	(5–76)	no cleavage	(5–76)	(5–67) (5–76)	(5–76)	(5–76)	(5–76)	(5–76)	no cleavage	(5–76)
CCL3 (1–70)	(1–47)	no cleavage	no cleavage	(9–63) (1–47)	(16–64)	(16–64)	no cleavage	(1–47)	no cleavage	no cleavage
CCL4 (1–69)	(16–62)	(6–44) (7–69)	(6–44)	(1–61) (6–69)	(7–69)	(14–61)	(7–69)	no cleavage	(6–44) (7–69)	(7–69)
CCL5 (1–69)	no cleavage	no cleavage	no cleavage	(1–65)	no cleavage	no cleavage	no cleavage	(5–69)	no cleavage	no cleavage
CCL7 (1–76)	(5–76)	(5–76)	(5–76)	(5–76) (7–76)	no cleavage	(5–76)	(5–76)	(5–76)	(5–76) (7–76) (9–76)	(5–76)
CCL8 (1–76)	(1–73) (5–73)	(1–73)	(5–76)	(1–73) (5–73)	(1–73) (5–73)	no cleavage	(7–76) (1–73) (5–73)	(1–73)	(6–76) (1–73) (5–73)	no cleavage
CCL11 (1–74)	no cleavage	(10–74)	no cleavage	(10–74)	no cleavage	no cleavage	(10–74)	no cleavage	no cleavage	no cleavage
CCL13 (1–75)	(8–72)	(5–75)	(4–75) (5–75)	(4–75) (6–75) (5–72)	(5–72)	(5–72)	(5–72)	(5–72)	(4–75)	(5–72)

CCL14 (1–74)	no cleavage	no cleavage	no cleavage	(1–70) (1–66)	no cleavage	no cleavage	(1–71) (4–72) (7–74)	no cleavage	no cleavage	no cleavage
CCL15 (1–92)	(25–92)	(14–92) (25–92) (27–92) (28–92)	(17–92) (25–92) (28–92)	(17–88) (17–92) (25–92)	(25–92) (28–92)	(25–92) (28–92)	(25–92) (28–92)	(25–92) (27–92) (28–92)	(14–92) (27–92)	(25–92) (27–92)
CCL16 (1–97)	(8–85)	(5–97) (8–77)	(8–85)	(8–85)	(8–85)	(8–85) (8–77)	(5–97) (8–85) (8–77)	(8–85) (8–77)	(5–97) (8–85)	(5–97)
CCL17 (1–71)	no cleavage	no cleavage	no cleavage	(4–69)	(9–71)	no cleavage	(4–69)	(4–71)	no cleavage	no cleavage
CCL23 (1–99)	(11–99) (17–99) (23–99) (26–99) (28–99) (30–99)	(11–99) (17–99) (26–99) (28–99)	(11–99) (21–99) (26–99) (1–90)	(11–99) (17–99) (23–99) (26–99) (1–90)	(9–71) (11–99) (17–99) (26–99)	(11–99) (28–99)	(11–99) (17–99) (23–99) (26–99)	(28–99) (1–90)	(14–99) (30–99)	(11–99) (21–99) (26–99)

The residue numbers are shown in brackets. Sites in black in the dark grey rectangles represent truncation sites and the light grey boxes indicate that the chemokine is not cleaved by that MMP.

The neutrophil influx is actively terminated by especially macrophage MMP12, which cleaves and inactivates every ELR motif containing CXCL chemokine (CXCL1, CXCL2, CXCL3, CXCL5, CXCL6, CXCL7, and CXCL8) that is a neutrophil chemoattractant [18]. More recently chemoattraction for monocytes/macrophages has been found to be enhanced by multiple MMPs cleaving and activating CCL15 and CCL23 to chemokines 10-fold more potent than CCL2 and CCL7 in chemoattraction activity for monocytes and macrophages [14]. The MCPs CCL2, CCL7, CCL8, and CCL13 are precisely processed at position 4↓5 or adjacent in MCP4, by multiple MMPs to inactivate their chemoattraction for macrophages [19] and especially efficiently by MMP12 [18]. In so doing, potent CC receptor (CCR) antagonists are generated that block all CCRs to terminate macrophage recruitment. Thus, each limb of the neutrophil and monocyte recruitment and termination responses is under control of MMP activity, especially by MMPs in the cells of the innate immune system that form positive and negative feedback loops. Likewise, lymphocytes are regulated by MMP cleavage of CXCL12 (SDFα and SDFβ) [20, 21], CX_3CL1 (fractalkine) [22], and CXCL11 [23]. Interestingly, MMP2 converts a cell membrane bound chemokine agonist, CX_3CL1, by two cleavages to a shed soluble form comprised of just the chemokine domain minus the cell membrane tethering stalk, but as an inactivated form. The first cleavage releases the chemokine from the stalk at the chemokine domain [24]AAA↓ [25]LTK and the second cleavage N terminally truncates the chemokine at position [1]QHLG↓ [5]MTK converting an agonist to a receptor antagonist.

During wound healing, the inflammatory response primarily acts as a gatekeeper to control/prevent the risk of infection [26]. Inflammatory cells support wound healing by releasing cytokines and growth factors to increase proliferation, migration, and chemotaxis for re-epithelialization and re-differentiation at the site of injury [27]. Vascular permeability is also increased, allowing serum proteins, for example, acute phase proteins and serum proteins including complement and antibodies, to penetrate to the site of injury. Among a multitude of proteins involved in inflammation and wound healing, several MMPs play active roles in many steps of these processes to check the balance between pro- and anti-inflammatory activities. This can be compared to the accelerator and brake pedals in a car: without the accelerator, it is impossible to gain speed but without the brake pedal, it is tricky to stop without causing an accident. Similarly, MMPs play critical roles during wound healing by mediating the activation and inactivation of various mediators in these responses: MMPs (i) process chemokines, activating or inactivating these proteins and converting these into agonists or antagonists to recruit immune cells to the site of injury and establish a chemotactic gradient as described above [14, 15, 18]; (ii) cleave cell–cell junctions and cell–matrix contacts to allow re-epithelialization [26, 28]; (iii) remodel the provisional matrix during the scarring process [26, 29]; (iv) promote the migration and proliferation of epithelial, endothelial, fibroblastic, and immune cells [28, 30, 31]; and (v) activate and inactivate the complement pathway to control the phagocytosis of pathogens or neutrophils at the site of injury and/or infection [32]. For example, in a model of 12-*O*-tetradecanoylphorbol 13-acetate (TPA) skin inflammation comparing *Mmp2*$^{-/-}$ mice to their wild type counterparts, serum and acute-phase proteins were reduced in the *Mmp2*$^{-/-}$ mice and the proteolytic networks were different due to less exudation in inflammation as shown by the analysis of the N-terminome by iTRAQ-TAILS [32]. In the wild-type animals, auf dem Keller et al [32] found that MMP2 precisely cuts serpin G1, also known as complement 1 (C1) inhibitor, at position ^{470}R↓S^{471}, which inactivated the serpin G1 *in vivo*. In the *Mmp2*$^{-/-}$ mice, the high levels of intact functional serpin G1 blocked complement activation *in vivo*. Thus, MMP2 dynamically regulates the levels of intact versus cleaved serpin G1 *in vivo* to control the complement cascade by forming the "metallo-serpin" switch. Serpin G1 also blocks plasma kallikrein cleavage of kininogen, which releases the vasoactive peptide bradykinin. In the *Mmp2*$^{-/-}$ mice the increase in the C1 inhibitor led to a decrease in release of bradykinin. Thus in inflammation in the skin of *Mmp2*$^{-/-}$ mice, the blood vessels exhibit an intact permeability barrier leading to reduced acute response proteins in the inflamed tissue, including reduced levels of complement proteins [32].

In the past decade, several groups have looked beyond the stereotypical roles of ECM degradation by MMPs and have demonstrated that they play key roles in multiple steps of wound healing and the immune/inflammatory responses. Are these observations of the novel roles of MMPs simply twisting a myth or has the dogma now been broken?

8.4 Enter the "omics" era: genomics, proteomics and degradomics

We have entered the "omics" era: the addition of "omics" to the end of a scientific word circumscribes a category of biological contexts, molecules, or relationships to study. Thus, degradomics, which was introduced in 2002, is defined as the characterization of all proteases, inhibitors, and protease substrates using both genomic and proteomic techniques [8].

Genomics continues to identify secrets kept silent within the 20, 135 human genes (or ~100,000 gene polymorphisms) through the development and application of high-throughput technologies in the fields of functional genomics, structural genomics, epigenomics, and metagenomics. Emerging technologies in the field of proteomics are being widely utilized to address the numerous challenges of protein identification in complex samples and the investigation of the roles of post-translation modifications (PTMs) in biological processes and diseases. The total number of different proteins in the human proteome far exceeds the number of genes due to multiple post-translational events. These include: (i) attachment of biochemical groups to proteins such as phosphate groups, carbohydrates, acetate, methyl, or lipids; (ii) chemical change of an amino acid such as citrullination; (iii) structural changes such as a disulfide bond formation; and (iv) proteolysis (processing). Although some of the estimated 300 PTMs are often rare events or are present in small amounts, they have indispensable roles in biology and human diseases. For example, every single protein in the human proteome will at some point be a "victim" of proteolysis whether it is: (i) removal of the initiating methionine in cytosolic proteins or the signal peptide during protein secretion; (ii) processing of a zymogens to remove the propeptide or prodomain; (iii) excision or disengagement of a specific domain from a multiple domain protein substrate in a pathway, for example in response to stress or pathogen such as in the blood coagulation cascade or complement activation system; (iv) during pathology as a bystander cleavage event by proteases during degradative events; and (v) degradation in the proteasome or lysosome.

It is challenging to detect some PTMs without a specific enrichment method (reviewed in [33, 34]). Proteolysis is no exception. Finding the cleaved neo N terminus is like finding a needle in a haystack, that is, one or a few neo N-terminal present as unique semi-tryptic peptides in a vast field of tryptic peptides as analyzed by proteomics. Several methods have been developed to study protease cleavages. Such methods specifically enrich for the neo-N-termini of proteins post-proteolysis and some also enrich the natural N-termini of proteins. For example, positive selection can be achieved by enzymatic biotinylation of N-terminal α-amines using subtiligase before release by the specific tobacco etch virus (TEV) protease [35]. Negative selection methods are advantageous as they enable full characterization of the original protein N-termini as well as the neo N-termini generated by proteolysis and include terminal amine isotopic labeling of substrates (TAILS) [33, 34, 36, 37] and N-terminal combined fractional diagonal chromatography (COFRADIC) [37]. These methods have been reviewed recently and so the specifics of these approaches will not be covered here [33, 34].

8.5 ECM versus Non-ECM MMP substrates

MMPs were discovered and initially described as matrix degrading enzymes and for several decades many studies have demonstrated key roles for MMPs in such events in human pathologies. However, using unbiased degradomics technologies that analyze the N-terminome and hence identify sometimes hundreds of cleavage sites in proteomes, MMPs are now recognized to process a plethora of substrates unrelated to the ECM [2, 38]. Still underappreciated in the literature and within the field, the finding of so many non-ECM substrates implies that one needs to restructure the current thinking on the roles of MMPs and break the ECM-centric dogma. These high-throughput proteomics experiments generate large datasets and bioinformatics technologies are essential for handling such big data. For proteases, the knowledgebase TopFIND is a useful tool to integrate information on protein N- and C-termini, amino acid modifications and proteolytic processing [39, 40]. Experimental evidence from degradomics techniques including TAILS and COFRADIC is presented from five different organisms (*Homo sapiens*, *Mus musculus*, *Escherichia coli*, *Saccharomyces cerevisia*, and *Arabidopsis thaliana*) [39, 40].

Using TOPFIND, we organized all known MMP substrates and cleavages into ECM versus non-ECM proteins (Table 8.2; Fig. 8.1(a,b)). Strikingly, ECM proteins comprise only 27% of all MMP substrates and so 73% are non-ECM proteins. MMP11 and MMP21 have no substrates that are ECM proteins, although MMP21 was discovered only recently and few reports have been published. Other more recently discovered MMPs (MMP23A/B, MMP27, and MMP28) have no validated substrates, but some have been postulated. MMP20 was discovered in dental matrix and the few studies reported have focused on identifying matrix substrates, therefore

Table 8.2 Reported ECM and non-ECM substrates for all 24 human MMPs taken from TopFIND [39].

Human MMP	Total number of reported substrates	ECM substrates	Non-ECM substrates	Percentage of ECM substrates (%)	Percentage of Non-ECM substrates (%)
MMP1	46	11	35	24	76
MMP2	94	27	67	29	71
MMP3	76	24	52	32	68
MMP7	72	14	58	18	81
MMP8	39	7	32	19	82
MMP9	79	17	62	22	78
MMP10	13	7	6	54	46
MMP11	18	0	18	0	100
MMP12	42	16	26	38	62
MMP13	44	23	50	32	68
MMP14	84	18	66	21	79
MMP15	8	4	4	50	50
MMP16	13	4	9	31	69
MMP17	9	3	6	33	67
MMP19	12	7	5	58	42
MMP20	10	9	1	90	10
MMP21	1	0	1	0	100
MMP23A/B	0	0	0	0	0
MMP24	4	0	4	0	100
MMP25	70	13	57	19	81
MMP26	10	5	5	50	50
MMP27	0	0	0	0	0
MMP28	0	0	0	0	0
Total number	744	208	536	27	73

90% of its known substrates are ECM proteins. Interestingly, using degradomics, MMP2, MMP9, MMP13, MMP14, and MMP25 are found to show a high percentage of non-ECM substrates, 71, 79, 62, 79 and 81%, respectively [15, 32, 41–44].

Two reports have demonstrated novel substrates of MMP2, one comparing *Mmp2*$^{-/-}$ fibroblast secretomes transfected with MMP2^{E375A} or MMP2 [45] and the other comparing recombinant MMP2 or buffer control added to *Mmp2*$^{-/-}$ fibroblast secretomes [42]. Both known ECM substrates (fibronectin, collagens, decorin) and hundreds of previously unknown non-ECM substrates (including CX_3CL1, galectin-1, hepatoma-derived growth factor, HSP90α, cystatin C, insulin growth factor receptor binding-protein-4 (IGFBP4) and IGFBP6) were discovered [22, 41]. In parallel, using iTRAQ-TAILS, seven new MMP9 substrates (one ECM (thrombospondin-2) and six non-ECM (cystatin C, galectin-1, IGFBP4, pyruvate kinase isozymes M1/M2, peptidyl-prolyl cis-trans isomerase A and Dickkopf-related protein 3)) were discovered and validated [42].

To investigate the stromal role and substrate repertoire of MMP13 in bone metastasis, the secretome from human breast cancer cell lines MDA-MB-231 or MDA-1833 co-cultured with MC3T3-E1 in a mineralized osteoid matrix was utilized in genomics and proteomics experiments [43]. The mRNA levels of five proteases

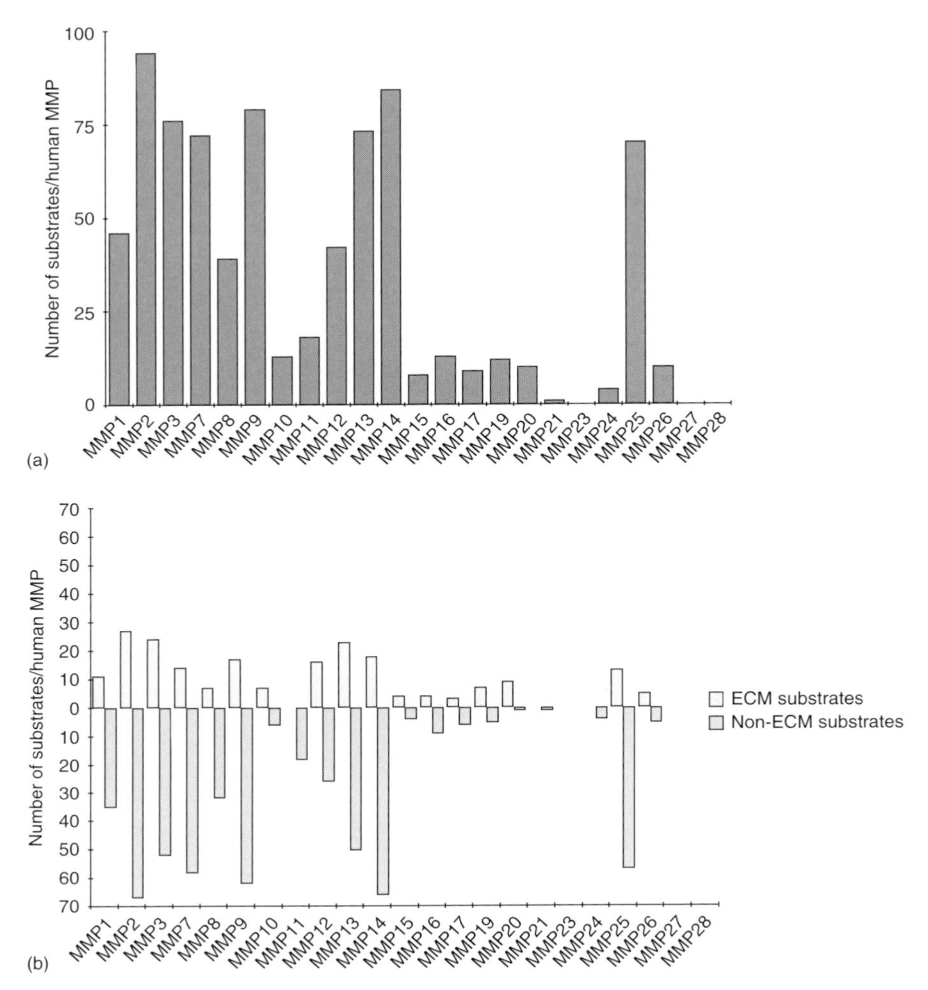

Figure 8.1 (a) All 773 reported human MMP substrates distributed for each of the 23 human MMPs. (b) All 773 reported human MMP substrates: the ECM substrates are shown in blue and the non-ECM substrates are shown in green. *(See insert for color representation of this figure.)*

were upregulated when cells were co-cultured: MMP13 (5.9-fold), proteasome catalytic subunit 2i (twofold), PHEX endopeptidase (twofold), carboxypeptidase X2 (twofold) and caspase-1 (twofold). Addition of recombinant MMP13 to these systems and analyzing these by iTRAQ proteomics identified 48 shed proteins from the cell membrane or pericellular matrix that were increased in the coculture supernatant. Novel substrates that were identified include platelet-derived growth factor (PDGF), which increased the phosphorylation of ERK1/2 upon cleavage, apolipoprotein SAA3, osteoprotegerin, antithrombin III which loses its inhibitory activity towards thrombin after MMP13 cleavage, and the protein CutA [43].

Isotope coded affinity tag (ICAT)-labeled conditioned media from MDA-MB-231 stably transfected with a vector encoding for MMP14 or MMP14^{E240A} as the control were analyzed by quantitative proteomics in the first degradomics paper reported in 2004 [46]. Of the novel substrates, several protease inhibitors (secretory leukocyte protease inhibitor (SLPI) and skin-derived antileukoproteinase) and chemokines/cytokines (IL-8, growth related oncogene (GRO)-α, GRO-γ, macrophage migration inhibitory factor, tumor necrosis factor-α, connective tissue growth factor) were identified as substrates, suggesting a role of MT1-MMP in controlling immune regulation and functions [46]. Several new ECM proteins were also identified as substrates of MT1-MMP, including fibronectin and epidermal growth factor containing fibulin-like extracellular matrix protein-1 [46], though it is clear that MMP14 is also a major collagenase with essential roles in collagen matrix degradation [47].

Substrates of the neutrophil-specific protease MMP25, also known as leukolysin, were identified using TAILS by adding recombinant MMP25 or $MMP25^{E234A}$ to human fetal lung fibroblast-1 (HFL-1) cells. 58 novel substrates were identified, 5 of which (cystatin C, IGFBP-7, galectin-1, vimentin, and SPARC) were validated biochemically [15]. Vimentin, which is displayed on the surface of apoptotic neutrophils, was cut by MMP25 at ten different sites. As demonstrated using a Transwell migration assay, THP-1 monocytic cells were attracted to full-length vimentin but not to MMP25 processed vimentin [15]. Nonetheless, the MMP25 cleaved form of vimentin increased phagocytosis by —two-fold in comparison with full length vimentin [15]. This may target macrophages to apoptotic neutrophils and then increase macrophage phagocytic activity to clear the end stage neutrophils. Thus, MMP25 is likely important in innate immunity by processing and degrading bioactive non-ECM proteins.

COFRADIC was recently used to map the *in vitro* secretome of gastric cancer-associated myofibroblasts collected from human patients: MMP1, MMP2 and MMP3 displayed higher activity in the cancer-associated myofibroblasts as compared to the adjacent tissue-derived myofibroblasts and the *in vivo* MMP activity was detected in a xenograft model of gastric cancer cells [48].

Before the use of global proteomics approaches, one assumed that most MMP substrates must be ECM proteins and often disregarded the non-ECM proteins as being "true" substrates. To date, 336 different substrates have been reported to be cleaved by the 23 human MMPs (Table 8.2 and Fig. 8.1(a)), some redundantly, others nonredundantly. Thus, many of these proteins are cleaved by more than one MMP and so in total, the 23 human MMPs cleave 776 substrates. As is evident in Table 8.2 and Fig. 8.1(b), 75% of known MMP substrates are not ECM proteins. As seen in Fig. 8.2(a), by compiling all gene ontology (GO) terms of these 246 non-ECM MMP substrates, the most highly enriched term was "inflammatory/immune response", accounting for 20% of the substrates, followed by chemokine (12%), cell migration (11%), cell proliferation (10%), cell signaling (9%), apoptosis (8%), blood coagulation (7%), calcium signaling (6%), cell adhesion (6%), wound healing (5%), angiogenesis (4%), and metalloprotease activity (3%). Furthermore, in Fig. 8.2(b), using pathway enrichment analysis, 55% substrates are involved in chemokine signaling, followed by complement and coagulation cascades (19%), NOD-like receptor signaling (11%), Toll-like receptor signaling (9%) and ECM-receptor (7%). Thus, from the nature of the substrates annotated in MEROPS, the most extensive roles of MMPs lie not in ECM degradation, but the modulation of inflammatory/immune responses by cleavage of signaling molecules.

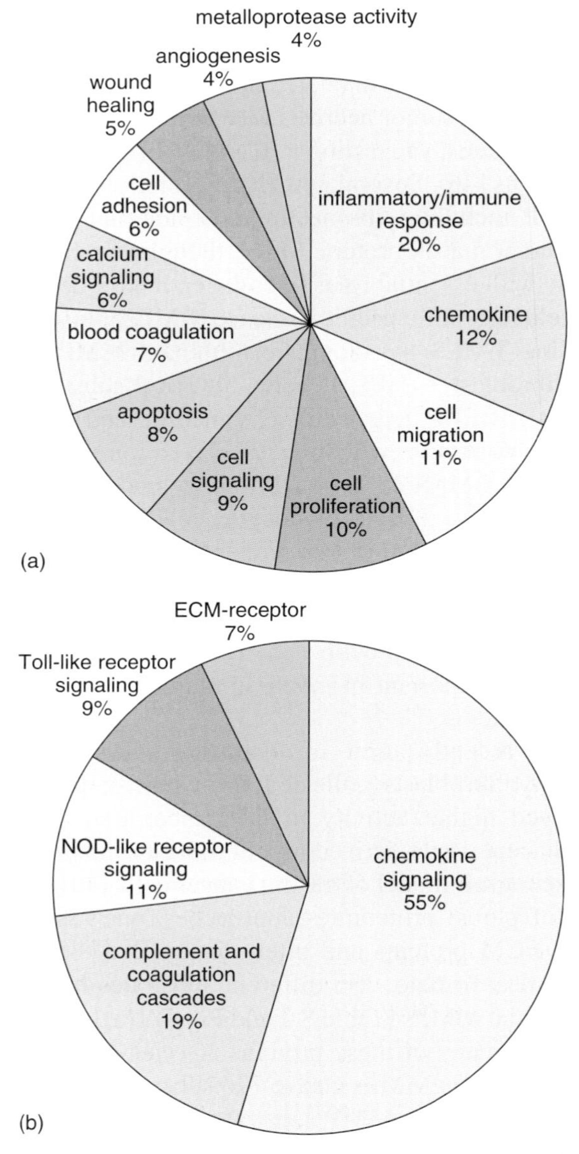

Figure 8.2 (a) Gene Ontology (GO) terms enrichment of all 246 reported non-ECM human MMP substrates. (b) Pathway enrichment analysis of the 246 reported non-ECM human MMP substrates. (*See insert for color representation of this figure.*)

8.6 Moonlighting protein substrates: intracellular proteins cleaved outside the cell

With new tools come new challenges, new observations, and new controversies in the quest for elucidating the functions of MMPs. The use of degradomics has revealed novel substrates that defy our current understanding of their roles and therefore, shifting gears into a second puzzle. Several novel observations have not only pointed toward an important contribution of MMPs in controlling immune processes, as described earlier, but also in cleaving intracellular substrates, once again defying the dogma that MMPs are only extracellular matrix-related enzymes. Whereas it has been thought that intracellular substrates of MMPs were in vitro phenomena due to cell death or cell lysis in culture and thus with little in vivo relevance, Cauwe and Opdenakker [4] proposed that such proteins had physiological relevance where MMPs are among the essential clearance systems to remove proteins resulting from escape during cell death.

Stress can also lead to the release of intracellular proteins [4] and proteomics analyses have consistently found some but not all intracellular proteins in the extracellular environment [2, 49]. MMPs would exert an essential clearing mechanism [4], but several intracellular proteins with *bona fide* extracellular roles have now been validated as being processed by MMPs, including peptidyl prolyl cis-trans isomerase A, DJ-1, hsp90α, adenylyl cyclase-associated protein1 and γ-enolase [2]. Such observations have promoted the hypothesis that by nonconventional secretion, intracellular proteins exit the cell and those with new extracellular roles functions and differing from their previously ascribed intracellular roles, increase the functional diversity of the proteome. These proteins are termed pleotropic proteins [49], also known as moonlighting proteins [2]. Since some moonlighting proteins are MMP substrates, then they must be considered to be physiologically relevant, just as those released upon cell death need to be cleared. The question is: Does MMP processing of these proteins alter their function yet again? As discussed earlier, the intracellular form of vimentin is chemoattractant for monocytic cells [15] with the MMP25 cleaved form of vimentin increasing phagocytosis by two-fold in comparison to full length vimentin [15]. High mobility group protein B1 (HMGB1) was discovered as a DNA-binding protein having DNA repair activity [50]. Interestingly, in arthritis [51], sepsis [52], and neurodegenerative diseases [53] where inflammation is increased, HMGB1 was demonstrated to activate coagulation and stimulate chemokines, thus inducing neutrophils recruitment [49]. In murine models of endotoxin challenge, inhibition of HMGB1 improved the outcome of the animals treated with HMBG1-specific antibodies [54, 55]. Importantly, secreted MMPs from inflammatory and immune cells can cleave and modulate the role of extracellular HMGB1 [2], thereby attracting other immune cells needed for clearance. MMPs can also cleave and modulate the functions of other damage-associated molecular pattern molecules (DAMPs), including peptidyl-prolyl cis-trans isomerase A (cyclophilin A), heat-shock protein-90α (HSP90α) and the calgranulins (S100 proteins) [56] and so regulate damage response through the amplification of the clearance signals and control the recruitment of immune cells to the site of injury [38]. Undoubtedly, many more exciting and well-documented examples of moonlighting protein substrates will be found.

8.7 Intracellular protein substrates cleaved inside the cell by MMPs

A number of reports indicate or make a claim for MMPs to be intracellular, although few highly convincing examples exist that unequivocally locate the enzyme past the lipid bilayer. Of course, as secreted proteins, MMPs exist intracellular but in discrete compartments associated with protein synthesis after crossing the ER lipid membrane bilayer through the translocon. Thereafter, secreted proteins transit the cell during maturation and sometimes storage before secretion in the golgi and secretory vesicles. Thus, immunolocalization will commonly detect secreted proteins in the cell, but within the protein synthesis and secretatory apparatus and thus segregated from the cytosol by a lipid bilayer. For example, in neutrophils, MMP8 and MMP9 are prestored together in granules and are released within minutes of pathogen detection; in macrophages, MMP9 is newly synthesized upon challenge, a process which happens after several hours, and is found in small golgi-derived cytoplasmic vesicles before being released [57]. MMP26 has an ER retention sequence and so it may cleave protein substrates within the ER. Indeed the prodomain sequence (P^{81}HCGVPD) differs from that of most MMPs (PRCGXPD); the histidine 81 residue is assumed to facilitate intracellular activation of MMP26 that occurs only during transient calcium influx [58]. Elevated levels of MMP26 are found in carcinoma cells, predominantly inside the cell and thus likely in the ER. When transfected into MCF7 breast carcinoma cells, MMP26 was mainly located in the intracellular milieu [58]. Whether or not MMP26 truly exists in the cytosol rather than functioning in the ER and golgi remains to be unequivocally shown.

A few reports suggest that MMPs actually are found in the cytosol or nucleus of cells, but these are often very controversial and incontrovertible evidence needs to be provided to support such claims. For example, MMP14 was demonstrated to be in the nucleus of macrophages and to impact their immune functions by dampening interleukin 12β [59]. In cardiomyocytes, MMP2, is located in the endoplasmic-reticulum and apparently in the cytosol, where it cleaves several substrates intracellularly, including troponin 1 (TN1), during hypoxia-reoxygenation injury [60]. In astrocytes, both MMP2 and MMP9 are found in different vesicles within the cytosol but also, it is claimed, in the cytoskeletal, membrane, and nuclear fractions [61]. After ischemic stroke, MMP2 and MMP9 cleave DNA repair proteins in neurons which promotes the accumulation of oxidative DNA damage in rat brains [62]. In the mouse striatum and using the *Hdh* striatal cell line, MMP10 was shown to inactivate huntingtin intracellularly by a cleavage at position ^{402}G↓Q, a key event in the pathogenesis of Huntington's disease [63]. However, until MMPs are accepted as having *bona fide* roles inside cells, evidence must be provided at many different levels to be convincing. For example, MMPs within intracellular vesicles or active in the golgi must be distinguished from MMPs in the cytosol without an intervening membrane barrier to substrate access.

MMP12 has very recently been convincingly reported in the nucleus of virus-infected cells [11]. Given the importance of IFNα in immunobiology, it is surprising that the intracellular signaling pathway for IFNα is not well established. In exploring this, Marchant et al. [11] obtained a striking result: The extracellular MMP12 translocates to the nucleus where it binds to the IκBα promoter where it was found to be essential for transcriptional up-regulation of IκBα. MMP12 does not bind the IκBα promoter before viral infection and only afterwards in the general transcriptional response to viral infection. In turn, IκBα was mechanistically linked to IFNα expression and secretion. In the absence of the protease, infected cells do not secrete alpha-interferon (IFNα), leading to more than 30% death rate in otherwise nonlethal viral infections. Thus, MMP12 is a moonlighting protease with *bona fide* intracellular nuclear roles markedly different from its conventional extracellular proteolytic actions that are absolutely essential for an effective anti-viral IFNα response. Indeed, MMP12 was also found to exercise a negative feedback loop by cleaving off the receptor (IFNαR2) binding site and thus terminating the IFNα pathway.

Marchant et al. [11] used multiple approaches to confirm the presence of intracellular and intranuclear MMP12 delivered *in trans* by the macrophage, including western blots of nuclear lysates using six different antibodies, confocal microscopy of different cells, and coculture experiments separated by cell impermeable membranes. Direct MMP12 binding to DNA was directly shown by electromobility shift assays and oligonucleotide binding was found to block MMP12 proteolytic activity. Finally, a truly unique set of substrates was described for MMP12 in the virus-infected cells. Using TAILS, more than 300 substrates were found for MMP12 and of these 250 were also bound by MMP12 in exons of their cognate genes. The authors showed that exon binding, MMP12 shut down transcription and formed a dual regulated substrate class. That is, MMP12 degrades a set of proteins and also reduces their transcription. Thus, within the cell, MMPs encounter a different palette of substrates compared to those secreted or displayed on the cell surface, thus expanding their biological roles. These observations further support that MMPs are more than matrix degrader enzymes.

8.8 Non-proteolytic roles of MMPs: missed in the myth?

Several zymogen forms of MMPs can also affect different biological processes, possibly contributing to disease progression without even being "active" and without cleaving substrates, but through protein–protein interactions or receptor binding. Many of these involve substrate exosites, which bind proteins to present substrate to the active center. However, as zymogens such sites are still available for protein interaction and so noncatalytic functions are possible [64] (Table 8.3).

For example, the binding of MMP14 to TIMP2 forms a receptor that facilitates proMMP2 activation by another MMP14 molecule [74]. Several proMMPs (MMP1, MMP2, MMP3, MMP9, MMP11, and MMP28) enhanced cell migration independent of their catalytic activities; TIMP-1, TIMP-2 or two broad spectrum hydroxamic acid-derived inhibitors were unable to inhibit the cell migration of COS-1 and MCF-7 cells [30, 73]. Mutations of the active site glutamic acid to an alanine which renders MMPs catalytically inactive had no inhibitory effect on this enhancement of cell motility [30]. In human neurons, proMMP1 can stimulate the dephosphorylation of Akt by binding to integrin $\alpha_2\beta_1$ and affect the cell survival through caspase activity [65]. Using a proteomic screen, MMP3 was found to bind through the hemopexin domain to heat-shock protein 90 β (HSP90β) and this interaction was critical for mammary epithelial invasion; blockade of this interaction with HSP90β antibodies inhibited invasion and branching morphogenesis [75].

MMP9 promotes B cell survival by docking via its hemopexin domain to CD44v and $\alpha_4\beta_1$ on the cell surface and promoting an intracellular signaling, thus activating

Table 8.3 Non-proteolytic roles of MMPs.

MMP	Non-proteolytic biological roles	Cell types and models	References
MMP1	– Dephosphorylates akt and increases cell death through caspase activation – Increases cell motility	– Cell culture (human neurons) – Cell culture (COS-1 and MCF7 cells transiently transfected with MMP1)	[30, 65]
MMP2	– Increases cell motility	– Cell culture (COS-1 and MCF7 cells transiently transfected with MMP2)	[30]
MMP3	– Increases cell motility	– Cell culture (COS-1 and MCF7 cells transiently transfected with MMP3)	[30]
MMP9	– Increases cell motility – Promotes B cell survival in chronic lymphocytic leukemia through Lyn and STAT3 activation – Induces signaling cascade through EGF/FAK/AKT/ERK by binding to CD44v – Induces breast cancer cell migration, proliferation, tumor growth, and metastasis	– Cell culture (COS-1 and MCF7 cells transiently transfected with MMP9) – Human B-CLL patients cells – Cell culture and *in vitro* biochemical assays, peptide mimicking the outer blades of MMP9 hemopexin domain – Cell culture, *in vitro* biochemical assays, and in mice treated with a small molecule binding the MMP9 hemopexin domain *in vivo*	[30, 66, 67]
MMP11	– Increases cell motility	– Cell culture (COS-1 and MCF7 cells transiently transfected with a MMP11)	[30]

MMP12	– The hemopexin domain displays antimicrobial properties against gram-positive and gram negative-bacteria (*Staphylococcus aureus* and *Escherichia coli*) – Transcriptional role of the catalytic domain of MMP12 by binding DNA and increasing transcription of Iκβα and repressing expression of 254 genes upon binding their exons	– *In vivo* (*MMP12*$^{-/-}$ mice) – Cell culture (HeLa, 1HAE and HL1 cells)	[11, 68]
MMP14	– The cytoplasmic tail regulates chemotaxis in macrophage – The cytoplasmic tail stimulates ATP production/glycoslysis – Controls inflammatory gene responses – Regulates the complement system by binding the globular domain of C1q – Induces breast cancer cell migration, angiogenesis, tumor growth and metastasis	– *In vivo* (*Mmp14*$^{-/-}$ mice) and *in vitro* chemotaxis assay – Cell culture and *in vitro* biochemical assays – Primary cells and peritoneal macrophages in cell culture and biochemical assays – Cell culture and *in vitro* biochemical assays – Cell culture, *in vitro* biochemical assays and *in vivo* (mice and chick embryo) treated with small molecule or peptide mimicking the outer blades of MMP14 hemopexin domain	[25, 59, 69–72]
MMP28	– Increases cell motility	– Cell culture (COS-1 and MCF7 transiently transfected with MMP28 cDNA)	[30, 73]

Lyn and STAT3, leading to a prevention of B cell apoptosis in chronic lymphocytic leukemia [66]. In breast cancer cells, MMP9 can also bind CD44v and induce a signaling cascade through the epidermal growth factor receptor/FAK/AKT/ERK pathway and peptides mimicking the outer blades of the hemopexin domain of MMP9 inhibited signaling and cell migration [67]. Small molecules, binding the central cavity of the MMP9 hemopexin domain, were also demonstrated to interfere with cell proliferation and migration both *in vitro* and *in vivo,* without blocking its MMP9 proteolytic activities [24].

MMP12 plays a critical role in the clearance of both Gram-positive and Gram-negative bacteria (*Staphylococcus aureus* and *Escherichia coli*) by adhering to the bacterial cell walls and disrupting cellular membrane integrity, resulting in bacterial cell death through a non-proteolytic mechanism; a four amino acids antimicrobial peptide based on the MMP12 hemopexin domain sequence was designed and showed efficacy in bacterial killing [68].

Independent of MMP14 proteolytic activity, by dampening the expression of interleukin 12β MMP14 was demonstrated to be involved in the control of immunoregulatory genes (Mi-2/NuRD) in macrophages. This facilitated nucleosome remodeling and histone acetylation at the cytokine promoter and in turn, triggered the activation of phosphoinositide 3-kinase δ (PPI3Kδ)/AKT/GSK3B signaling cascade [59]. Thus, MMP14 is involved in the regulation of immunity by the downregulation of proinflammatory genes and upregulation of anti-inflammatory cytokines. The 20 amino acid cytoplasmic tail of MMP14 also has a role in regulating the migration of macrophages to the sites of inflammation. This was shown in *Mmp14*$^{-/-}$mice using deletion constructs [69]. The cytoplasmic tail of MMP14 also stimulates the production of ATP in macrophages by interacting with Factor Inhibiting HIF-1 (FIH-1) and amyloid beta A4 precursor protein-binding family A member 3 (Mint3), thus, alleviating transcriptional repression and enhancing the transcriptional activity of hypoxia-inducible factor 1α (HIF-1α) to stimulate glycolysis [25]. Both the catalytically active and inactive form of MMP14 interact with the globular domain of C1q component of the complement system, suggesting that MMP14 would regulate the levels of C1q and interfere during complement activation since C1q was resistant to MMP14 cleavage [70]. MMP14 peptides mimicking the outer blades of the hemopexin domain were tested *in vivo* in a breast cancer mouse xenograft model and a chick embryo chorioallantoic membrane angiogenic assay, where the peptides were demonstrated to inhibit cancer cell migration, tumor growth, metastasis, and angiogenesis [71]. Another study utilized a virtual ligand screening of small molecules binding to the center of the hemopexin domain of MMP14, similar to the ones demonstrated to bind MMP9, which were shown to inhibit tumor growth by causing a fibrotic phenotype in a breast cancer mouse xenograft model [72]. Targeting exosites such as those on the hemopexin domain of MMP14 could potentially reduce its binding capacity to collagen and impair its role as a collagen receptor for adhesion [76].

Overall, studying the non-proteolytic functions of MMPs has revealed unexpected sites that could be used to develop blocking monoclonal antibodies, peptides, or small molecules to control their biological roles and especially their detrimental roles in diseases. By increasing selectivity, one decreases the chance of inhibiting the beneficial roles of MMPs that were unknown at the time that MMP inhibitors entered the cancer clinical trials. Another potential therapeutic option is to inhibit the downstream pathways affected by non-proteolytic roles of MMPs.

8.9 The fairy TAIL of a frog Has an unexpected ending

The seminal discovery of MMPs being matrix degraders was empirical evidence at the time but it appears to have transcended into *a priori* knowledge. In understanding the roles of MMPs in biology, one cannot accept *a priori* knowledge without critical thinking and analysis, but one must constantly defy preconceived notions using new evidence being collected with new technologies such as genomics, proteomics, and degradomics. However, these recent observations also need to be carefully validated both *in vitro* and *in vivo*, not only in rodents but also in humans, before being accepted as *knowledge*.

Numerous studies demonstrate various roles of MMPs in matrix remodeling events in several diseases but it is now time to investigate the 75% majority of MMP substrates in order to reveal new exciting roles in biological processes and diseases (Fig. 8.1). In order to understand the diverse roles of MMPs in pathobiology, it is critical to globally integrate all the players. Genomics and proteomics studies have not yet been conducted on all MMPs and thus the field is still open for new discoveries. Efficacious bioinformatics methods will be critical to facilitate the intricate analysis of these complex genomics and proteomics *in vivo* samples, especially when enrichment methods are utilized. These novel discoveries in the field of MMPs will renew our understanding and will modify our approaches for effective drug developments in various diseases caused by MMPs.

Acknowledgements

We thank Georgina Butler for her insightful suggestions and editing of the manuscript.

References

1. Sodek, J. & Overall, C.M. (1988) Matrix degradation in hard and soft connective tissues. In: Davidovitch, Z. (ed), *The biological mechanisms of tooth eruption and root resorption*. EBSCO Media, Birmingham, AL, pp. 303–311.
2. Butler, G.S. & Overall, C.M. (2009) Updated biological roles for matrix metalloproteinases and new 'Intracellular' substrates revealed by degradomics. *Biochemistry*, 48, 10830–10845.
3. Iyer, R.P., Patterson, N.L., Fields, G.B. & Lindsey, M.L. (2012) The history of matrix metalloproteinases: milestones, myths, and misperceptions. *American Journal of Physiology - Heart and Circulatory Physiology*, 303, H919–H930.
4. Cauwe, B. & Opdenakker, G. (2010) Intracellular substrate cleavage: a novel dimension in the biochemistry, biology and pathology of matrix metalloproteinases. *Critical Reviews in Biochemistry and Molecular Biology*, 45, 351–423.
5. Mason, S.D. & Joyce, J.A. (2011) Proteolytic networks in cancer. *Trends in Cell Biology*, 21, 228–237.
6. Gross, J.J. & Lapiere, C.M.C. (1962) Collagenolytic activity in amphibian tissues: a tissue culture assay. *Proceedings of the National academy of Sciences of the United States of America*, 48, 1014–1022.
7. Overall, C.M. & Kleifeld, O. (2006) Tumour microenvironment – Opinion: Validating matrix metalloproteinases as drug targets and anti-targets for cancer therapy. *Nature Reviews. Cancer*, 6, 227–239.

8. López-Otín, C. & Overall, C.M. (2002) Protease degradomics: a new challenge for proteomics. *Nature Reviews. Molecular Cell Biology*, 3, 509–519.
9. Visse, R.R. & Nagase, H.H. (2003) Matrix metalloproteinases and tissue inhibitors of metalloproteinases: structure, function, and biochemistry. *Circulation Research*, 92, 827–839.
10. Nagase, H.H., Visse, R.R. & Murphy, G.G. (2006) Structure and function of matrix metalloproteinases and TIMPs. *Cardiovascular Research*, 69, 12–12.
11. Marchant, D.J., Bellac, C.L., Moraes T.J. *et al.* (2014) A new transcriptional role for matrix metalloproteinase-12 in antiviral immunity. *Nature Medicine*, 20, 493–502.
12. Wilson, C.L., Ouellette, A.J., Satchell, D.P. *et al.* (2000) Regulation of intestinal alpha-defensin activation by the metalloproteinase matrilysin in innate host defense. *Science*, 286, 113–117.
13. McQuibban, G.A., Gong, J.H., Tam, E.M. *et al.* (2000) Inflammation dampened by gelatinase A cleavage of monocyte chemoattractant protein-3. *Science*, 289, 1202–1206.
14. Starr, A.E., Dufour, A., Maier, J. & Overall, C.M. (2012) Biochemical analysis of matrix metalloproteinase activation of chemokines CCL15 and CCL23 and increased glycosaminoglycan binding of CCL16. *The Journal of Biological Chemistry*, 287, 5848–5860.
15. Starr, A.E., Bellac, C.L., Dufour, A., Goebeler, V. & Overall, C.M. (2012) Biochemical characterization and N-terminomics analysis of leukolysin, the membrane-type 6 matrix metalloprotease (MMP25): chemokine and vimentin cleavages enhance cell migration and macrophage phagocytic activities. *The Journal of Biological Chemistry*, 287, 13382–13395.
16. Tester, A.M., Cox J.H., Connor, A.R. *et al.* (2007) LPS responsiveness and neutrophil chemotaxis in vivo require PMN MMP-8 activity. *PLoS ONE*, 2, e312.
17. Fortelny, N., Cox, J.H., Kappelhoff, R. *et al.* (2014) Network analyses reveal pervasive functional regulation between proteases in the human protease web. *PLOS Biology*, 12, e1001869. doi:10.1371–journal.pbio.1001869.
18. Dean, R.A., Cox, J.H., Bellac, C.L. *et al.* (2008) Macrophage-specific metalloelastase (MMP-12) truncates and inactivates ELR+ CXC chemokines and generates CCL2, -7, -8, and -13 antagonists: potential role of the macrophage in terminating polymorphonuclear leukocyte influx. *Blood*, 112, 3455–3464.
19. McQuibban, G.A., Gong, J.H., Wong, J.P. *et al.* (2002) Matrix metalloproteinase processing of monocyte chemoattractant proteins generates CC chemokine receptor antagonists with anti-inflammatory properties in vivo. *Blood*, 100, 1160–1167.
20. McQuibban, G.A., Butler, G.S., Gong, J.H. *et al.* (2001) Matrix metalloproteinase activity inactivates the CXC chemokine stromal cell-derived factor-1. *Journal of Biological Chemistry*, 276, 43503–43508.
21. Zhang, K., McQuibban, G.A., Silva, C. *et al.* (2003) HIV-induced metalloproteinase processing of the chemokine stromal cell derived factor-1 causes neurodegeneration. *Nature Neuroscience*, 6, 1064–1071.
22. Dean, R.A. & Overall, C.M. (2007) Proteomics discovery of metalloproteinase substrates in the cellular context by iTRAQ labeling reveals a diverse MMP-2 substrate degradome. *Molecular and Cellular Proteomics*, 6, 611–623.
23. Cox, J.H. & Overall, C.M. (2008) Cytokine substrates: MMP regulation of inflammatory signaling molecules. In: *The Cancer Degradome*. Springer, New York, pp. 519–539. DOI:10.1007/978-0-387-69057-5_26.
24. Dufour, A., Sampson, N.S., Li, J. *et al.* (2011) Small-molecule anticancer compounds selectively target the hemopexin domain of matrix metalloproteinase-9. *Cancer Research*, 71, 4977–4988.
25. Sakamoto, T. & Seiki, M. (2010) A membrane protease regulates energy production in macrophages by activating hypoxia-inducible factor-1 via a non-proteolytic mechanism. *The Journal of Biological Chemistry*, 285, 29951–29964.
26. Gill, S.E. & Parks, W.C. (2008) Metalloproteinases and their inhibitors: regulators of wound healing. *The International Journal of Biochemistry & Cell Biology*, 40, 1334–1347.
27. Martin, P. & Leibovich, S.J. (2005) Inflammatory cells during wound repair: the good, the bad and the ugly. *Trends in Cell Biology*, 15, 599–607.
28. Kessenbrock, K., Plaks, V. & Werb, Z. (2010) Matrix metalloproteinases: regulators of the tumor microenvironment. *Cell*, 141, 52–67.
29. Sternlicht, M.D. & Werb, Z. (2001) How matrix metalloproteinases regulate cell behavior. *Annual Review of Cell and Developmental Biology*, 17, 463–516.
30. Dufour, A., Sampson, N.S., Zucker, S. & Cao, J. (2008) Role of the hemopexin domain of matrix metalloproteinases in cell migration. *Journal of Cellular Physiology*, 217, 643–651.
31. Coussens, L.M., Fingleton, B. & Matrisian, L.M. (2002) Matrix metalloproteinase inhibitors and cancer: trials and tribulations. *Science*, 295, 2387–2392.
32. auf dem Keller, U., Prudova, A., Eckhard, U., Fingleton, B. & Overall, C.M. (2013) Systems-level analysis of proteolytic events in increased vascular permeability and complement activation in skin inflammation. *Science Signaling*, 6, rs2.

33. Rogers, L. & Overall, C.M. (2013) Proteolytic post translational modification ofproteins: proteomic tools and methodology. *Molecular and Cellular Proteomics*. doi:10.1074/mcp.M113.031310
34. Huesgen, P.F. & Overall, C.M. (2012) N- and C-terminal degradomics: new approaches to reveal biological roles for plant proteases from substrate identification. *Physiologia Plantarum*, 145, 5–17.
35. Mahrus, S., Trinidad, J.C., Barkan, D.T. *et al.* (2008) Global sequencing of proteolytic cleavage sites in apoptosis by specific labeling of protein N termini. *Cell*, 134, 866–876.
36. Kleifeld, O., Doucet, A., auf dem Keller, U. *et al.* (2010) Isotopic labeling of terminal amines in complex samples identifies protein N-termini and protease cleavage products. *Nature Biotechnology*, 28, 281–288.
37. Gevaert, K., Goethals, M., Martens, L. *et al.* (2003) Exploring proteomes and analyzing protein processing by mass spectrometric identification of sorted N-terminal peptides. *Nature Biotechnology*, 21, 566–569.
38. Dufour, A. & Overall, C.M. (2013) Missing the target: matrix metalloproteinase antitargets in inflammation and cancer. *Trends in Pharmacological Sciences*, 34, 233–242.
39. Lange, P.F. & Overall, C.M. (2011) TopFIND, a knowledgebase linking protein termini with function. *Nature Methods*, 8, 703–704.
40. Lange, P.F., Huesgen, P.F. & Overall, C.M. (2012) TopFIND 2.0—linking protein termini with proteolytic processing and modifications altering protein function. *Nucleic Acid Research*.
41. auf dem Keller, U., Prudova, A., Gioia, M., Butler, G.S. & Overall, C.M. (2010) A statistics-based platform for quantitative N-terminome analysis and identification of protease cleavage products. *Molecular and Cellular Proteomics*, 9, 912–927.
42. Prudova, A., auf dem Keller, U., Butler, G.S. & Overall, C.M. (2010) Multiplex N-terminome analysis of MMP-2 and MMP-9 substrate degradomes by iTRAQ-TAILS quantitative proteomics. *Molecular and Cellular Proteomics*, 9, 894–911.
43. Morrison, C., Mancini, S., Cipollone, J. *et al.* (2011) Microarray and proteomic analysis of breast cancer cell and osteoblast co-cultures: role of osteoblast matrix metalloproteinase (MMP)-13 in bone metastasis. *Journal of Biological Chemistry*, 286, 34271–34285.
44. Butler, G.S., Dean, R.A., Tam, E.M. & Overall, C.M. (2008) Pharmacoproteomics of a metalloproteinase hydroxamate inhibitor in breast cancer cells: dynamics of membrane type 1 matrix metalloproteinase–mediated membrane protein shedding. *Molecular and Cellular Biology*, 28, 4896–4914.
45. Dean, R.A., Butler, G.S., Hamma-Kourbali, Y. *et al.* (2007) Identification of candidate angiogenic inhibitors processed by matrix metalloproteinase 2 (MMP-2) in cell-based proteomic screens: disruption of vascular endothelial growth factor (VEGF)/heparin affin regulatory peptide (pleiotrophin) and VEGF/Connective tissue growth factor angiogenic inhibitory complexes by MMP-2 proteolysis. *Molecular and Cellular Biology*, 27, 8454–8465.
46. Tam, E.M., Morrison, C.J., Wu, Y.I., Stack, M.S. & Overall, C.M. (2004) Membrane protease proteomics: Isotope-coded affinity tag MS identification of undescribed MT1-matrix metalloproteinase substrates. *Proceedings of the National academy of Sciences of the United States of America*, 101, 6917–6922.
47. Sabeh, F., Shimizu-Hirota, R. & Weiss, S.J. (2009) Protease–dependent versus -independent cancer cell invasion programs: three-dimensional amoeboid movement revisited. *The Journal of Cell Biology*, 185, 11–19.
48. Holmberg, C., Ghesquière, B., Impens, F. *et al.* (2013) Mapping Proteolytic Processing in the Secretome of Gastric Cancer-Associated Myofibroblasts Reveals Activation of MMP-1, MMP-2, and MMP-3. *Journal of Proteome Research*, 12, 3413–3422.
49. Butler, G.S. & Overall, C.M. (2009) Proteomic identification of multitasking proteins in unexpected locations complicates drug targeting. *Nature Reviews. Drug Discovery*, 8, 935–948.
50. Lange, S.S., Mitchell, D.L. & Vasquez, K.M. (2008) High mobility group protein B1 enhances DNA repair and chromatin modification after DNA damage. *Proceedings of the National academy of Sciences of the United States of America*, 105, 10320–10325.
51. Pisetsky, D.S., Erlandsson-Harris, H. & Andersson, U. (2008) High-mobility group box protein 1 (HMGB1): an alarmin mediating the pathogenesis of rheumatic disease. *Arthritis Research and Therapy*, 10, 209.
52. Wang, H., Yang, H. & Tracey, K.J. (2004) Extracellular role of HMGB1 in inflammation and sepsis. *Journal of Internal Medicine*, 255, 320–331.
53. Fossati, S. & Chiarugi, A. (2007) Relevance of high-mobility group protein box 1 to neurodegeneration. *International Review of Neurobiology*, 82, 137–148.
54. Yang, H., Ochani, M., Li, J. *et al.* (2004) Reversing established sepsis with antagonists of endogenous high-mobility group box 1. *Proceedings of the National academy of Sciences of the United States of America*, 101, 296–301.
55. Abraham, E., Arcaroli, J., Carmody, A., Wang, H. & Tracey, K.J. (2000) Cutting edge: HMG-1 as a mediator of acute lung inflammation. *The Journal of Immunology*, 165, 2950–2954.

56. Butler, G.S. & Overall, C.M. (2013) Matrix metalloproteinase processing of signaling molecules to regulate inflammation. *Periodontology 2000*, 2000 (63), 123–148.
57. Vandooren, J., Van den Steen, P.E. & Opdenakker, G. (2013) Biochemistry and molecular biology of gelatinase B or matrix metalloproteinase-9 (MMP-9): The next decade. *Critical Reviews in Biochemistry and Molecular Biology*. doi:10.3109/10409238.2013.770819.
58. Marchenko, N.D., Marchenko, G.N., Weinreb, R.N. *et al.* (2004) Beta-catenin regulates the gene of MMP-26, a novel metalloproteinase expressed both in carcinomas and normal epithelial cells. *The International Journal of Biochemistry & Cell Biology*, 36, 942–956.
59. Shimizu-Hirota, R., Xiong, W., Baxter, B.T. *et al.* (2012) MT1-MMP regulates the PI3Kδ·Mi-2/NuRD-dependent control of macrophage immune function. *Genes & Development*, 26, 395–413.
60. Ali, M.A., Chow, A.K., Kandasamy, A.D. *et al.* (2012) Mechanisms of cytosolic targeting of matrix metalloproteinase-2. *Journal of Cellular Physiology*, 227, 3397–3404.
61. Sbai, O., Ould-Yahoui, A., Ferhat, L. *et al.* (2010) Differential vesicular distribution and trafficking of MMP-2, MMP-9, and their inhibitors in astrocytes. *Glia*, 58, 344–366.
62. Hill, J.W., Poddar, R., Thompson, J.F., Rosenberg, G.A. & Yang, Y. (2012) Intranuclear matrix metalloproteinases promote DNA damage and apoptosis induced by oxygen-glucose deprivation in neurons. *Neuroscience*, 220, 277–290.
63. Miller, J.P., Holcomb, J., Al-Ramahi, I. *et al.* (2010) Matrix metalloproteinases are modifiers of huntingtin proteolysis and toxicity in Huntington's disease. *Neuron*, 67, 199–212.
64. Overall, C.M. & Lopez-Otin, C. (2002) Strategies for MMP inhibition in cancer: Innovations for the post-trial era. *Nature Reviews. Cancer*, 2, 657–672.
65. Conant, K., St Hillaire, C., Nagase, H. *et al.* (2004) Matrix metalloproteinase 1 interacts with neuronal integrins and stimulates dephosphorylation of Akt. *The Journal of Biological Chemistry*, 279, 8056–8062.
66. Redondo-Muñoz, J., Ugarte-Berzal, E., Terol, M.J. *et al.* (2010) Matrix Metalloproteinase-9 Promotes Chronic Lymphocytic Leukemia B Cell Survival through Its Hemopexin Domain. *Cancer Cell*, 17, 13–13.
67. Dufour, A., Zucker, S., Sampson, N.S., Kuscu, C. & Cao, J. (2010) Role of matrix metalloproteinase-9 dimers in cell migration: design of inhibitory peptides. *Journal of Biological Chemistry*, 285, 35944–35956.
68. Houghton, A.M.A., Hartzell, W.O.W., Robbins, C.S.C., Gomis-Rüth, F.X.F. & Shapiro, S.D.S. (2009) Macrophage elastase kills bacteria within murine macrophages. *Nature*, 460, 637–641.
69. Sakamoto, T. & Seiki, M. (2009) Cytoplasmic tail of MT1-MMP regulates macrophage motility independently from its protease activity. *Genes to Cells*, 14, 617–626.
70. Rozanov, D.V., Sikora, S., Godzik A. *et al.* (2004) Non-proteolytic, receptor/ligand interactions associate cellular membrane type-1 matrix metalloproteinase with the complement component C1q. *The Journal of Biological Chemistry*, 279, 50321–50328.
71. Zarrabi, K., Dufour, A., Li, J. *et al.* (2011) Inhibition of matrix metalloproteinase 14 (MMP-14)-mediated cancer cell migration. *The Journal of Biological Chemistry*, 286, 33167–33177.
72. Remacle, A.G., Golubkov, V.S., Shiryaev, S.A. *et al.* (2012) Novel MT1-MMP small-molecule inhibitors based on insights into hemopexin domain function in tumor growth. *Cancer Research*, 72, 2339–2349.
73. Pavlaki, M., Zucker, S., Dufour, A. *et al.* (2011) Furin functions as a nonproteolytic chaperone for matrix metalloproteinase-28: MMP-28 propeptide sequence requirement. *Biochemistry Research International*, 2011, 630319.
74. Strongin, A.Y., Collier, I., Bannikov, G. *et al.* (1995) Mechanism of cell surface activation of 72-kDa type IV collagenase. Isolation of the activated form of the membrane metalloprotease. *The Journal of Biological Chemistry*, 270, 5331–5338.
75. Correia, A.L., Mori, H., Chen, E.I., Schmitt, F.C. & Bissell, M.J. (2013) The hemopexin domain of MMP3 is responsible for mammary epithelial invasion and morphogenesis through extracellular interaction with HSP90β. *Genes & Development*, 27, 805–817.
76. Tam, E.M., Wu, Y.I., Butler, G.S. *et al.* (2002) Collagen Binding Properties of the Membrane Type-1 Matrix Metalloproteinase (MT1-MMP) Hemopexin C Domain. The ectodomain of the 44-kda autocatalytic product of MT1-MMP inhibits cell invasion by disrupting native type I collagen cleavage. *Journal of Biological Chemistry*, 277, 39005–39014.

9 MMPs: From Genomics to Degradomics

Barbara Grünwald[1], Pascal Schlage[2], Achim Krüger[1], and Ulrich auf dem Keller[2]

[1] *Institute for Experimental Oncology and Therapy Research, Klinikum rechts der Isar, Technische Universität München, Munich, Germany*
[2] *ETH Zurich, Department of Biology, Institute of Molecular Health Sciences, Zurich, Switzerland*

9.1 Introduction

Since their discovery in the 1980s, Matrix metalloproteinases (MMPs) have been well characterized as central drivers of a variety of pathological processes such as inflammation and cancer [1–4]. Increasing knowledge about the pathological relevance of MMPs in cancer quickly led to the development of synthetic inhibitors. Failure of these compounds in several phase III clinical trials due to therapeutic inefficiency and adverse effects hit the field by surprise. This chapter discusses how the drawbacks of a conventional genomics approach led to the misevaluation of MMPs as therapeutic targets. While representing state-of-the-art of molecular-biological research in the last decade of the 20th century, genomics alone could only give rise to overestimated expectations due to missing and misinterpreted context information. As a result, the premature introduction of MMP inhibitors into the clinic led to the seminal failure of this therapeutic approach. Here, we summarize recent advancements in our understanding of MMPs as master regulators of tissue homeostasis, and point to specific challenges of using MMPs as therapeutic targets. This knowledge, also based on genomics but combined with the newly emerging approaches of systems biology, should help explain the complex biology of individual MMPs and facilitate our judgment on whether some of them can be considered as targets for therapeutic intervention.

In the second part of this chapter we introduce the concepts of degradomics-based systems biology and demonstrate how its application to MMP research deepened our understanding of the interconnected activities of members of this protease class within the protease web. The rapid development of systems biology approaches enabled global analysis of MMP expression as well as activity. Driven by mass spectrometry-based proteomics, the specificity profiles for individual MMPs can be recorded as a prerequisite for the development of specific inhibitors. Novel quantitative proteomics techniques for the system-wide analysis of protease substrates in complex proteomes identified hundreds of new MMP substrates. This

Matrix Metalloproteinase Biology, First Edition. Edited by Irit Sagi and Jean P. Gaffney.

extends our knowledge about their multiple roles in health and disease. With the help of integrative data repositories, we will gain decisive systems-level insight into MMP biology, enabling us to design novel strategies to counteract their activities in disease.

9.1.1 Genomics: general aspects

The pathological impact of MMPs ranges from inflammation [5], auto-immune disease, and cardiovascular disorders [6] to cancer, angiogenesis, and tumor metastasis [1–4], all of which represent rather complex conditions. While inherited disorders like cystic fibrosis and phenylketonuria are caused by single genes, malignant or cardiovascular diseases typically arise as result of a combination of genetic and environmental factors. In the 20th century, scientific study of such complex disorders was typically approached *via* genomics, a more recent field of genetics that employs the analysis of all of the genes of an individual, the genome [7]. The field of genomics employs the very powerful analytical approach of "Methodological Reduction". This concept is based on the idea that a complex system represents the sum of its parts and can therefore be understood by analysis of its simpler components [8]. Accordingly, the co-operative and comprehensive function of all parts in the complex system can be explained and predicted after revelation of their individual structures and functions [9]. This reductionist approach is the basis of the central dogma of molecular biology: genes are transcribed into mRNAs, which are then translated into proteins [10]. Based on this mechanistic relation between DNA and proteins, all phenotypic features of an organism could be explained by genetic determination according to the following chain of causation: *gene [leads to] protein [leads to] function [leads to] phenotype*. Thus, the experimental study of a phenotype or a genetic disease is considered most productive if it is aimed at revealing the underlying genetic cause [8]. Consequently, the field of genomics has put large effort into sequencing, assembly, and analysis of genes in the last decades, culminating in several whole genome sequencing projects launched in the 1990s [11]. The resulting great amount of knowledge paved the way for *functional genomics*, a field of genomics that makes use of genome-wide sequence data to annotate and manipulate functions and interactions [12]. By now the functions of a vast majority of the human genes have been identified and genomic mapping has been further supplemented by differential gene expression patterns from various physiological and pathological conditions [7]. Especially in cancer, this approach has led to the identification of many promising therapeutic targets such as MMPs [13]. However, from the failure of the clinical trials on MMP inhibitors we have learned that essential information on MMPs as therapeutic targets in cancer treatment was still missing or misinterpreted. Accordingly, the functions of MMPs – as the individual part – have not yet been sufficiently revealed in order to predict their co-operative and comprehensive functions within a system as complex as cancer [14]. We next review the results obtained from the genomics approach to MMP function in detail, pointing out valuable information as well as former missing links, those which eventually resulted in the misled predictions and expectations of the effects of therapeutic intervention with MMPs [15].

9.1.2 The genomics approach to MMP function in cancer

In the past, the reductionist-genomic approach to cancer as a genetically determined disease was proved to be successful in many regards. Cancer biology reached a milestone when the first human oncogene *ras* was discovered by the basic examination of DNA homologies between retroviral oncogenes and transforming sequences [16]. Soon, attention was drawn to down-stream effector proteins, among which proteases were found to be induced upon malignant transformation [17] and to be more abundantly expressed in malignant than in benign tumors [18]. Subsequent purification and characterization of these downstream effectors of oncogenes led to their identification as MMPs [19, 20] and demonstrated their direct involvement in tumor invasion and metastasis [21]. Two major corner stones of cancer biology were laid by this observation: enzymatic break-down of the basement membrane had already been established as a cardinal feature of malignant disease [22] and could now be shown to occur downstream of an oncogene. Further biochemical studies broadened the knowledge on MMP substrates showing that MMPs are capable of degrading virtually every component of the extracellular matrix (ECM) [19, 23–25]. Accumulating evidence from *in vitro* studies further supported the notion that deregulated MMP expression correlates with the progression of the transformed phenotype of tumor cells [26]. Taken together, these correlative studies strongly pointed to MMP-mediated tissue breakdown as a major determinant of cancer progression. Thus, the logical next step was to test whether the causal chain, *MMP deregulation [leads to] matrix degradation [leads to] cancer progression* will hold true in the complex system. This is typically realized via two approaches: (i) basic correlation of the component's *presence*/expression levels with clinicopathologic parameters of disease progression in human patients and (ii) correlation of the component's *absence* with disease progression in animal disease models. The former was soon realized by numerous studies, most of which indeed showed a positive correlation of MMP *presence* with cancer progression in human patients. For instance, collagenolytic activity was shown to correspond to histological de-differentiation of tumors [27] and tumoral MMP-2/MMP-3 levels to malignant behavior [28]. Initial studies on the correlation of MMP-2 serum levels with disease progression were controversial; while some groups reported a positive correlation with the presence of distant metastasis and response failure [29], others found no increase of serum MMP-2 in cancer patients [30]. However, by 1997, expression of MMP-1, MMP-2, MMP-3, and MMP-9 had repeatedly been reported to be of prognostic value for the patients' survival in various types of cancer [31–34]. In turn, the *absence* of MMPs was shown to correlate with reduced cancer progression in functional genetic mouse models, corroborating the notion that MMPs could be promising therapeutic targets. MMP-7 knock-out mice, for instance, showed reduced spontaneous intestinal tumorigenesis [35], MMP-11-defficient mice suffered less chemically-induced carcinogenesis [36] and MMP-2 knock-out led to impaired angiogenesis and tumor progression [37]. Furthermore, deficiency in MMP-9 resulted in reduced experimental metastasis formation [38] and reduced carcinogenesis in skin [39], while pharmacological MMP-9 inhibition in a genetic mouse model of spontaneous pancreatic carcinogenesis was shown to reduce the angiogenic switch [40]. Taken together, *absence* of MMP expression in functional genetic mouse models appeared to be beneficial for the host, implying a decisive role of MMPs in malignant progression. In accordance, abundant experimental evidence had shown that the *presence* of MMPs strongly correlates with cancer progression in human patients. These insights, in combination with increasing knowledge about the enzymatic properties of MMPs, have led to the following four principal assumptions on the role of MMPs in cancer [41]:

i) MMPs are produced by tumors
ii) MMPs degrade ECM
iii) MMPs promote tumor progression
iv) MMPs contribute to invasion and metastasis

Accordingly, all currently available knowledge has resulted in the logical – but as we now know over-simplified –concept that cancer invasion and metastasis in humans is driven by cancer cell-released MMPs whose predominant role is to digest surrounding connective tissues [42].

9.1.3 Taking first steps towards MMP inhibition in cancer therapy

According to the presumably essential function of MMPs in the invasive growth of cancer, great expectations were placed on therapeutic interference with MMPs. The targeted inhibition of cancer invasion was expected to block local as well as distant organ infiltration [43]. The resulting therapeutic spectrum would have represented a tremendous clinical progress, as the inability to control metastasis still accounts for approximately 90% of deaths in cancer patients. Interestingly, MMPs had already received attention as mediators of joint degeneration in rheumatoid arthritis [44, 45], before their recognition as potential targets in cancer therapy. Therefore, many companies had launched research programs on development of various pharmacological MMP inhibitors in the early 1980s. Theoretically, selective MMP inhibitors would have provided greater specificity and more safety than broad-spectrum inhibition due to an increased therapeutic index. On the other hand, different MMPs are often co-expressed in various types of cancer [28], rendering identification of a single MMP crucial for the disease process very difficult. As a result, initial concepts were based on effective but unselective inhibitors, in order to achieve maximal repression of tumor cell spread by inhibition of all member of the MMP family. The enzymatic mechanism of MMPs is prone to be targeted by such a broad-spectrum approach due to the highly conserved structure of their zinc-dependent catalytic domain [46]. First generation MMP inhibitors were hydroxamate-based derivates that specifically mimic the peptide residues of a principal cleavage site. While their peptide backbone mediates a tight interaction, the hydroxamate group chelates the zinc atom in the active site, resulting in potent but reversible inhibition of the enzymatic activity [47]. Several hydroxamate derivatives have been developed, such as batimastat and marimastat (the orally bioavailable follow-up compound), GM6001 [48], MMI270

[49], and GI129471 [50]. Batimastat was the first broad-spectrum MMP inhibitor that entered a clinical trial in cancer patients and will be used in the following as an example to demonstrate the variety of observations that supported or contradicted the expectations on MMP inhibition as a therapeutic approach. Initial *in vitro* evaluation of the compound was entirely in-line with these expectations: Efficient inhibition of extracellular matrix degradation and invasion was observed [47], along with a weak cytostatic activity but no cytotoxic activity on cancer cell lines [51–53]. When batimastat entered a phase I clinical trial for oral administration in 1991, it exhibited very poor bioavailability [47]. However, soon after, intra-peritoneal application of batimastat was shown to resolve ascites derived from ovarian carcinoma *xenografts* in mice which resulted in a highly significant prolongation of survival [52]. Batimastat entered a second clinical trial instantaneously as treatment for malignant ascites in early 1993 [47]. An impressive therapeutic benefit from intra-peritoneal batimastat application was meanwhile demonstrated in two murine *xenograft* models of colorectal carcinoma: orthotopic tumor growth and spontaneous metastasis formation [54] and experimental metastasis formation to the liver and lung [55] were efficiently inhibited upon batimastat treatment. Similar observations were reported in a *syngeneic* melanoma model [51]. In sum, various lines of experimental evidence had substantiated the notion that broad-spectrum MMP inhibition might indeed be the key to effective inhibition of invasive growth:

- Expression of various MMPs had repeatedly been reported to be of prognostic value for patient survival in various types of cancer [31–33].
- Absence of MMP expression in functional genetic mouse models appeared to be beneficial for the host [35–40].
- Efficacy of pharmacological broad-spectrum inhibition of MMPs had been demonstrated in pre-clinical studies employing experimental and spontaneous murine tumor models [47, 51, 52, 54, 55].

Based on this compelling data, it seemed reasonable at that time to rapidly route MMP inhibitors into trials with human cancer patients.

9.1.4 Lessons from the failure of unselective MMP inhibition

In the meantime, however, a considerable amount of information had accumulated which suggested that MMP biology was far more complex than initially assumed. In contrast to the assumption that MMPs are produced by tumor cells, it was shown that in several cancer types MMPs are abundantly and sometimes exclusively expressed by stromal cells in the tumor microenvironment [34, 56–58]. Though MMPs are secreted as soluble zymogens and activated in the extracellular space [59], their expression by stromal cells surrounding invasive tumor cells was still in agreement with their postulated role in degradation of the ECM. Moreover, the targeting of components produced by non-malignant cells in the tumor is therapeutically very attractive, because these cells will presumably not mutate and develop drug resistance [60]. Another interesting aspect of MMP biology received increasing awareness at about the same time: the action of individual proteases is embedded in complex interaction networks. For example the MMP system was shown to co-operate closely with the plasmin/plasminogen activators system in ECM degradation during tumor progression [43, 61]. This brought up the notion that simultaneous targeting of several protease systems might be necessary to achieve full inhibition of invasive behavior [62]. Comparison of *in vitro* effects of concomitant inhibition of two cooperating protease systems with invasive behavior *in vivo* pointed out another obstacle for assessment of the effects of MMP inhibition: while the uPA inhibitor aprotinin alone and in combination with batimastat almost entirely abolished degradation of casein and collagen type IV *in vitro*, no inhibition of tumor cell invasion was observed *in vivo* [62]. Batimastat, on the other hand, showed only marginal effectiveness on casein and collagen type IV degradation *in vitro* but inhibited intraperitoneal tumor growth [62], suggesting that the inhibition pattern observed *in vitro* does not necessarily reflect the *in vivo* situation.

Nevertheless, the failure of early clinical trials due to inefficacy and unacceptable side effects was not to be expected at that time. Various factors have contributed to this, which must be considered before an ultimate conclusion about the therapeutic benefit of MMP inhibition in cancer can be drawn. For instance, the broad-spectrum approach turned out to be a major conceptual limitation in terms of unexpected side effects as information about physiological roles of MMPs was still missing at that time. In early phase I clinical trials (dose escalation studies designed to evaluate safety) prolonged treatment of human patients with MMP inhibitors was found to cause pain and inflammation of skeletal muscles. This revealed that MMP activity is not limited to malignant disease but plays a physiological role in normal joint function [59]. The conditions were reversible upon short treatment breaks but limited the administrable dose for subsequent trials [63]. MMP-1 was expected to be responsible for these arthralgic side effects, so large effort was put into the development of MMP inhibitors with enhanced selectivity. Compounds selective for the deep-pocket (MMP-2, MMP-3, MMP-8, MMP-9, MMP-13, and MT1-MMP) over the shallow-pocket enzymes MMP-1 and MMP-7 were designed, which indeed showed more favorable side-effect profiles. After this it was realized that the MMP family contains anti-targets which cause unexpected side effects, while other MMPs might still be valid targets for anti-cancer therapy [3]. Also, new members of the MMP family were discovered only after initiation of the clinical trials, such as the class of *A disintegrin and metalloproteinases* (ADAMs), which have important cellular functions independent of cancer progression [64].

Additionally, insufficient interpretation of pre-clinical studies resulted in unsuitable study design [63]. MMP inhibitors were tested in phase II/III combination trials that evaluated efficacy in direct comparison to standard treatments in patients with late stage cancer [63]. However, it is important to note that in the pre-clinical animal models, MMP inhibitor treatment had generally started at early time-points and maintained during disease progression. Moreover, it was directly shown that initiation of MMP inhibitor treatment at a minimal tumor load had a much more profound effect on growth inhibition than initiation of treatment at later stages [51, 65]. A vivid comparison depicted the significance of evaluating MMP inhibitors in patients with late stage cancers as "locking the barn door after the horse has bolted out of the stable"

[66]. As a result, the potential effectiveness of MMP inhibition as treatment in early stage cancer has still not been addressed. All these observations provide explanations for the inefficiency of MMP inhibitors in trials on advanced human malignancies but also show that the roles of MMPs expand much further than promotion of invasive growth and metastasis. Moreover, it became evident that unselective MMP inhibition in cancer was not only ineffective but can even cause adverse effects, when some studies had to be terminated due to a significant decrease in survival of MMP inhibitor treated patients [67]. At that time, only a very small number of publications had shown a limited effectiveness of these compounds [68] but adverse effects of MMP inhibition in cancer were yet unknown. However, one study reported the startling observation that batimastat inhibited tumor growth but promoted liver metastasis formation [62]. The mechanism behind this metastasis-promoting effect of batimastat was soon thereafter elucidated: Broad-spectrum inhibition of MMPs gave rise to a metastasis-promoting environment in the liver, characterized by an increased expression of MMP-2 and MMP-9, hepatocyte growth factor (HGF), basic fibroblast growth factor (bFGF), angiogenin, and caspase-1 [69]. This observation suggested for the first time, that the role of MMPs in tissue homeostasis has to be considered as a major determinant of metastatic progression [70].

In addition to the pre-clinical testing of MMP inhibitors, the formerly compelling straightforward data obtained from functional genetic mouse models had to undergo revision when several studies reported that the effects upon ablation of individual MMPs were context-dependent. For instance, a mouse strain-dependent effect of MMP-9 ablation on the anti-metastatic outcome was demonstrated in a mammary tumor model [71]. This suggests that responses to MMP inhibition are controlled by very short genetic distances [71]. The effects of MMP-7 ablation in mice turned out to be organ-specific: while formation and growth of intestinal tumors was significantly reduced by MMP-7 deficiency [35], MMP-7 knock-out was shown to be irrelevant during mammary tumor progression [71]. Eventually, more information pointed to the context-dependency of the action of individual MMPs. MMP-8 was identified as the first specific MMP with an even protective function in cancer [72]. Genetic ablation of MMP-8 led to an increase in skin tumor formation in male mice due to the function of MMP-8 in the immune system. Interestingly, this role of MMP-8 during carcinogenesis of the skin was shown to be gender-specific: the susceptibility to tumors was induced in female MMP-8 knock-out mice only after removal of their ovaries or treatment with the estrogen-receptor antagonist tamoxifen [72]. This suggests that the sex hormone system plays a significant role in altered homeostasis after manipulation of the MMP system [73]. As a result, the information on MMP function derived from the genomics-based approach was misinterpreted due to missing knowledge about the potential context-dependency of any observations made. Therefore, unselective MMP inhibition in cancer was wrongly predicted as accurate therapeutic rationale for cancer treatment. These very intriguing observations, however, shed more light on the particular nature of MMP biology only after the clinical trials had failed. In the following paragraph, we address why major characteristics of MMP biology cannot be addressed by a genomics approach alone and how these challenges can be met by newly developing approaches of systems biology.

9.1.5 Limitations of the genomic approach to MMP function

As discussed above, the genomic approach is valid for the description of biological systems that can be described by the following chain of causation: *gene [leads to] protein [leads to] function [leads to] phenotype.* However, a significant challenge to the reductionist-genomic concept stems from the fact that higher level biological features can be realized by different single molecular kinds, so that *many* molecular kinds can correspond to *one* higher level kind [8]. For example, excessive cleavage of a substrate in a disease state can be mediated by various different proteases with over-lapping substrate specificity. *Vice versa*, the reductionist-genomic approach reaches its limitations as soon as the effects of molecular processes depend on the context in which they occur, so that *one* molecular kind can have *many* higher level features [8]. This is the case when a protease has multiple substrates so that its actual activity is dependent on the spatio–temporal distribution pattern of its substrates. Both of these scenarios – in which a genomic approach remains to depict an incomplete description of the system – are major characteristics of MMP biology [74].

Compared to other proteases, MMPs exhibit a relatively high redundancy in function [75]. The lack of one MMP can therefore be compensated by increased expression of another [76]. For example, MMP-13 knock-out mice show enhanced MMP-8 expression during wound healing [77], while MMP-8 deficiency in wound repair led to an increased MMP-9 expression [78]. The notion that ablation of one MMP leads to a counter-regulation by other MMPs [43] is further supported by the fact that most MMP knock-out models lack significant phenotypes [13] although the MMP system has been recognized as a major determinant of organ homeostasis [73]. Hence, systemic ablation of an MMP is compensated by the establishment of a new homoeostasis [15] but with sometimes unforeseeable effects regarding disease susceptibility [73]. Knock-out of MMP-9, a well-described tumor promoter [79, 80], drastically induced invasion of tumor cells to the liver in an experimental colorectal carcinoma model [73]. This was caused by MMP-9 deficiency-induced changes in the local proteolytic network of the bone marrow, resulting in a strong induction of interleukin-6 (IL6) in the circulation. Over a distance, re-adaptation of the organism upon manipulation of MMP-9 induced a pro-metastatic microenvironment in the liver [73]. This example illustrates – along with unexpected compensatory effects upon systemic ablation of an MMP – that different local proteolytic networks within the body can effectively communicate *via* signaling molecules during the re-establishment of a new homeostasis.

As mentioned above, another major obstacle for a genomic approach is the biological case of *one* molecular kind having *many* higher level features [8], which is given by the broad substrate diversity of single MMPs and the MMP familiy in general [74]. Although MMPs are largely responsible for the turnover and degradation of the extracellular matrix, recent data indicates that this is neither the sole nor the main function. The emerging knowledge about the broad substrate repertoire of MMPs has considerably influenced the understanding of MMPs. More than being path-clearing enzymes for invading tumor cells, MMPs produce a large number of bioactive degradation products with signaling functions involved in cell growth [81], adhesion and migration [82], and angiogenic switching [40, 83]. In fact, the majority of MMP substrates are non-matrix molecules [75]. MMPs have been shown to act in an activating as well as an in-activating manner on various cytokines and chemokines [84], thereby affecting most diverse aspects of inflammation and innate immunity [5, 85]. Moreover, another recently identified function of MMPs is the mobilization of growth factors and cytokines – not only by cleaving the ECM molecules they might be bound to as originally proposed – but by releasing masking carrier proteins such as CTGF, pleotrophin, follistatin and IGFBPs from their cognate cytokines and growth factors [75]. Considering the biological activity of these newly identified substrates, MMPs function as key regulators of tumor extracellular environment in terms of both matrix turnover and the signaling milieu controlling cell function (Fig. 9.1).

9.1.6 Approaching proteolysis as a system

The first step in the protease-mediated induction of a signaling pathway is the activation of the respective protease. Proteolytic activity is tightly controlled at the transcriptional level by differential expression and at the protein level by activation of inactive zymogens and/or binding of inhibitors or cofactors. Activation can be either auto-catalytic or catalyzed by other proteases [86] and is thereby embedded in a complex network [74]. Once activated, the actual signaling capacity of a certain protease

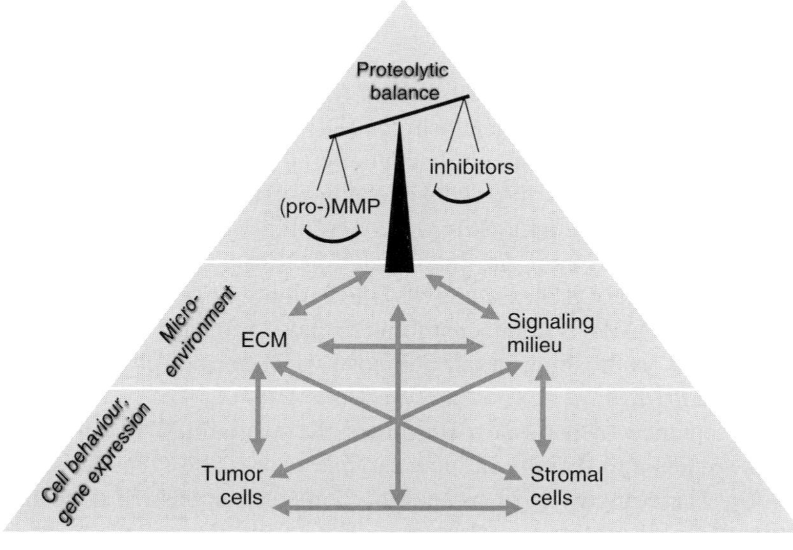

Figure 9.1 Biological activity of an individual MMP within a local tumor microenvironment. MMPs are central regulators of tumor extracellular environment in terms of both extracellular matrix (ECM) turnover and the signaling milieu controlling cell function. Proteolytic balance is tightly controlled at the protein level by activation of individual MMPs from inactive zymogens (proMMPs) and by the binding of inhibitors. Upon activation, each MMP mediates specific effects on the local microenvironment, dependent on its substrate repertoire. These effects derive either down-stream of the MMPs individual ECM substrates or via activation and/or inactivation of signaling molecules, such as cytokines and growth factors. In consequence, the proteolytic balance influences gene expression and behavior of cancer as well as stromal cells, which in turn are major determinants of the proteolytic balance, the local ECM composition and signaling milieu. (*See insert for color representation of this figure.*)

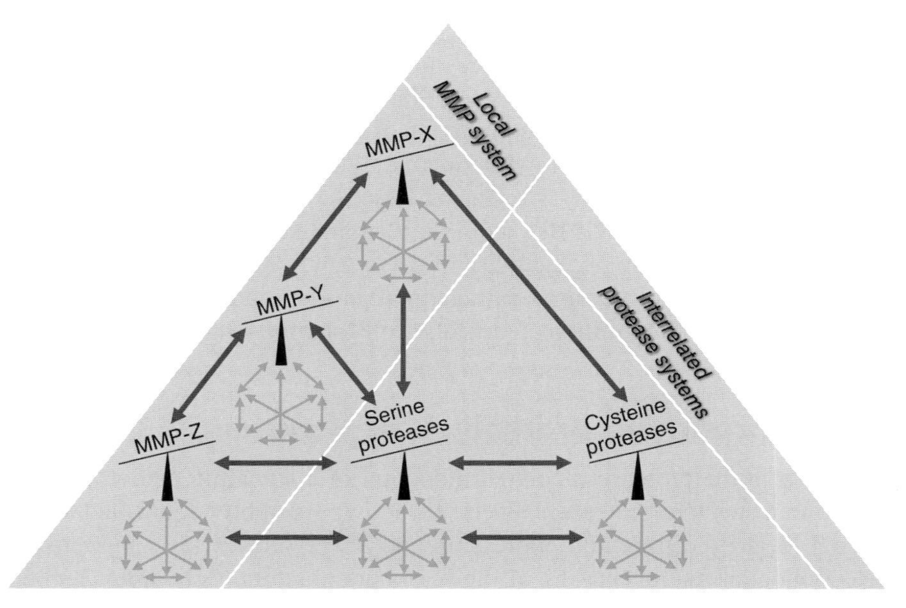

Figure 9.2 Local proteolytic network consisting of interrelated protease systems. The biological effects of an individual MMP (as depicted in Fig. 9.1.) are embedded in the interaction with other MMPs that exhibit partially over-lapping but also distinct substrate specificities, forming the local MMP system. Individual proteolytic systems are interrelated, mutually influencing each other in the modulation of protease activity, substrate availability, and action within a tissue. (*See insert for color representation of this figure.*)

is determined by its substrate repertoire and thereby dependent on the availability and functionality of its respective substrates [87]. This renders the investigation of protease-mediated signaling particularly challenging: In addition to immediate effects of the protease on its substrate, the resultant changes in the substrate's biological activity have to be taken into account [86]. Moreover, individual proteolytic systems, like the MMP and the uPA system, are interrelated [43, 61], mutually influencing each other in the modulation of protease activity and action within a tissue. This local proteolytic network of MMPs is connected to interrelated proteases, their inhibitors, and substrates, thereby establishing organ homeostasis in normal physiology as well as during disease (Fig. 9.2).

Proteolytic balance of a tissue determines the availability of all ECM-derived extracellular signals and thereby influences gene expression signatures and cellular behavior [70]. The complexity of proteolytic homeostasis has led us to believe that investigation of single proteolysis pathways is not a fruitful approach to determine the function of a respective MMP within a tissue or even a disease [70]. Consideration of the interconnectivity of a specific proteolytic pathway with other protease systems and the resulting net effects within the protease web will strongly improve the evaluation of the role of a protease, for example, in a certain disease state [74]. Consequently, in the past years, various systems biology approaches have been developed, suitable to elucidate the specific role of individual proteases as well as their involvement in higher orders of complexity [6, 88]. In light of such more holistic perspectives, it can be appreciated how information flows in proteolytic networks, depending on its organization either in linear pathways and amplification cascades (one way), or in regulatory circuits (feed-back loop) [74]. Upon stimulation, the availability of interacting proteins, inhibitors, substrates and biological activities of cleavage products altogether determine the output signal [74]. Accordingly, the net activity of a protease depends on the activities of many proteases and inhibitors. Due to constant challenges and re-adaptations, a proteolytic network is not in equilibrium but in constant flux [74]. MMPs can be considered as the key nodal proteases of the local protease networks, forming many critical cross-class and protease family connections [74]. Different local proteolytic tissue networks communicate with each other in the organism over a distance *via* the circulatory system, forming the proteolytic internet [73]. Information is transmitted *via* systemic up-regulation or down-regulation of soluble factors such as cytokines and hormones [73], as well as secreted protease inhibitors such as TIMP-1 [15]. The status of homeostasis in the regional proteolytic network of an organ is thereby reported to other tissues in the body [70]. Consequently, manipulation of a member of a regional proteolytic network can be of systemic impact due to the interconnectivity of the protease web. The resulting re-adaptation during establishment of a new systemic homeostasis determines susceptibility of the organism to disease (Fig. 9.3) [73].

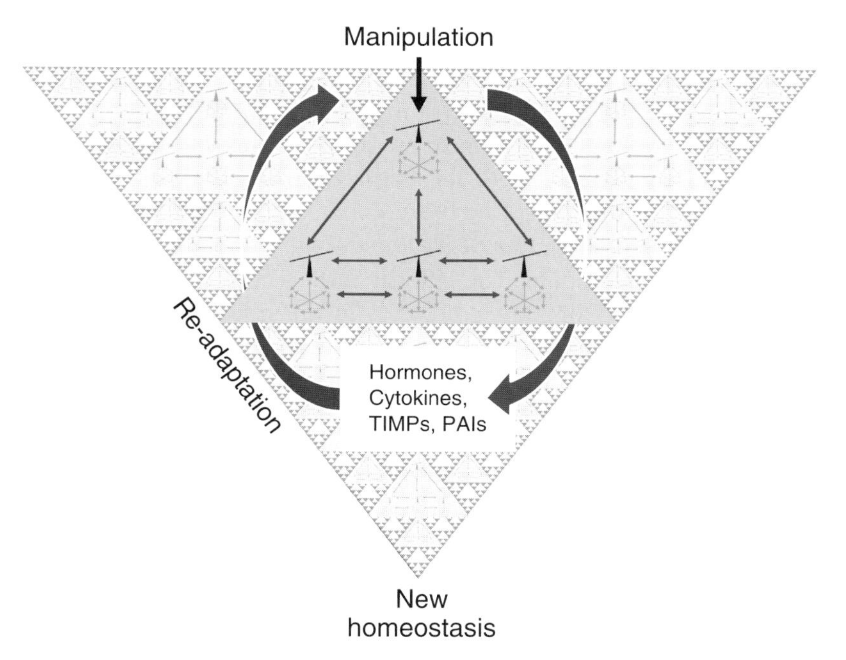

Figure 9.3 Interconnectivity of local proteolytic networks within an organism. Local proteolytic tissue networks (as depicted in Fig.9.2.) within an organism communicate with each other over a distance via the circulatory system, forming the proteolytic internet. Information is transmitted systemically via up-regulation or down-regulation of soluble factors such as cytokines, hormones, as well as secreted protease inhibitors such as TIMP-1 and PAI-1. The status of homeostasis in the regional proteolytic network of an organ is thereby reported to other tissues in the body. Accordingly, any manipulation of a single member of the proteolytic network results in a re-adaptation. This process is subject to a multitude of net effects that altogether impact on the formation of a new homeostasis, which determines the susceptibility of the organism to disease. (*See insert for color representation of this figure.*)

In case of unselective MMP inhibition in cancer, the induced changes in homeostasis even increased the susceptibility of the organism to the disease that was expected to be treated. From that we have learned that predictions about the nature of a new homeostasis can only be valid if they consider the system-wide net effects that are induced upon manipulation of one (or a group of) member(s) of the proteolytic network. This complexity represents the principal challenge to evaluating MMPs as therapeutic targets. In the following sections we discuss how newly emerging approaches of systems biology enable us to improve our understanding of the complex biology of individual MMPs and facilitate our judgment on whether some of them can be considered as targets for therapeutic intervention.

9.2 Degradomics – an overview

The fundamental change in the understanding of proteases as mediators of specific proteolytic cleavages rather than unspecific degraders founded the field of degradomics [87]. In MMP biology, this revolution was mainly driven by the seminal finding that MMPs, through specific processing, convert chemokines from agonists to antagonists with a profound impact on inflammatory responses [89, 90]. Degradomics aims to identify and functionally characterize all components and interactions of a proteolytic system to finally define the protease web and its disturbances in a perturbed state, such as in disease [91]. Following the central dogma of molecular biology, the degradome of an organism comprises the entire repertoire of proteolytic genes encoded in a genome [92]. These are all proteases and protease-homologues that can be produced giving rise to the degradome of a cell or tissue in a particular state, that is, all proteases, which are expressed in this condition. In view of post-transcriptional regulation, this degradome is sub-divided into the transcriptional and the translational degradome. The latter is further controlled at the activity level, presenting the actual set of active proteolytic enzymes, the activity degradome [93]. Each active protease cleaves multiple substrate proteins that all together form the substrate degradome of the individual protease. Ultimately, the interconnected functions of these components at each level of complexity define the state of the protease web and thus the proteolytic potential of the system (Fig. 9.4) [3, 74, 93].

The rapid progress of degradomics within recent years was facilitated by the development of an array of systems biology techniques suitable for the comprehensive analysis of genomes, RNA, and proteins. A prerequisite for most of these approaches was the availability of complete genome sequences at the beginning of the 21st century [94, 95]. They helped to define the human and the mouse degradome, which could be used to develop tools for the specific assessment of protease transcripts [92]. Moreover, it became possible to establish comprehensive sequence databases that are pivotal to deconvolute data from analysis of highly complex analyte mixtures. This was particularly important for mass spectrometry-based proteomics, which is mostly based on the reconstruction of proteins from measurements of peptide fragments [96]. High confidence in assignment of fragment spectra to correct peptides is achieved by use of specific digesting proteases that generate peptides from proteins with defined N and C termini. Therefore, semi-specific peptides derived from proteolytic

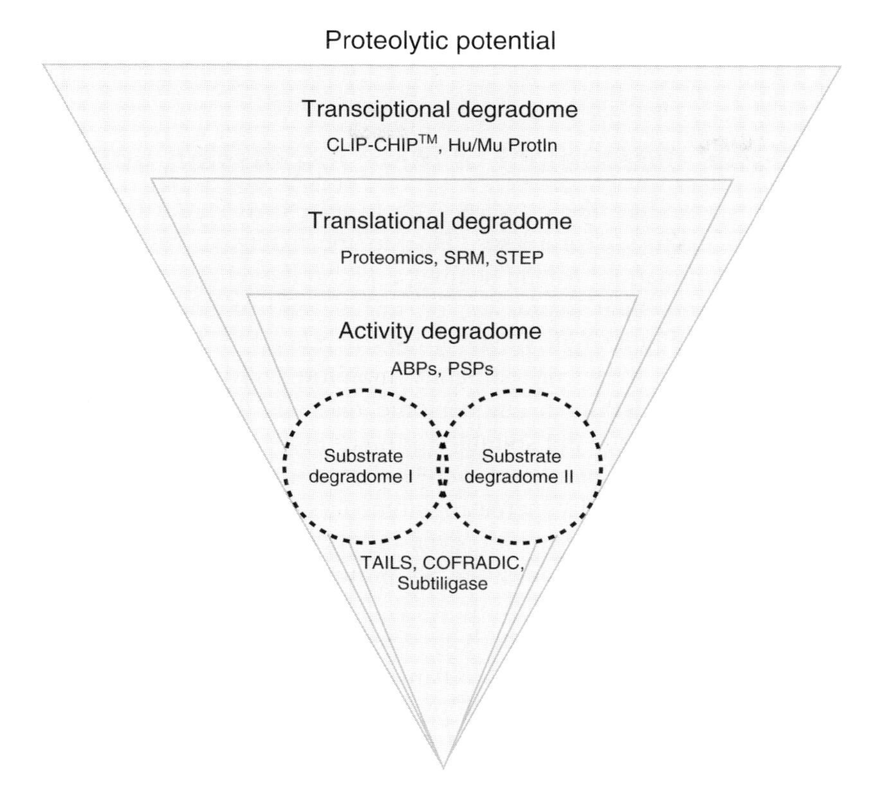

Figure 9.4 Degradomes and degradomics approaches. The transcriptional degradome defines the translational degradome, of which the activity degradome represents the active proteases. Individual active proteases give rise to partial overlapping substrate degradomes. All together, they define the proteolytic potential of a system. For each level of complexity powerful degradomics techniques have been developed. CLIP-CHIP^{TM}, Hu/Mu ProtIn, dedicated protease microarrays; SRM, selected reaction monitoring; STEP, STandard of Expressed Protein peptides; ABPs, activity-based probes; PSPs, proteolytic signature peptides; TAILS, terminal amine isotopic labeling of substrates; COFRADIC, combined diagonal fractional chromatography; Subtiligase, engineered peptide ligase for modification of protein N termini. (*See insert for color representation of this figure.*)

cleavage events with unknown specificity were disregarded, and proteolysis as post-translational modification was mostly neglected. Even with the establishment of quantitative proteomics, the assessment of protease activity remained for a long time restricted to the observation of reduction in protein abundance by complete degradation. Later, these datasets were exploited to map peptides with differential abundance to different regions of the same protein, indicating specific cleavage events [97]. The further development of these approaches culminated in a set of powerful proteomics techniques, which directly and quantitatively measure protein N and/or C termini in complex proteomes that had been exposed to a protease of interest [98–100]. With these achievements we are getting closer to deciphering substrate degradomes of individual proteases and their interconnected activities within the protease web.

In this second part of the chapter, we describe the development of degradomics techniques, their current applications in MMP research, and their future perspectives. Furthermore, we discuss strategies for data dissemination and how integrated data interpretation will help to dissect complex MMP-dependent proteolytic pathways and networks.

9.2.1 Global assessment of MMP expression and activity

The first step in assessing the protease web in a defined state of a system is the determination of expressed proteases. Due to the extensive research on MMPs, particularly in inflammation, wound healing, and carcinogenesis, a multitude of data is available measuring their expression on mRNA and protein level. However, in most studies, either only a single MMP or closely related MMPs, such as MMP-14, MMP-2, and MMP-9, were concomitantly monitored [101]. This limitation was mainly due to the restriction to cumbersome methods developed for analysis of single gene products that are not suitable for recording expression profiles of whole gene families in parallel. Such tasks were highly simplified with the invention of cDNA microarrays with rapidly increasing numbers of probes covering thousands of transcripts in parallel [102]. However, even the most comprehensive genome-wide microarrays do not include probes for all proteases and protease-related genes. Especially, inactive homologues are often overlooked, which might have important roles in the regulation of protease activity despite their lack of proteolytic activity. To address this limitation, Overall and co-workers introduced the CLIP-CHIP™, a dedicated degradome microarray with 1561 probes for protease-related genes, based on a comparative genomics survey of protease genes in mouse and men [92, 103, 104]. This focused array uses 70mer oligonucleotides as probes with high specificity and is printed on glass slides in a modular design, allowing concomitant analysis of known non-protease gene signatures that are characteristic for the physiological process under study, for example, breast carcinogenesis. The CLIP-CHIP™ comes in a murine and in a human version and was successfully employed to assess changes to the degradome by lack of MMP-8 in a mouse model of inflammatory arthritis [105] and to identify MMP-13 as an important factor in the interaction of metastatic human breast cancer cells with osteoblasts [106]. Illustrating the strong interdependence of protease expression within the protease web, the CLIP-CHIP™ identified changes in the expression of more than 82 proteases, including MMP-11 and MMP-28 in inflamed mouse skin upon loss of MMP-2 [107]. A similar tool for the global analysis of protease transcripts is the Hu/Mu ProtIn oligonucleotide microarray, which was specifically designed to analyze samples from xenografted human tumor cells in mice [108]. This is particularly important for discerning the cellular origin of proteases, especially MMPs that might be either released from the tumor or in many cases from the host stroma [81]. Using the Hu/Mu ProtIn array, a protective role for stroma-derived MMP-12 in lung cancer was revealed [109], and Sinnamon et al. [110] identified mast cells as protective host cells in intestinal tumorigenesis through their specific protease expression pattern. While microarrays provide massive parallel analysis of a multitude of mRNA transcripts, they usually require validation and are limited in their dynamic range [111]. Optimal sensitivity and specificity for analysis of expression on the RNA level is achieved by quantitative real-time polymerase chain reaction (qRT-PCR) that has been extensively used in MMP research. For this purpose, Dylan Edwards' laboratory has designed and evaluated primers and probes for all human and mouse MMPs, TIMPs, and ADAMs [112]. These were applied in a series of studies to comprehensively assess MMP expression in several human cancer types [113, 114], host-pathogen interactions [115], and preclinical tumor models [116]. They provide an invaluable resource for the validation of high-throughput data and the focused analysis of MMP mRNA levels with high sensitivity and accuracy. In recent years, unprecedented coverage, sensitivity, and accuracy in transcriptome profiling has been achieved by massive parallel sequencing of cDNA libraries in RNA-Seq experiments [117]. This method allows, in addition to the identification of fully transcribed gene products, the evaluation of isoforms and splice variants. In the beginning hampered by the complicated data analysis, RNA-Seq is now becoming a standard technology that is commonly applied in transcriptomics. Consequently, it will also become the method of choice for the global assessment of transcriptional degradomes. As examples, RNA-Seq was used to evaluate MMP expression in megakaryocytes, which transfer MMP mRNAs to platelets that play important roles in remodeling and inflammation [118] and to identify MMPs and their related genes as contributors to integrity and damage in the cochlear sensory epithelium [119]. Through its ability to characterize 5' boundaries of transcripts, RNA-Seq will also play an important role in defining, if protein N termini that are not compliant with known gene structures are derived from alternative splicing or proteolysis. Discussing the role of microRNAs in the regulation of protease-related transcripts would be beyond the scope of this chapter, but it should be noted that with microRNA arrays and RNA-Seq these regulatory elements are accessible and will gain increasing attention in the field of degradomics.

The global assessment of the transcriptional degradome defines the sample space for the translational degradome. As described in the first part of this chapter, many studies aimed to analyze MMP protein expression in various experimental systems. This resulted in the development and validation of specific and sensitive antibodies to most MMPs that can be used for the determination of relative protein levels of individual family members. However, these antibodies are not always commercially available, and a general distribution in large quantities to the public is not feasible for research laboratories. In a global antibody-based approach, the Human Protein Atlas project strives to evaluate and develop specific antibodies to virtually all human proteins [120]. Currently, this resource provides information for 16 MMPs with expression data in cells and normal and cancer tissues, which ideally will be extended to a full set of standardized antibodies for MMP research. If specific antibodies are available, monitoring expression of individual MMPs by immunoblotting and/or immunohistochemistry is highly sensitive and has the unique advantage to also provide spatial information in tissues. However, this approach is less suited for the system-wide analysis of translational degradomes, even if it is restricted to MMPs and their inhibitors. A powerful method for the multiplexed quantitative analysis of MMP protein expression is the Luminex assay system that uses antibodies coupled to magnetic beads and for which assays for the concomitant analysis of 9 MMPs are available [121]. With the rapid technical advancements in recent years, mass spectrometry-based proteomics theoretically allows the comprehensive assessment of translational degradomes in complex samples. However, MMPs are rather low in abundance and in many cases only transiently expressed. As a consequence, according to PeptideAtlas, only 13 human MMPs have been detected so far (PeptideAtlas Build Human 2013-08) [122]. This limitation might be overcome by development of STandard of Expressed Protein (STEP) peptides [123] and targeted proteomics assays for selected reaction monitoring (SRM) [124]. First assays for the identification and accurate quantification of MMPs in complex samples have already been established [125], and with initiatives, such as the collaboration between OriGene Technologies, Inc. and the Institute of Systems Biology in Seattle (http://www.origene.com/protein/Mass_Spectrometry.aspx), it is expected that assays for many more MMPs and related proteins will be available soon.

The ultimate step in defining the actual degradome that shapes the proteome and thereby modulates signaling pathways within cells and tissues is the assessment of protease activity. In recent years, activity-based probes (ABPs) have emerged as powerful tools to determine active proteases in complex proteomes [100, 126]. This approach was originally explored for cysteine proteases, which form transient covalent acyl-enzyme intermediates that can be exploited for the development of covalent binders. These in combination with a common affinity tag, such as biotin, can then be used to extract active proteases from cell or tissue lysates or even whole organisms that are subsequently identified by mass spectrometry-based proteomics [127]. The non-covalent mechanism of matrix metalloproteinase catalysis prevents the direct applicability of activity-based protease proteomics and requires, for example, an additional photoactivatable moiety covalently attaching the probe to the enzyme outside of the active site. Using this strategy, Cravatt and co-workers analyzed more than 20 metalloproteases, including six MMPs and two ADAMs in murine cancer cell lines [128]. An alternative approach to directly target MMPs by ABPs was described by Matt Bogyo's laboratory, which, by protein engineering introduced a cysteine residue near the active site for covalent attack by ABPs. Using this method, they monitored activity of engineered MMP-12 and MMP-14 in mammalian cells and zebrafish embryos [129]. However, the need of engineered proteases precludes analysis of endogenous MMPs unless the mutant enzyme is introduced into the genome using knock-in strategies. Despite this limitation, this technique has great potential for temporal and spatial recording of individual MMP activity in complex systems. Very recently, Fahlman et al. [123] presented proteolytic signature peptides (PSPs), which are spiked into complex lysates to detect active proteases by monitoring removal of propeptides from zymogens. This allows not only detecting many active MMPs in parallel but also determining the absolute and relative amounts of the inactive and active forms of the proteases and thus the degree of their activity. Such an approach is a big step forward toward the assessment of the complete MMP activity degradome in complex systems with high sensitivity and accuracy.

9.2.2 Defining MMP active site specificity

In view of the increasing number of examples for target and anti-target MMPs in cancer and other diseases [3, 130], the identification of novel and specific inhibitors became a major challenge in current MMP research. This might be facilitated by a better understanding of non-catalytic substrate interactions and by a comprehensive in-depth characterization of cleavage site specificities. Many approaches used random cleavage site sequences that were presented to the test protease by phage display, in the form of peptide microarrays or as mixture-based oriented or positional scanning peptide libraries [131–135]. All these powerful approaches helped to define cleavage site specificities for many proteases but are cumbersome and rely on artificial rather than natural peptide substrates.

These limitations were overcome by the introduction of the proteomic identification of protease cleavage sites (PICS) method that uses natural peptide libraries, which are generated from the proteome of interest [136]. Peptide libraries are then digested with the test protease generating new N termini (neo-N termini), which are subsequently specifically enriched by biotin affinity tags and analyzed by mass spectrometry-based proteomics. As sequences of proteins initially used for library generation are known from sequence databases, both prime- and nonprime-side specificities can be determined in the same experiment. Employing PICS, cleavage specificities for proteases of various classes including several MMPs have been explored, revealing differences that might be exploited for inhibitor design [136, 137]. Due to the broad applicability and the high-throughput capabilities of PICS, detailed specificity profiles for all MMPs might be available in the near future. Originally developed as a qualitative technique, PICS has now been modified by use of tandem mass tags enabling simultaneous analysis of multiple proteases and/or conditions within a single experiment [138]. It can be expected that this improvement will even shorten the time, until a comprehensive and family-wide dataset of MMP specificities can be provided as an invaluable resource for MMP researchers. Moreover, the general idea of using natural peptide libraries for protease specificity profiling has also been extended to the analysis of carboxypeptidases [139], demonstrating the impact of this concept on current degradomics method development.

9.2.3 MMP substrate degradomics

Most important for the function of a protease is its substrate degradome, that is, all substrate proteins processed under physiological conditions. Through many excellent studies, it could be established that MMPs can cleave virtually all components of the extracellular matrix [140]. As a consequence, focused substrate surveys for newly identified members of the MMP family concentrated on this substrate class for a long time. Moreover, protein cleavage was mostly determined by direct incubation of substrate and protease in the absence of natural interactors. This drastically changed with the failure of MMP inhibitors in clinical cancer trials and the identification of novel substrate classes, such as chemokines [74, 141, 142]. Importantly, the latter was driven by a modified yeast two-hybrid assay using the non-catalytic hemopexin domain of MMP-2 as bait [90]. However, ideally, the activity of a protease is analyzed while catalytically acting on a substrate protein under physiological conditions. With its rapid development and the ability to monitor thousands of peptides and their corresponding proteins in complex mixtures, mass spectrometry-based proteomics became the method of choice for substrate degradomics [87]. Introducing isotopic labels through chemical modifications provides quantitative information and thus allows assessing changes in protein abundances in the presence and absence of a protease. Using this strategy, several new MMP-14 substrates shed from the membrane could be identified by quantitative comparison of cell culture supernatants from MMP-14-transfected and control breast cancer cells that had been differentially labeled with isotope-coded affinity tags (ICAT) [143, 144]. The same approach was applied to the identification of MMP-2 substrates using MMP-2-deficient murine embryonic fibroblasts and controls rescued by MMP-2 transfection [145]. Later, by use of isobaric tags for relative and absolute quantitation (iTRAQ) and a comparable experimental system, coverage of the MMP-2 substrate degradome could be expanded [97]. In the same experiment, regions for cleavage sites were determined by assigning peptides with differential abundance to different domains within the same protein in a process termed 'peptide mapping'. As an alternative to isotopic labeling, label-free quantitative proteomics revealed novel MMP-9 substrates in prostate cancer cells and macrophages [146, 147]. Complementing these solution-based techniques, several groups employed gel-based proteomics approaches to identify MMP-14 targets in human plasma and MMP-9 substrates in monocytes [148, 149].

Global quantitative proteomics approaches identify protease substrates by monitoring changes to protein abundances in response to the active protease. However, ideally, the cleavage site should be also revealed to gain information on cleavage specificity and putative modulations of biological functions. This can be achieved by direct analysis of protein neo-N termini generated by a test protease in a complex proteome. Thereby, neo-N-terminal peptides are specifically enriched either by positive or negative selection. Several positive enrichment strategies have been developed that employ amine reactive affinity tags to extract protein neo-N termini from protease-treated samples [150–152]. In most cases these tags do not allow introducing isotopic labels and thus effectively discriminate between basal proteolysis within the sample and the activity of the test protease. These limitations are overcome by negative selection approaches for N-terminal peptides that in combination with metabolic or isotopic labeling specifically assess cleavage events mediated by the activity of the test protease. A widely applied substrate degradomics method is combined fractional diagonal chromatography (COFRADIC), which uses differential labeling of primary amines in full-length proteins and tryptic peptides to chromatographically separate N-terminal from internal peptides [153]. In combination with metabolic stable isotope labeling, COFRADIC revealed the substrate degradomes of several proteases [154, 155], but has not been applied in MMP research. A more recently introduced technique based on negative selection of protein N termini is terminal amine isotopic labeling of substrates (TAILS) [156, 157]. TAILS uses a unique combination of isotopic labeling and a highly efficient negative selection strategy to detect protease substrates and their cleavage sites in complex proteomes. In a first step, proteomes exposed to the test protease and control samples are differentially labeled at the protein level. Next, labeled proteins from both samples are combined and subjected to trypsin digestion. This generates internal tryptic peptides with primary α-amines, which are selectively removed by covalent binding to a hyper-branched aldehyde functionalized polyglycerol polymer (HPG-ALD). Protein N termini blocked by isotopic labeling or natural N-terminal modification (e.g., acetylation) do not bind to the polymer and are enriched upon polymer removal. Finally, quantitative comparison of peptides in protease treated and control samples filters out neo-N termini that had been released by the test protease and thus are only present in the protease-treated sample. By use of iTRAQ-labels, TAILS was extended to a robust multiplex quantitative proteomics analysis platform for the high-throughput assessment of N-terminomes and protease substrate degradomes [158, 159]. CLIPPER, a dedicated bioinformatics analysis platform for TAILS data, provides algorithms for the statistical evaluation of protease cleavage events and the automated annotation of protein N termini within substrate proteins [160]. TAILS had a profound impact on MMP substrate discovery and was applied to characterize the MMP-2 substrate degradome in mouse fibroblast secretomes [157, 159], to reveal new substrates for MMP-11 and MMP-26 [157, 161], and to concomitantly study cleavage events mediated by the closely related MMP-2 and MMP-9 in a single experiment [158]. In its first application to *in vivo* MMP substrate degradomics, TAILS identified the complement 1 (C1) inhibitor as an MMP-2 substrate that when cleaved loses inhibitory activity against plasma kallikrein and complement C1, leading to increased vascular permeability and complement activation in skin inflammation [107]. Moreover, this is an example of interclass protease cross talk, where a metalloproteinase controls serine protease activity by proteolytic inactivation of an inhibitor. TAILS has also been employed to elucidate substrates of meprin and cathepsin proteases *in vitro* and *in vivo,* demonstrating the wide applicability of this approach to the comprehensive analysis of substrate degradomes in complex systems [162–166]. Most recently, TAILS was further extended to an 8plex degradomics platform and applied to the time-resolved analysis of the MMP-10 substrate degradome [167]. This study not only identified several novel MMP-10 substrates but also sub-classified them by primary cleavage specificity and structural accessibility. As not all neo-N-terminal peptides are accessible to mass spectrometry, monitoring protein C termini can greatly enhance the number of identified substrates. For this purpose and to specifically record cleavage events mediated by carboxypeptidases, Schilling et al. [168] introduced a modified version of TAILS that follows the same principle but negatively enriches C-terminal peptides from protease-exposed and control samples. In the future, complementary assessment of both N- and C-terminomes will expand the accessible fraction of substrate degradomes and further deepen our understanding of MMP function.

9.2.4 Targeted degradomics

Despite the enormous progress in substrate degradomics, these techniques are still limited in their ability to specifically and directly detect substrates *in vivo*. This is reflected by the relatively low number of MMP-2 substrates identified by TAILS in murine skin compared to cell-based experiments [107, 159]. One reason for this discrepancy might be the effect of redundancy and compensation in highly complex multi-cellular environments that complicate the direct relation of cleavage events to a protease of interest, which has been deleted from the genome [73]. Furthermore, proteolytic processing could be locally restricted to small cell populations within the complex tissue architecture, preventing detection of these cleavage events. On the other hand, *in vitro* assays produce large datasets of cleavage sites that might be contaminated by bystander substrates not processed under more physiological conditions.

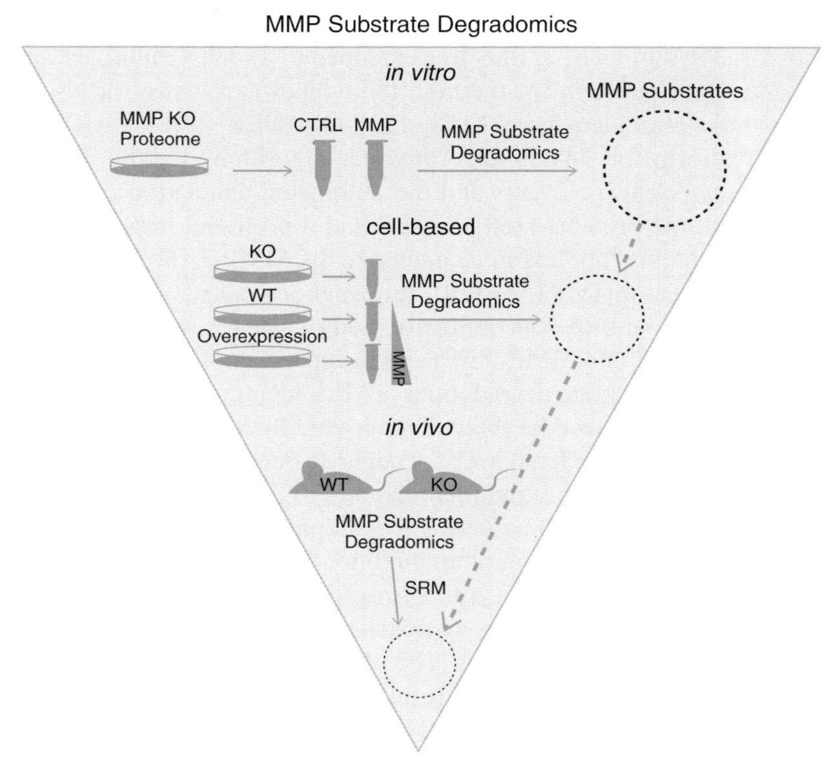

Figure 9.5 Integrated strategy to elucidate physiological MMP substrates. Multiple candidate substrates from unbiased *in vitro* and cell-based experiments serve as templates for the development of targeted SRM assays that are applied in appropriate *in vivo* models. KO, knockout; WT, wild-type; SRM, selected reaction monitoring. (*See insert for color representation of this figure.*)

Still, these datasets contain *bona fide in vivo* substrates of the test protease, which are not directly accessible by analysis of *in vivo* samples. Hence, a major challenge of current substrate degradomics is the efficient discrimination of critical from bystander substrates and their *in vivo* validation.

We propose an integrated strategy that narrows in on physiological substrates of a test protease by first assessing cleavage events in *in vitro* assays monitoring MMP activity on cell culture supernatants (Fig. 9.5). Using appropriate cell lines, a more comprehensive proteome reflecting the cellular composition of the target tissue is covered. By quantifying protease-generated neo-N termini at multiple time points of incubation, bystander substrates that are cleaved with low efficiency might be discarded [169]. This dataset is complemented by cell-based assays comparing N-terminomes of MMP-deficient, wild-type and/or cells transfected with the active protease or cultured in the presence of the active enzyme in the medium. This could be particularly important for detecting MMP-mediated membrane shedding events. Identified cleavages serve as templates for the development of targeted proteomics assays employing SRM to specifically monitor cleavage events in samples from MMP knock-out mice and wild-type controls. Basing these measurements on *in vitro* data with more defined MMP substrate relationships will also help to distinguish direct from indirect cleavages, which might be mediated by systemic alterations to the protease web resulting from genetic depletion of one component [73]. In contrast to unbiased shotgun proteomics, SRM specifically measures peptides and their fragments from individual proteins, significantly increasing sensitivity in detection and accuracy in quantification [124]. Designing assays for quantitative measurements of neo-N-terminal peptides and internal tryptic peptides spanning the cleavage site will allow monitoring cleavage events with high reliability and estimating ratios of cleaved and non-cleaved substrate proteins. Moreover, with zymogen removal as an example it has already been demonstrated that use of spiked-in isotopically labeled marker peptides even enables absolute quantification of each form in complex proteomes [123]. Combining this approach with SRM will further enhance sensitivity and specificity in monitoring MMP-dependent processing events *in vivo* with the possibility to record tens to hundreds of cleavages in parallel.

9.2.5 Data integration and repositories

A major challenge in degradomics is the interpretation of large datasets from high-throughput analyses of complex samples. Commercially available software packages are designed for identification and quantification of complete proteins rather than cleavage fragments. Powerful tools, such as the CLIPPER analysis pipeline [160], have been developed and are currently being further improved for statistical evaluation and automated annotation of N-terminomics datasets. However, for a comprehensive picture of MMP activities and their roles within the protease web, data from multiple experiments acquired in different laboratories have to be combined and integrated within a commonly accessible framework. Several public data repositories for proteolytic events have been launched, which provide annotated information on proteases, substrates, and cleavage sites. Covering proteases of all species, substrates, cleavage sites, and inhibitors, MEROPS represents the most comprehensive resource for protease research [170]. Although providing dumps of point releases for local installation, MEROPS has some limitations in automated data queries for meta-analyses. This has been addressed by the Termini-oriented protein function inferred database (TopFIND), an integrative resource of MEROPS and UniProtKB data that features an advanced programming interface based on the popular PSI common query interface (PSICQUIC) [171, 172]. There are similar resources, such as the DegraBase [173] and The Online Protein Resource (TOPPR) [174], which also include original N-terminomics data but are currently limited to specific cellular processes or technologies. The ultimate goal of integrative degradomics data interpretation will be the creation of interdependent proteolytic cascades, pathways, circuits, and finally networks that all together build the protease web. In the long term, such a resource will enable the design of computational models with predictive power for consequences of perturbations of proteolytic pathways in disease and thus the development of new drugs to counteract these detrimental disturbances. First steps in formalizing the representation of interactions of proteolytic mediators have been made in the form of "The Proteolysis Map" at the Center on Proteolytic Pathways [175]. Moreover, 'truncation' has been implemented as a specific and decisive inter- and intra-molecular interaction in the Systems Biology Markup

Language and Systems Biology Graphical Notation systems [176, 177] to facilitate integration of proteolysis as an additional layer of control in cellular behavior.

9.3 Conclusions

MMPs are central regulators of systemic homeostasis, and the MMP system connects various local proteolytic networks within an organism's protease web. As a consequence, deregulated MMP activity promotes various complex pathological processes, such as inflammation, cancer, autoimmune, and cardiovascular diseases. Conventional genomics approaches are limited to unravel MMP function, as the interconnectivity of proteolytic networks with MMPs as signaling hubs requires system-wide technologies to decipher their complex activities. The rapid progress in mass spectrometry-based proteomics allows assessing specificity profiles of individual MMPs, facilitating development of more specific inhibitors. Newly developed quantitative proteomics techniques enable system-wide analysis of MMP substrates in complex proteomes, providing valuable insights into their multitude of biological functions. Global assessment of MMP expression and, more importantly, activity will bring us closer to depicting complete MMP activity degradomes in complex systems. In combination with integrative data analysis, the resulting knowledge will provide decisive systems-level insight into MMP biology. Instead of peeking through the keyhole, we will soon be able to obtain a global view of the MMP system, helping us to finally unravel individual functions of single MMPs within the protease web in order to design novel strategies to therapeutically target MMPs in disease.

Acknowledgments

Our work is supported by grants from the Swiss National Science Foundation (31003A_140726 to U.a.d.K.) and from the European Commission (FP7-PEOPLE-2010-RG/SkiNterminomics to U.a.d.K.; FP7-NMP-2010-263307/SaveMe to A.K.).

References

1. Hu, J.L., Van den Steen, P.E., Sang, Q.X.A. & Opdenakker, G. (2007) Matrix metalloproteinase inhibitors as therapy for inflammatory and vascular diseases. *Nature Reviews. Drug Discovery*, 6, 480–498.
2. Lopez-Otin, C. & Matrisian, L.M. (2007) Emerging roles of proteases in tumour suppression. *Nature Reviews. Cancer*, 7, 800–808.
3. Overall, C.M. & Kleifeld, O. (2006) Tumour microenvironment – opinion: validating matrix metalloproteinases as drug targets and anti-targets for cancer therapy. *Nature Reviews. Cancer*, 6, 227–239.
4. Sternlicht, M.D. & Werb, Z. (2001) How matrix metalloproteinases regulate cell behavior. *Annual Review of Cell and Developmental Biology*, 17, 463–516.
5. Parks, W.C., Wilson, C.L. & López-Boado, Y.S. (2004) Matrix metalloproteinases as modulators of inflammation and innate immunity. *Nature Reviews. Immunology*, 4, 617–629.
6. Sela-Passwell, N., Rosenblum, G., Shoham, T. & Sagi, I. (2010) Structural and functional bases for allosteric control of MMP activities: can it pave the path for selective inhibition? *Biochimica et Biophysica Acta*, 1803, 29–38.
7. Lockhart, D.J. & Winzeler, E.A. (2000) Genomics, gene expression and DNA arrays. *Nature*, 405, 827–836.
8. Brigandt, I. & Love, A. (2008) Reductionism in biology. In: Edward N. Zalta (ed.) *Stanford Encyclopedia of Philosophy*. http://plato.stanford.edu/archives/sum2012/entries/reduction-biology/

9. Peacocke, A. (1985) *Reductionism in Academic Disciplines*. Society for Research into Higher Education & NFER-NELSON, Guildford, Surrey, UK.
10. Crick, F.H. (1958). On protein synthesis. Paper presented at: Symposia of the Society for Experimental Biology.
11. Collins, F., Lander, E., Rogers, J., Waterston, R. & Conso, I. (2004) Finishing the euchromatic sequence of the human genome. *Nature*, 431, 931–945.
12. Hieter, P. & Boguski, M. (1997) Functional genomics: it's all how you read it. *Science*, 278, 601–602.
13. Fingleton, B. (2007) Matrix metalloproteinases as valid clinical target. *Current Pharmaceutical Design*, 13, 333–346.
14. Folkman, J., Hahnfeldt, P. & Hlatky, L. (2000) Cancer: looking outside the genome. *Nature Reviews. Molecular Cell Biology*, 1, 76–79.
15. Kopitz, C. & Krüger, A. (2008) Janus-faced effects of broad-spectrum and specific MMP inhibition on metastasis. In: *The Cancer Degradome*. Springer, New York, USA, pp. 495–517.
16. Parada, L.F., Tabin, C.J., Shih, C. & Weinberg, R.A. (1982) Human EJ bladder carcinoma oncogene is homologue of Harvey sarcoma virus ras gene. *Nature*, 297, 474–478.
17. Matrisian, L., Glaichenhaus, N., Gesnel, M. & Breathnach, R. (1985) Epidermal growth factor and oncogenes induce transcription of the same cellular mRNA in rat fibroblasts. *The EMBO Journal*, 4, 1435.
18. Matrisian, L.M., Bowden, G.T., Krieg, P. *et al.* (1986) The mRNA coding for the secreted protease transin is expressed more abundantly in malignant than in benign tumors. *Proceedings of the National Academy of Sciences of the United States of America*, 83, 9413–9417.
19. Gunja-Smith, Z., Nagase, H. & Woessner, J. (1989) Purification of the neutral proteoglycan-degrading metalloproteinase from human articular cartilage tissue and its identification as stromelysin matrix metalloproteinase-3. *Biochemical Journal*, 258, 115–119.
20. Sanchez-Lopez, R., Nicholson, R., Gesnel, M.C., Matrisian, L.M. & Breathnach, R. (1988) Structure-function relationships in the collagenase family member transin. *The Journal of Biological Chemistry*, 263, 11892–11899.
21. Ostrowski, L.E., Finch, J., Krieg, P. *et al.* (1988) Expression pattern of a gene for a secreted metalloproteinase during late stages of tumor progression. *Molecular Carcinogenesis*, 1, 13–19.
22. Liotta, L., Tryggvason, K., Garbisa, S., Hart, I., Foltz, C. & Shafie, S. (1980) Metastatic potential correlates with enzymatic degradation of basement membrane collagen. *Nature*, 284, 67–68.
23. Baricos, W.H., Murphy, G., Zhou, Y., Nguyen, H.H. & Shah, S.V. (1988) Degradation of glomerular basement membrane by purified mammalian metalloproteinases. *Biochemical Journal*, 254, 609–612.
24. Bejarano, P.A., Noelken, M.E., Suzuki, K., Hudson, B. & Nagase, H. (1988) Degradation of basement membranes by human matrix metalloproteinase 3 (stromelysin). *Biochemical Journal*, 256, 413–419.
25. Okada, Y., Nagase, H. & Harris, E. (1986) A metalloproteinase from human rheumatoid synovial fibroblasts that digests connective tissue matrix components. Purification and characterization. *The Journal of Biological Chemistry*, 261, 14245–14255.
26. Matrisian, L.M., Rautmann, G. & Breathnach, R. (1988) Differential expression of preproenkephalin and transin mRNAs following oncogenic transformation: evidence for two classes of oncogene induced genes. *Oncogene Res*, 23, 251–262.
27. van der Stappen, J.W., Hendriks, T. & Wobbes, T. (1990) Correlation between collagenolytic activity and grade of histological differentiation in colorectal tumors. *International Journal of Cancer*, 45, 1071–1078.
28. Shima, I., Sasaguri, Y., Kusukawa, J. *et al.* (1992) Production of matrix metalloproteinase-2 and metalloproteinase-3 related to malignant behavior of esophageal carcinoma. A clinicopathologic study. *Cancer*, 70, 2747–2753.
29. Garbisa, S., Scagliotti, G., Masiero, L. *et al.* (1992) Correlation of serum metalloproteinase levels with lung cancer metastasis and response to therapy. *Cancer Research*, 52, 4548–4549.
30. Zucker, S., Lysik, R.M., Zarrabi, M.H. *et al.* (1992) Type IV collagenase/gelatinase (MMP-2) is not increased in plasma of patients with cancer. *Cancer Epidemiology, Biomarkers & Prevention*, 1, 475–479.
31. Chenard, M.P., O'Siorain, L., Shering, S. *et al.* (1996) High levels of stromelysin-3 correlate with poor prognosis in patients with breast carcinoma. *International Journal of Cancer*, 69, 448–451.
32. Kodate, M., Kasai, T., Hashirnoto, H., Yasumoto, K., Iwata, Y. & Manabe, H. (1997) Expression of matrix metalloproteinase (gelatinase) in T1 adenocarcinoma of the lung. *Pathology International*, 47, 461–469.
33. Liabakk, N.-B., Talbot, I., Smith, R.A., Wilkinson, K. & Balkwill, F. (1996) Matrix metalloprotease 2 (MMP-2) and matrix metalloprotease 9 (MMP-9) type IV collagenases in colorectal cancer. *Cancer Research*, 56, 190–196.
34. Zeng, Z.S. & Guillem, J.G. (1996) Colocalisation of matrix metalloproteinase-9-mRNA and protein in human colorectal cancer stromal cells. *British Journal of Cancer*, 74, 1161–1167.

35. Wilson, C.L., Heppner, K.J., Labosky, P.A., Hogan, B.L. & Matrisian, L.M. (1997) Intestinal tumorigenesis is suppressed in mice lacking the metalloproteinase matrilysin. *Proceedings of the National Academy of Sciences of the United States of America*, 94, 1402–1407.
36. Masson, R., Lefebvre, O., Noël, A. *et al.* (1998) In vivo evidence that the stromelysin-3 metalloproteinase contributes in a paracrine manner to epithelial cell malignancy. *The Journal of Cell Biology*, 140, 1535–1541.
37. Itoh, T., Tanioka, M., Yoshida, H., Yoshioka, T., Nishimoto, H. & Itohara, S. (1998) Reduced angiogenesis and tumor progression in gelatinase A-deficient mice. *Cancer Research*, 58, 1048–1051.
38. Itoh, T., Tanioka, M., Matsuda, H. *et al.* (1999) Experimental metastasis is suppressed in MMP-9-deficient mice. *Clinical and Experimental Metastasis*, 17, 177–181.
39. Coussens, L.M., Tinkle, C.L., Hanahan, D. & Werb, Z. (2000) MMP-9 supplied by bone marrow–derived cells contributes to skin carcinogenesis. *Cell*, 103, 481–490.
40. Bergers, G., Brekken, R., McMahon, G. *et al.* (2000) Matrix metalloproteinase-9 triggers the angiogenic switch during carcinogenesis. *Nature Cell Biology*, 2, 737–744.
41. Verhagen, A.M. & Lock, P. (2002) Revealing the intricacies of cancer. *Genome Biology*, 3.reports4015
42. Bolon, I., Devouassoux, M., Robert, C., Moro, D., Brambilla, C. & Brambilla, E. (1997) Expression of urokinase-type plasminogen activator, stromelysin 1, stromelysin 3, and matrilysin genes in lung carcinomas. *The American Journal of Pathology*, 150, 1619.
43. DeClerck, Y.A., Imren, S., Montgomery, A.M., Mueller, B.M., Reisfeld, R.A. & Laug, W.E. (1997) Proteases and protease inhibitors in tumor progression. In: *Chemistry and Biology of Serpins*. Springer, pp. 89–97.
44. Murphy, G., Cambray, G., Virani, N., Page-Thomas, D. & Reynolds, J. (1981) The production in culture of metalloproteinases and an inhibitor by joint tissues from normal rabbits, and from rabbits with a model arthritis. *Rheumatology International*, 1, 17–20.
45. Werb, Z. & Reynolds, J.J. (1975) Purification and properties of a specific collagenase from rabbit synovial fibroblasts. *Biochemical Journal*, 151, 645–653.
46. Bode, W. & Maskos, K. (2003) Structural basis of the matrix metalloproteinases and their physiological inhibitors, the tissue inhibitors of metalloproteinases. *Biological Chemistry*, 384, 863–872.
47. Brown, P. & Giavazzi, R. (1995) Matrix metalloproteinase inhibition: A review of anti-tumour activity Matrix metalloproteinase inhibition: A review of anti-tumour activity. *Annals of Oncology*, 6, 967–974.
48. Pavlaki, M. & Zucker, S. (2003) Matrix metalloproteinase inhibitors (MMPIs): the beginning of phase I or the termination of phase III clinical trials. *Cancer and Metastasis Reviews*, 22, 177–203.
49. Levitt, N.C., Eskens, F.A., O'Byrne, K.J. *et al.* (2001) Phase I and pharmacological study of the oral matrix metalloproteinase inhibitor, MMI270 (CGS27023A), in patients with advanced solid cancer. *Clinical Cancer Research*, 7, 1912–1922.
50. Maquoi, E., Frankenne, F., Noël, A., Krell, H.-W., Grams, F. & Foidart, J.-M. (2000) Type IV collagen induces matrix metalloproteinase 2 activation in HT1080 fibrosarcoma cells. *Experimental Cell Research*, 261, 348–359.
51. Chirvi, R.G., Garofalo, A., Crimmin, M.J. *et al.* (1994) Inhibition of the metastatic spread and growth of B16-BL6 murine melanoma by a synthetic matrix metalloproteinase inhibitor. *International Journal of Cancer*, 58, 460–464.
52. Davies, B., Brown, P.D., East, N., Crimmin, M.J. & Balkwill, F.R. (1993) A synthetic matrix metalloproteinase inhibitor decreases tumor burden and prolongs survival of mice bearing human ovarian carcinoma xenografts. *Cancer Research*, 53, 2087–2091.
53. Taraboletti, G., Garofalo, A., Belotti, D. *et al.* (1995) Inhibition of angiogenesis and murine hemangioma growth by batimastat, a synthetic inhibitor of matrix metalloproteinases. *Journal of the National Cancer Institute*, 87, 293–298.
54. Wang, X., Fu, X., Brown, P.D., Crimmin, M.J. & Hoffman, R.M. (1994) Matrix metalloproteinase inhibitor BB-94 (batimastat) inhibits human colon tumor growth and spread in a patient-like orthotopic model in nude mice. *Cancer Research*, 54, 4726–4728.
55. Watson, S.A., Morris, T.M., Robinson, G., Crimmin, M.J., Brown, P.D. & Hardcastle, J.D. (1995) Inhibition of organ invasion by the matrix metalloproteinase inhibitor batimastat (BB-94) in two human colon carcinoma metastasis models. *Cancer Research*, 55, 3629–3633.
56. Karelina, T.V., Hruza, G.J., Goldberg, G.I. & Eisen, A.Z. (1993) Localization of 92-kDa type IV collagenase in human skin tumors: comparison with normal human fetal and adult skin. *The Journal of Investigative Dermatology*, 100, 159–165.
57. Poulsom, R., Hanby, A.M., Pignatelli, M. *et al.* (1993) Expression of gelatinase A and TIMP-2 mRNAs in desmoplastic fibroblasts in both mammary carcinomas and basal cell carcinomas of the skin. *Journal of Clinical Pathology*, 46, 429–436.

58. Poulsom, R., Pignatelli, M., Stetler-Stevenson, W.G. *et al.* (1992) Stromal expression of 72 kda type IV collagenase (MMP-2) and TIMP-2 mRNAs in colorectal neoplasia. *The American Journal of Pathology*, 141, 389–396.
59. Brinckerhoff, C.E. & Matrisian, L.M. (2002) Matrix metalloproteinases: a tail of a frog that became a prince. *Nature Reviews. Molecular Cell Biology*, 3, 207–214.
60. McCawley, L.J. & Matrisian, L.M. (2000) Matrix metalloproteinases: multifunctional contributors to tumor progression. *Molecular Medicine Today*, 6, 149–156.
61. DeClerck, Y. & Laug, W. (1995) Cooperation between matrix metalloproteinases and the plasminogen activator-plasmin system in tumor progression. *Enzyme and Protein*, 49, 72–84.
62. Della Porta, P., Soeltl, R., Krell, H.-W. *et al.* (1999) Combined treatment with serine protease inhibitor aprotinin and matrix metalloproteinase inhibitor Batimastat (BB-94) does not prevent invasion of human esophageal and ovarian carcinoma cells in vivo. *Anticancer Research*, 19, 3809–3816.
63. Coussens, L.M., Fingleton, B. & Matrisian, L.M. (2002) Matrix metalloproteinase inhibitors and cancer—trials and tribulations. *Science*, 295, 2387–2392.
64. Schlondorff, J. & Blobel, C.P. (1999) Metalloprotease-disintegrins: modular proteins capable of promoting cell-cell interactions and triggering signals by protein-ectodomain shedding. *Journal of Cell Science*, 112, 3603–3617.
65. Eccles, S.A., Box, G.M., Bone, E.A., Thomas, W. & Brown, P.D. (1996) Control of Lymphatic and Hematogenous Metastasis of a Rat Mammary Carcinoma by the Matrix Metalloproteinas Inhibitor Batimastat (BB-94). *Cancer Research*, 56, 2815–2822.
66. Zucker, S., Cao, J. & Chen, W.T. (2000) Critical appraisal of the use of matrix metalloproteinase inhibitors in cancer treatment. *Oncogene*, 19, 6642–6650.
67. Moore, M., Hamm, J., Dancey, J. *et al.* (2003) Comparison of gemcitabine versus the matrix metalloproteinase inhibitor BAY 12-9566 in patients with advanced or metastatic adenocarcinoma of the pancreas: a phase III trial of the National Cancer Institute of Canada Clinical Trials Group. *Journal of Clinical Oncology*, 21, 3296–3302.
68. Wylie, S., MacDonald, I.C., Varghese, H.J. *et al.* (1999) The matrix metalloproteinase inhibitor batimastat inhibits angiogenesis in liver metastases of B16F1 melanoma cells. *Clinical and Experimental Metastasis*, 17, 111–117.
69. Krüger, A., Soeltl, R., Sopov, I. *et al.* (2001) Hydroxamate-type matrix metalloproteinase inhibitor batimastat promotes liver metastasis. *Cancer Research*, 61, 1272–1275.
70. Krüger, A., Kates, R.E. & Edwards, D.R. (2010) Avoiding spam in the proteolytic internet: Future strategies for anti-metastatic MMP inhibition. *Biochimica et Biophysica Acta*, 1803, 95–102.
71. Martin, M.D., Carter, K.J., Jean-Philippe, S.R. *et al.* (2008) Effect of ablation or inhibition of stromal matrix metalloproteinase-9 on lung metastasis in a breast cancer model is dependent on genetic background. *Cancer Research*, 68, 6251–6259.
72. Balbín, M., Fueyo, A., Tester, A.M. *et al.* (2003) Loss of collagenase-2 confers increased skin tumor susceptibility to male mice. *Nature Genetics*, 35, 252–257.
73. Krüger, A. (2009) Functional genetic mouse models: promising tools for investigation of the proteolytic internet. *Biological Chemistry*, 390, 91–97.
74. Overall, C.M. & Dean, R.A. (2006) Degradomics: systems biology of the protease web. Pleiotropic roles of MMPs in cancer. *Cancer and Metastasis Reviews*, 25, 69–75.
75. Rodríguez, D., Morrison, C.J. & Overall, C.M. (2010) Matrix metalloproteinases: what do they not do? New substrates and biological roles identified by murine models and proteomics. *Biochimica et Biophysica Acta*, 1803, 39–54.
76. Woessner, J. (1998) The matrix metalloproteinase family. In: Parks, W.C. & Mecham, R.P. (eds), *Matrix Metalloproteinases*. Acadamic Press, San Deigo, pp. 1–14.
77. Hartenstein, B., Dittrich, B.T., Stickens, D. *et al.* (2005) Epidermal development and wound healing in matrix metalloproteinase 13-deficient mice. *The Journal of Investigative Dermatology*, 126, 486–496.
78. Gutiérrez-Fernández, A., Inada, M., Balbín, M. *et al.* (2007) Increased inflammation delays wound healing in mice deficient in collagenase-2 (MMP-8). *The FASEB Journal*, 21, 2580–2591.
79. Himelstein, B., Canete-Soler, R., Bernhard, E., Dilks, D. & Muschel, R. (1994) Metalloproteinases in tumor progression: the contribution of MMP-9. *Invasion and Metastasis*, 14, 246.
80. Masson, V., de la Ballina, L.R., Munaut, C. *et al.* (2005) Contribution of host MMP-2 and MMP-9 to promote tumor vascularization and invasion of malignant keratinocytes. *The FASEB Journal*, 19, 234–236.
81. Egeblad, M. & Werb, Z. (2002) New functions for the matrix metalloproteinases in cancer progression. *Nature Reviews. Cancer*, 2, 161–174.
82. Belkin, A.M., Akimov, S.S., Zaritskaya, L.S., Ratnikov, B.I., Deryugina, E.I. & Strongin, A.Y. (2001) Matrix-dependent proteolysis of surface transglutaminase by membrane-type metalloproteinase regulates cancer cell adhesion and locomotion. *The Journal of Biological Chemistry*, 276, 18415–18422.

83. Johnson, C., Sung, H.-J., Lessner, S.M., Fini, M.E. & Galis, Z.S. (2004) Matrix Metalloproteinase-9 Is Required for Adequate Angiogenic Revascularization of Ischemic Tissues Potential Role in Capillary Branching. *Circulation Research*, 94, 262–268.
84. Van Lint, P. & Libert, C. (2007) Chemokine and cytokine processing by matrix metalloproteinases and its effect on leukocyte migration and inflammation. *Journal of Leukocyte Biology*, 82, 1375–1381.
85. Pelus, L.M., Bian, H., King, A.G. & Fukuda, S. (2004) Neutrophil-derived MMP-9 mediates synergistic mobilization of hematopoietic stem and progenitor cells by the combination of G-CSF and the chemokines GROβ/CXCL2 and GROβT/CXCL2δ4. *Blood*, 103, 110–119.
86. Turk, B. (2006) Targeting proteases: successes, failures and future prospects. *Nature Reviews. Drug Discovery*, 5, 785–799.
87. Lopez-Otin, C. & Overall, C.M. (2002) Protease degradomics: a new challenge for proteomics. *Nature Reviews. Molecular Cell Biology*, 3, 509–519.
88. Overall, C. & Kleifeld, O. (2006) Towards third generation matrix metalloproteinase inhibitors for cancer therapy. *British Journal of Cancer*, 94, 941–946.
89. McQuibban, G.A., Gong, J.H., Wong, J.P., Wallace, J.L., Clark-Lewis, I. & Overall, C.M. (2002) Matrix metalloproteinase processing of monocyte chemoattractant proteins generates CC chemokine receptor antagonists with anti-inflammatory properties in vivo. *Blood*, 100, 1160–1167.
90. McQuibban, G.A., Gong, J.H., Tam, E.M., McCulloch, C.A., Clark-Lewis, I. & Overall, C.M. (2000) Inflammation dampened by gelatinase A cleavage of monocyte chemoattractant protein-3. *Science*, 289, 1202–1206.
91. Overall, C.M. & Blobel, C.P. (2007) In search of partners: linking extracellular proteases to substrates. *Nature Reviews. Molecular Cell Biology*, 8, 245–257.
92. Puente, X.S., Sánchez, L.M., Overall, C.M. & López-Otín, C. (2003) Human and mouse proteases: a comparative genomic approach. *Nature Reviews. Genetics*, 4, 544–558.
93. auf dem Keller, U., Doucet, A. & Overall, C.M. (2007) Protease research in the era of systems biology. *Biological Chemistry*, 388, 1159–1162.
94. Venter, J.C., Adams, M.D., Myers, E.W. *et al.* (2001) The sequence of the human genome. *Science*, 291, 1304–1351.
95. Lander, E.S., Linton, L.M., Birren, B. *et al.* (2001) Initial sequencing and analysis of the human genome. *Nature*, 409, 860–921.
96. Aebersold, R. & Mann, M. (2003) Mass spectrometry-based proteomics. *Nature*, 422, 198–207.
97. Dean, R.A. & Overall, C.M. (2007) Proteomics discovery of metalloproteinase substrates in the cellular context by iTRAQ labeling reveals a diverse MMP-2 substrate degradome. *Molecular and Cellular Proteomics*, 6, 611–623.
98. Rogers, L. & Overall, C.M. (2013) Proteolytic post translational modification ofproteins: proteomic tools and methodology. *Mol Cell Proteomics*. 12, 3532–3542. doi:10.1074/mcp.M1113.031310.
99. Plasman, K., Van Damme, P. & Gevaert, K. (2013) Contemporary positional proteomics strategies to study protein processing. *Current Opinion in Chemical Biology*, 17, 66–72.
100. auf dem Keller, U. & Schilling, O. (2010) Proteomic techniques and activity-based probes for the system-wide study of proteolysis. *Biochimie*, 92, 1705–1714.
101. Maatta, M., Soini, Y., Liakka, A. & Autio-Harmainen, H. (2000) Differential expression of matrix metalloproteinase (MMP)-2, MMP-9, and membrane type 1-MMP in hepatocellular and pancreatic adenocarcinoma: implications for tumor progression and clinical prognosis. *Clinical Cancer Research*, 6, 2726–2734.
102. Schena, M., Shalon, D., Heller, R., Chai, A., Brown, P. *et al.* (1996) Parallel human genome analysis: microarray-based expression monitoring of 1000 genes. *Proceedings of the National Academy of Sciences of the United States of America*, 93, 10614–10619.
103. Kappelhoff, R., auf dem Keller, U. & Overall, C.M. (2010) Analysis of the degradome with the CLIP-CHIP microarray. *Methods in Molecular Biology*, 622, 175–193.
104. Overall, C.M., Tam, E.M., Kappelhoff, R., Connor, A., Ewart, T. *et al.* (2004) Protease degradomics: mass spectrometry discovery of protease substrates and the CLIP-CHIP, a dedicated DNA microarray of all human proteases and inhibitors. *Biological Chemistry*, 385, 493–504.
105. Cox, J.H., Starr, A.E., Kappelhoff, R., Yan, R., Roberts, C.R. & Overall, C.M. (2010) Matrix metalloproteinase 8 deficiency in mice exacerbates inflammatory arthritis through delayed neutrophil apoptosis and reduced caspase 11 expression. *Arthritis and Rheumatism*, 62, 3645–3655.
106. Morrison, C., Mancini, S., Cipollone, J., Kappelhoff, R., Roskelley, C. & Overall, C. (2011) Microarray and proteomic analysis of breast cancer cell and osteoblast co-cultures: role of osteoblast matrix metalloproteinase (MMP)-13 in bone metastasis. *The Journal of Biological Chemistry*, 286, 34271–34285.

107. auf dem Keller, U., Prudova, A., Eckhard, U., Fingleton, B. & Overall, C.M. (2013) Systems-level analysis of proteolytic events in increased vascular permeability and complement activation in skin inflammation. *Science Signaling*, 6, rs2.
108. Schwartz, D.R., Moin, K., Yao, B. *et al.* (2007) Hu/Mu ProtIn oligonucleotide microarray: dual-species array for profiling protease and protease inhibitor gene expression in tumors and their microenvironment. *Molecular Cancer Research*, 5, 443–454.
109. Acuff, H.B., Sinnamon, M., Fingleton, B. *et al.* (2006) Analysis of host- and tumor-derived proteinases using a custom dual species microarray reveals a protective role for stromal matrix metalloproteinase-12 in non-small cell lung cancer. *Cancer Research*, 66, 7968–7975.
110. Sinnamon, M.J., Carter, K.J., Sims, L.P., Lafleur, B., Fingleton, B. & Matrisian, L.M. (2008) A protective role of mast cells in intestinal tumorigenesis. *Carcinogenesis*, 29, 880–886.
111. Allanach, K., Mengel, M., Einecke, G. *et al.* (2008) Comparing microarray versus RT-PCR assessment of renal allograft biopsies: similar performance despite different dynamic ranges. *American Journal of Transplantation*, 8, 1006–1015.
112. Pennington, C.J. & Edwards, D.R. (2010) Real-time PCR expression profiling of MMPs and TIMPs. *Methods in Molecular Biology*, 622, 159–173.
113. Baren, J.P., Stewart, G.D., Stokes, A. *et al.* (2012) mRNA profiling of the cancer degradome in oesophago-gastric adenocarcinoma. *British Journal of Cancer*, 107, 143–149.
114. Stokes, A., Joutsa, J., Ala-Aho, R. *et al.* (2010) Expression profiles and clinical correlations of degradome components in the tumor microenvironment of head and neck squamous cell carcinoma. *Clinical Cancer Research*, 16, 2022–2035.
115. Green, J.A., Elkington, P.T., Pennington, C.J. *et al.* (2010) Mycobacterium tuberculosis upregulates microglial matrix metalloproteinase-1 and -3 expression and secretion via NF-kappaB- and Activator Protein-1-dependent monocyte networks. *The Journal of Immunology*, 184, 6492–6503.
116. Atkinson, J.M., Falconer, R.A., Edwards, D.R. *et al.* (2010) Development of a novel tumor-targeted vascular disrupting agent activated by membrane-type matrix metalloproteinases. *Cancer Research*, 70, 6902–6912.
117. Wang, Z., Gerstein, M. & Snyder, M. (2009) RNA-Seq: a revolutionary tool for transcriptomics. *Nature Reviews. Genetics*, 10, 57–63.
118. Cecchetti, L., Tolley, N.D., Michetti, N., Bury, L., Weyrich, A.S. & Gresele, P. (2011) Megakaryocytes differentially sort mRNAs for matrix metalloproteinases and their inhibitors into platelets: a mechanism for regulating synthetic events. *Blood*, 118, 1903–1911.
119. Hu, B.H., Cai, Q., Hu, Z. *et al.* (2012) Metalloproteinases and their associated genes contribute to the functional integrity and noise-induced damage in the cochlear sensory epithelium. *The Journal of Neuroscience*, 32, 14927–14941.
120. Uhlen, M., Oksvold, P., Fagerberg, L. *et al.* (2010) Towards a knowledge-based Human Protein Atlas. *Nature Biotechnology*, 28, 1248–1250.
121. Thrailkill, K.M., Moreau, C.S., Cockrell, G. *et al.* (2005) Physiological matrix metalloproteinase concentrations in serum during childhood and adolescence, using Luminex Multiplex technology. *Clinical Chemistry and Laboratory Medicine*, 43, 1392–1399.
122. Farrah, T., Deutsch, E.W., Hoopmann, M.R. *et al.* (2013) The state of the human proteome in 2012 as viewed through PeptideAtlas. *Journal of Proteome Research*, 12, 162–171.
123. Fahlman, R.P., Chen, W. & Overall, C.M. (2013) *Absolute Proteomic Quantification of the Activity State of Proteases and Proteolytic Cleavages Using Proteolytic Signature Peptides and Isobaric Tags.* J Proteomics.
124. Picotti, P. & Aebersold, R. (2012) Selected reaction monitoring-based proteomics: workflows, potential, pitfalls and future directions. *Nature Methods*, 9, 555–566.
125. Ocana, M.F. & Neubert, H. (2010) An immunoaffinity liquid chromatography-tandem mass spectrometry assay for the quantitation of matrix metalloproteinase 9 in mouse serum. *Analytical Biochemistry*, 399, 202–210.
126. Deu, E., Verdoes, M. & Bogyo, M. (2012) New approaches for dissecting protease functions to improve probe development and drug discovery. *Nature Structural and Molecular Biology*, 19, 9–16.
127. Fonović, M. & Bogyo, M. (2008) Activity-based probes as a tool for functional proteomic analysis of proteases. *Expert Review of Proteomics*, 5, 721–730.
128. Sieber, S.A., Niessen, S., Hoover, H.S. & Cravatt, B.F. (2006) Proteomic profiling of metalloprotease activities with cocktails of active-site probes. *Nature Chemical Biology*, 2, 274–281.
129. Xiao, J., Broz, P., Puri, A.W. *et al.* (2013) A coupled protein and probe engineering approach for selective inhibition and activity-based probe labeling of the caspases. *Journal of the American Chemical Society*, 135, 9130–9138.

130. Kessenbrock, K., Plaks, V. & Werb, Z. (2010) Matrix metalloproteinases: regulators of the tumor microenvironment. *Cell*, 141, 52–67.
131. Matthews, D.J. & Wells, J.A. (1993) Substrate phage: selection of protease substrates by monovalent phage display. *Science*, 260, 1113–1117.
132. Boulware, K.T. & Daugherty, P.S. (2006) Protease specificity determination by using cellular libraries of peptide substrates (CLiPS). *Proceedings of the National Academy of Sciences of the United States of America*, 103, 7583–7588.
133. Salisbury, C.M., Maly, D.J. & Ellman, J.A. (2002) Peptide microarrays for the determination of protease substrate specificity. *Journal of the American Chemical Society*, 124, 14868–14870.
134. Rosse, G., Kueng, E., Page, M.G. *et al.* (2000) Rapid identification of substrates for novel proteases using a combinatorial peptide library. *Journal of Combinatorial Chemistry*, 2, 461–466.
135. Turk, B.E., Huang, L.L., Piro, E.T. & Cantley, L.C. (2001) Determination of protease cleavage site motifs using mixture-based oriented peptide libraries. *Nature Biotechnology*, 19, 661–667.
136. Schilling, O. & Overall, C.M. (2008) Proteome-derived, database-searchable peptide libraries for identifying protease cleavage sites. *Nature Biotechnology*, 26, 685–694.
137. Schilling, O., Huesgen, P.F., Barré, O., auf dem Keller, U. & Overall, C.M. (2011) Characterization of the prime and non-prime active site specificities of proteases by proteome-derived peptide libraries and tandem mass spectrometry. *Nature Protocols*, 6, 111–120.
138. Jakoby, T., van den Berg, B.H. & Tholey, A. (2012) Quantitative protease cleavage site profiling using tandem-mass-tag labeling and LC-MALDI-TOF/TOF MS/MS analysis. *Journal of Proteome Research*, 11, 1812–1820.
139. Van Damme, P., Evjenth, R., Foyn, H. *et al.* (2011) Proteome-derived peptide libraries allow detailed analysis of the substrate specificities of N{alpha}-acetyltransferases and point to hNaa10p as the post-translational actin N{alpha}-acetyltransferase. *Molecular and Cellular Proteomics*, 10(M110), 004580.
140. Lynch, C.C. & Matrisian, L.M. (2002) Matrix metalloproteinases in tumor-host cell communication. *Differentiation*, 70, 561–573.
141. Khokha, R., Murthy, A. & Weiss, A. (2013) Metalloproteinases and their natural inhibitors in inflammation and immunity. *Nature Reviews. Immunology*, 13, 649–665.
142. Page-McCaw, A., Ewald, A.J. & Werb, Z. (2007) Matrix metalloproteinases and the regulation of tissue remodelling. *Nature Reviews. Molecular Cell Biology*, 8, 221–233.
143. Butler, G.S., Dean, R.A., Tam, E.M. & Overall, C.M. (2008) Pharmacoproteomics of a metalloproteinase hydroxamate inhibitor in breast cancer cells: dynamics of membrane type 1 matrix metalloproteinase-mediated membrane protein shedding. *Molecular and Cellular Biology*, 28, 4896–4914.
144. Tam, E.M., Morrison, C.J., Wu, Y.I., Stack, M.S. & Overall, C.M. (2004) Membrane protease proteomics: Isotope-coded affinity tag MS identification of undescribed MT1-matrix metalloproteinase substrates. *Proceedings of the National Academy of Sciences of the United States of America*, 101, 6917–6922.
145. Dean, R.A., Butler, G.S., Hamma-Kourbali, Y. *et al.* (2007) Identification of candidate angiogenic inhibitors processed by matrix metalloproteinase 2 (MMP-2) in cell-based proteomic screens: disruption of vascular endothelial growth factor (VEGF)/heparin affin regulatory peptide (pleiotrophin) and VEGF/Connective tissue growth factor angiogenic inhibitory complexes by MMP-2 proteolysis. *Molecular and Cellular Biology*, 27, 8454–8465.
146. Vaisar, T., Kassim, S.Y., Gomez, I.G. *et al.* (2009) MMP-9 sheds the beta2 integrin subunit (CD18) from macrophages. *Molecular and Cellular Proteomics*, 8, 1044–1060.
147. Xu, D., Suenaga, N., Edelmann, M., Fridman, R., Muschel, R. & Kessler, B. (2008) Novel MMP-9 substrates in cancer cells revealed by a label-free quantitative proteomics approach. *Molecular and Cellular Proteomics*, 7, 2215.
148. Hwang, I.K., Park, S.M., Kim, S.Y. & Lee, S.T. (2004) A proteomic approach to identify substrates of matrix metalloproteinase-14 in human plasma. *Biochimica et Biophysica Acta*, 1702, 79–87.
149. Cauwe, B., Martens, E., Proost, P. & Opdenakker, G. (2009) Multidimensional degradomics identifies systemic autoantigens and intracellular matrix proteins as novel gelatinase B/MMP-9 substrates. *Integrative Biology*, 1, 404–426.
150. Xu, G., Shin, S.B. & Jaffrey, S.R. (2009) Global profiling of protease cleavage sites by chemoselective labeling of protein N-termini. *Proceedings of the National Academy of Sciences of the United States of America*, 106, 19310–19315.
151. Mahrus, S., Trinidad, J.C., Barkan, D.T., Sali, A., Burlingame, A.L. & Wells, J.A. (2008) Global sequencing of proteolytic cleavage sites in apoptosis by specific labeling of protein N termini. *Cell*, 134, 866–876.
152. Timmer, J.C., Enoksson, M., Wildfang, E. *et al.* (2007) Profiling constitutive proteolytic events in vivo. *Biochemical Journal*, 407, 41–48.

153. Gevaert, K., Goethals, M., Martens, L. *et al.* (2003) Exploring proteomes and analyzing protein processing by mass spectrometric identification of sorted N-terminal peptides. *Nature Biotechnology*, 21, 566–569.
154. Van Damme, P., Maurer-Stroh, S., Plasman, K. *et al.* (2009) Analysis of protein processing by N-terminal proteomics reveals novel species-specific substrate determinants of granzyme B orthologs. *Molecular and Cellular Proteomics*, 8, 258–272.
155. Demon, D., Van Damme, P., Vanden Berghe, T. *et al.* (2009) Proteome-wide substrate analysis indicates substrate exclusion as a mechanism to generate caspase-7 versus caspase-3 specificity. *Molecular and Cellular Proteomics*, 8, 2700–2714.
156. Kleifeld, O., Doucet, A., Prudova, A. *et al.* (2011) Identifying and quantifying proteolytic events and the natural N terminome by terminal amine isotopic labeling of substrates. *Nature Protocols*, 6, 1578–1611.
157. Kleifeld, O., Doucet, A., auf dem Keller, U. *et al.* (2010) Isotopic labeling of terminal amines in complex samples identifies protein N-termini and protease cleavage products. *Nature Biotechnology*, 28, 281–288.
158. Prudova, A., auf dem Keller, U., Butler, G.S. & Overall, C.M. (2010) Multiplex N-terminome analysis of MMP-2 and MMP-9 substrate degradomes by iTRAQ-TAILS quantitative proteomics. *Molecular and Cellular Proteomics*, 9, 894–911.
159. auf dem Keller, U., Prudova, A., Gioia, M., Butler, G.S. & Overall, C.M. (2010) A statistics-based platform for quantitative N-terminome analysis and identification of protease cleavage products. *Molecular and Cellular Proteomics*, 9, 912–927.
160. auf dem Keller, U. & Overall, C.M. (2012) CLIPPER-An add-on to the Trans-Proteomic Pipeline for the automated analysis of TAILS N-terminomics data. *Biological Chemistry*, 393, 1477–1483.
161. Starr, A.E., Bellac, C.L., Dufour, A., Goebeler, V. & Overall, C.M. (2012) Biochemical characterization and N-terminomics analysis of leukolysin, the membrane-type 6 matrix metalloprotease (MMP25): chemokine and vimentin cleavages enhance cell migration and macrophage phagocytic activities. *The Journal of Biological Chemistry*, 287, 13382–13395.
162. Tholen, S., Biniossek, M.L., Gansz, M. *et al.* (2013) Deletion of cysteine cathepsins B or L yields differential impacts on murine skin proteome and degradome. *Molecular and Cellular Proteomics*, 12, 611–625.
163. Jefferson, T., auf dem Keller, U., Bellac, C. *et al.* (2013) The substrate degradome of meprin metalloproteases reveals an unexpected proteolytic link between meprin beta and ADAM10. *Cellular and Molecular Life Sciences*, 70, 309–333.
164. Tholen, S., Biniossek, M.L., Gessler, A.L. *et al.* (2011) Contribution of cathepsin L to secretome composition and cleavage pattern of mouse embryonic fibroblasts. *Biological Chemistry*, 392, 961–971.
165. Jefferson, T., Čaušević, M., auf dem Keller, U. *et al.* (2011) Metalloprotease meprin beta generates nontoxic N-terminal amyloid precursor protein fragments in vivo. *The Journal of Biological Chemistry*, 286, 27741–27750.
166. Becker-Pauly, C., Barré, O., Schilling, O. *et al.* (2011) Proteomic analyses reveal an acidic prime side specificity for the astacin metalloprotease family reflected by physiological substrates. *Molecular and Cellular Proteomics*, 10. doi:M111.009233.
167. Schlage, P., Egli, F.E., Nanni, P. *et al.* (2013) Time-resolved analysis of the matrix metalloproteinase 10 substrate degradome. *Molecular and Cellular Proteomics*. doi:10.1074/mcp.M1113.035139.
168. Schilling, O., Barré, O., Huesgen, P.F. & Overall, C.M. (2010) Proteome-wide analysis of protein carboxy termini: C terminomics. *Nature Methods*, 7, 508–511.
169. Plasman, K., Van Damme, P., Kaiserman, D. *et al.* (2011) Probing the efficiency of proteolytic events by positional proteomics. *Molecular and Cellular Proteomics*, 10. doi:10.1074/mcp.M1110.003301.
170. Rawlings, N.D., Barrett, A.J. & Bateman, A. (2012) MEROPS: the database of proteolytic enzymes, their substrates and inhibitors. *Nucleic Acids Research*, 40, D343–350.
171. Lange, P.F., Huesgen, P.F. & Overall, C.M. (2012) TopFIND 2.0—Linking protein termini with proteolytic processing and modifications altering protein function. *Nucleic Acids Research*, 40, D351–361.
172. Lange, P.F. & Overall, C.M. (2011) TopFIND, a knowledgebase linking protein termini with function. *Nature Methods*, 8, 703–704.
173. Crawford, E.D., Seaman, J.E., Agard, N. *et al.* (2013) The DegraBase: a database of proteolysis in healthy and apoptotic human cells. *Molecular and Cellular Proteomics*, 12, 813–824.
174. Colaert, N., Maddelein, D., Impens, F. *et al.* (2013) The Online Protein Processing Resource (TOPPR): a database and analysis platform for protein processing events. *Nucleic Acids Research*, 41, D333–337.
175. Igarashi, Y., Heureux, E., Doctor, K.S. *et al.* (2009) PMAP: databases for analyzing proteolytic events and pathways. *Nucleic Acids Research*, 37, D611–618.
176. Le Novere, N., Hucka, M., Mi, H. *et al.* (2009) The Systems Biology Graphical Notation. *Nature Biotechnology*, 27, 735–741.
177. Hucka, M., Finney, A., Sauro, H.M. *et al.* (2003) The systems biology markup language (SBML): a medium for representation and exchange of biochemical network models. *Bioinformatics*, 19, 524–531.

10 MMPs in Biology and Medicine

Di Jia[1] *, Roopali Roy[1] *, and Marsha A. Moses[1]

[1] *Vascular Biology Program and Department of Surgery, Boston Children's Hospital and Harvard Medical School, Boston, MA, USA*

10.1 Introduction

Matrix metalloproteases (MMP) are a family of zinc- and calcium-dependent proteolytic enzymes that play diverse roles in physiological and pathological processes. The majority of the twenty-three characterized human MMPs are secreted save for six members which are membrane-anchored (MT-MMP). The basic domain structure of MMP family members is provided in Fig. 10.1 [1]. All MMPs include an amino-terminal signal peptide which directs them for secretion, a pro-domain that confers latency via a cysteine-switch mechanism and a Zn^{2+}-dependent catalytic domain that is responsible for enzymatic function. A majority of MMPs also contain a C-terminal hemopexin-like domain that provides substrate specificity. MMP activity can collectively degrade most extracellular matrix (ECM) proteins as well as regulate the activity of other proteases, growth factors, cytokines, and cell-surface ligands and receptors. MMP activity may be regulated at four different levels: (i) gene expression via transcriptional and post-transcriptional regulation, (ii) extracellular localization, (iii) pro-enzyme activation by removal of the pro-domain, and (iv) direct inhibition of enzymatic function by tissue inhibitors of MMPs [2–5]. MMP activity is inhibited specifically and reversibly by a group of structurally related, endogenous inhibitors known as TIMPs (Tissue Inhibitors of Metalloproteases). To date, four TIMPs have been identified: TIMP-1, TIMP-2, TIMP-3, and TIMP-4 [6–9]. The role of MMPs and TIMPs in tumor growth, metastasis and angiogenesis has been widely investigated. We refer the reader to a number of comprehensive reviews on this topic, as well as to reviews discussing the general biochemistry of the MMP family and its regulation [10–14]. MMP-mediated activities that have important physiological implications include cell migration and invasion, differentiation, proliferation, angiogenesis, apoptosis, inflammation, and platelet aggregation, which collectively contribute to diverse biological functions under normal and pathological conditions.

ADAMs (A Disintegrin And Metalloprotease) comprise a family of integral membrane and secreted glycoproteins with conserved protein domains that consist of two subgroups: the membrane-anchored ADAMs [15–18] and the secreted members

*Authors contributed equally to this work.

Matrix Metalloproteinase Biology, First Edition. Edited by Irit Sagi and Jean P. Gaffney.

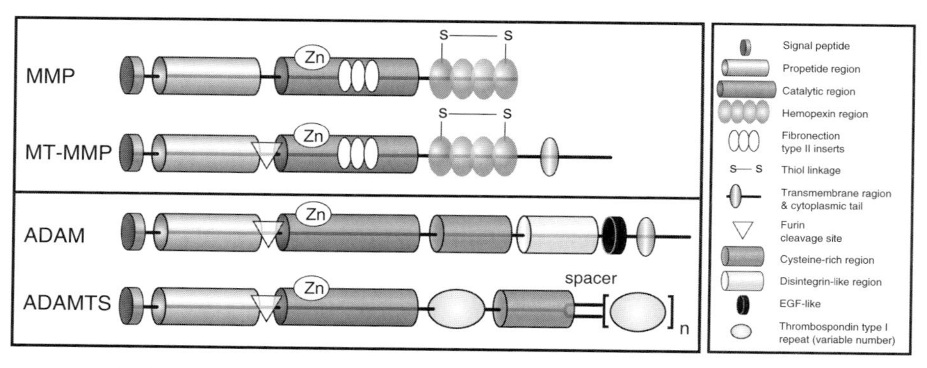

Figure 10.1 Basic domain structure of MMP and ADAM family members. The characteristic domain structure of MMPs includes (i) the signal peptide domain, which guides the enzyme into the rough endoplasmic reticulum during synthesis, (ii) the propeptide domain, which sustains the latency of these enzymes until it is removed or disrupted, (iii) the catalytic domain, which houses the highly conserved Zn^{2+} binding region and is responsible for enzyme activity, (iv) the hemopexin domain, which determines the substrate specificity of MMPs, and (v) a small hinge region, which enables the hemopexin region to present substrate to the active core of the catalytic domain. The subfamily of membrane-type MMPs (MT-MMPs) possesses an additional transmembrane domain and an intracellular domain. MMPs are produced in a latent form and most are activated by extracellular proteolytic cleavage of the propeptide. MT-MMPs also contain a cleavage site for furin proteases, providing the basis for furin-dependent activation of latent MT-MMPs prior to secretion. ADAMs are multidomain proteins composed of propeptide, metalloprotease, disintegrin-like, cysteine-rich, and epidermal growth factor-like domains. Membrane-anchored ADAMs contain a transmembrane and cytoplasmic domain. ADAMTSs have at least one Thrombospondin type I Sequence Repeat (TSR) motif [1]. (Reprinted with permission © (2009) American Society of Clinical Oncology. All rights reserved). (*See insert for color representation of this figure*).

with multiple thrombospondin repeats referred to as ADAMTSs [19, 20] (Fig. 10.1). Analogous to the MMPs, ADAMs comprise the following domains: signal peptide, pro-domain and a Zn^{2+}-dependent metalloprotease domain; additionally, ADAMs contain disintegrin, cysteine-rich, and EGF-like domains, a transmembrane region (for membrane-anchored ADAMs), and a cytoplasmic tail (Fig. 10.1). Of the 21 ADAMs identified in the human genome, only 12 have an intact catalytic domain and are therefore proteolytically active. ADAMs are multifunctional proteins involved in cell–cell/matrix interactions, cell signaling, ectodomain shedding, and regulation of growth factor availability in normal physiology, as well as in pathological conditions such as inflammatory diseases, atherosclerosis, and tumorigenesis.

10.2 Functional roles of MMPs and ADAMs

10.2.1 ECM remodeling

The extracellular matrix (ECM) is an acellular component of all tissues and organs and provides the necessary physical scaffolding for cells in addition to serving as a source of crucial biochemical and biomechanical cues that can, in turn, regulate tissue morphogenesis, differentiation, and homeostasis [21]. The ECM is comprised of fibrous proteins such as collagens, laminins, and entactin that serve a structural role, as well as proteoglycans that occupy the extracellular interstitial space within tissues. Aberrant ECM synthesis and/or degradation are hallmarks of many diseases. Collectively, MMP/ADAM family members can degrade the majority of ECM proteins and

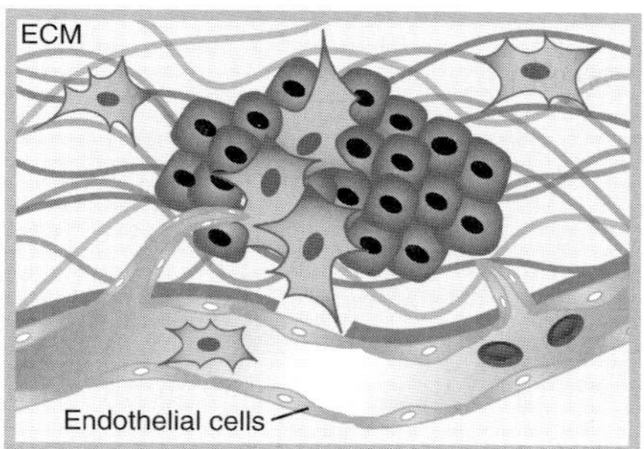

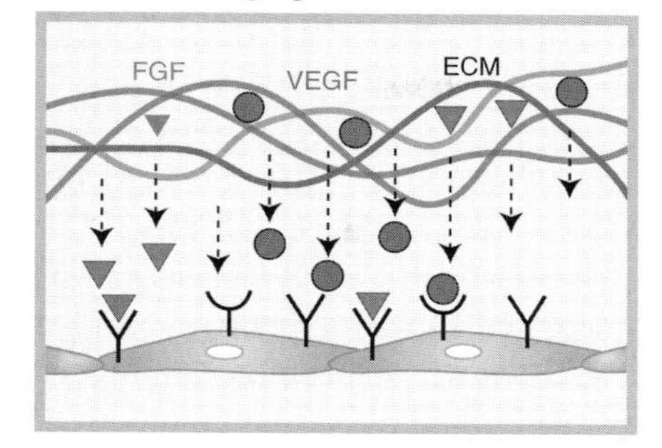

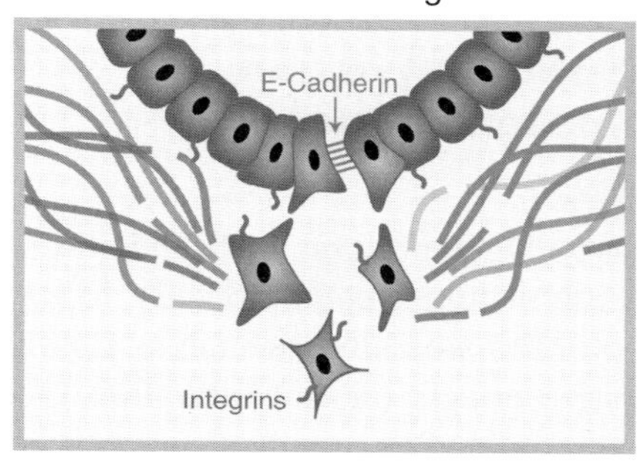

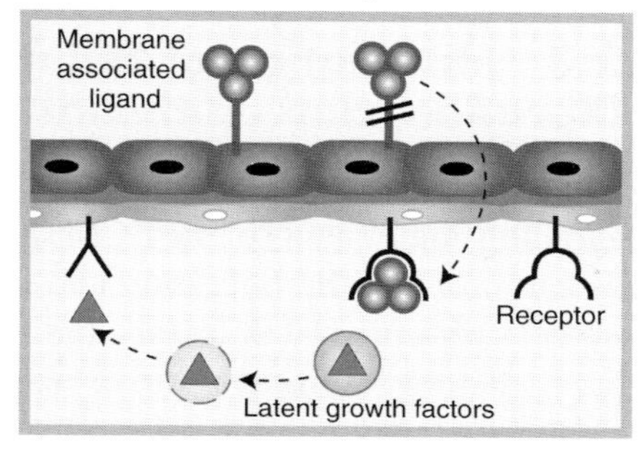

Figure 10.2 Multiple functions of MMPs in cancer progression. (Counterclockwise) MMPs degrade components of ECM, facilitating angiogenesis, tumor cell invasion and metastasis. MMPs modulate the interactions between tumor cells by cleaving E-cadherin, and between tumor cells and ECM by processing integrins, which also enhances the invasiveness of tumor cells. MMPs also process and activate signaling molecules, including growth factors and cytokines, making these factors more accessible to target cells by either liberating them from the ECM (e.g., VEGF and bFGF) and inhibitory complexes (e.g., TGF-β), or by shedding them from cell surface (e.g., HB-EGF) [1]. (Reprinted with permission © (2009) American Society of Clinical Oncology. All rights reserved). (*See insert for color representation of this figure*).

thereby contribute to a variety of biological activities, including cell migration, cell invasion, differentiation, proliferation, apoptosis, inflammatory reactions, and angiogenesis [5, 10, 16, 22]. MMP proteolytic activity plays a role in the local expansion of primary tumors as well as intravasation of cancer cells into nearby blood vessels and subsequent extravasation and invasion at a distant location (Fig.10.2).

10.2.2 Processing of growth factors and receptors

MMPs and ADAMs can modulate the bioavailability of growth factors as well as the function of cell-surface receptors. Ligands for several growth factor receptors are processed by MMPs. The most prominent among them are the epidermal growth factor receptor (EGFR) ligands: EGF, HB-EGF, TGFα, amphiregulin, betacellulin, and epiregulin. In general, signaling via the EGFR pathway is tightly regulated; however, under pathological conditions, increased MMP/ADAM activity results in increased shedding of active EGFR ligands and the induction of constitutively active

EGFR kinases can result in aberrantly upregulated signaling pathways, which in turn stimulate uncontrolled cell proliferation, migration, and survival (Fig. 10.2). For example, MMP-3, MMP-7, ADAM10, ADAM12, and ADAM17 have all been implicated in shedding HB-EGF and TGFα [23–26]. Similarly, the bioavailability of insulin-like growth factor (IGF) is mainly regulated by the cleavage of IGF binding proteins (IGFBP) mediated by MMP-1, MMP-2, MMP-3, ADAM12, and ADAM28, which cleave IGFPB-3 [27–29] or MMP-11 which cleaves IGFBP-1 [30]. The MMP-mediated cleavage of IGFBPs subsequently enables the activation of IGF-mediated signaling pathways and stimulates cell proliferation [31]. MMPs have also been shown to promote angiogenesis via the release of angiogenic mitogens sequestered in the ECM including vascular endothelial growth factor (VEGF)[32, 33] and basic fibroblast growth factor (bFGF) [34].

10.2.3 Modulation of cell migration, invasion, proliferation, and epithelial to mesenchymal transition (EMT)

By degrading components of the ECM such as collagen IV, laminin-5, and fibronectin, MMPs can promote the migration and invasion of tumor cells [35–37]. MMPs can also process certain growth factors and receptor tyrosine kinases such as EphA2 and CXCL12, thereby exerting positive or negative effects on cell migration [38, 39]. MMP-mediated degradation of cell surface molecules involved in cell-cell or cell-ECM interaction such as E-cadherin, integrins, and CD44 can also contribute to the stimulation of cell migration and invasion [40–45]. In addition to MMPs, specific ADAMs such as ADAM10 and the secreted form of ADAM12 (ADAM12-S) have also been shown to increase cell migration and invasion via their proteolytic activities [46, 47]. Recent studies have shown that MMPs can also promote cell migration independent of their proteolytic activity. One such example is MT1-MMP, which promotes macrophage migration by potentiating the ATP production in these cells [48].

Perturbation of cell-cell contact regulation can lead not only to changes in cell migration, but also to altered cell proliferation. Hence, certain MMPs such as MMP-7 [49], MMP-9 and MMP-12 [50] increase cell proliferation by cleavage of E-cadherin and N-cadherin which, in turn, leads to increased β-catenin signaling and cyclin D1 levels. MT1-MMP has been shown to process cell surface protein syndecan-1, thereby transforming this transmembrane protein into a diffusible factor that stimulates breast cancer cell proliferation [51, 52].

By processing growth factors and cleaving cell adhesion molecules, MMPs serve as important regulators of EMT, an important process by which malignant tumor cells activate invasion and metastasis, one of the hallmarks of cancer [53]. For example, MMP-28-mediated proteolytic processing of latent TGF-β leads to EMT in lung carcinoma cells [54]. MMP-3 expression in mammary epithelial cells leads to EMT through a series of downstream cascades, including cleavage of E-cadherin, increased expression of vimentin, and an alternatively spliced form of Rac-1, and increased cellular reactive oxygen species [55, 56].

10.2.4 Regulation of angiogenesis

Angiogenesis, the formation of new blood vessels from preexisting ones, is a key process under both normal physiological conditions such as embryonic development and pathological conditions such as cancer initiation, progression, and metastasis. Due to their activities in ECM degradation, MMPs are utilized both by tumor cells and by endothelial cells to digest the ECM and the vascular basement membrane, thereby facilitating the invasion of tumor cells into the stroma and migration of the endothelial cells toward the tumor [11, 57, 58]. MMPs and ADAMS can also activate signaling pathways in endothelial cells. For example, MT1-MMP releases bioactive TGF-β via its proteolytic activity, which in turn activates the TGF-β/Smad signaling pathway in endothelial cells [59]. PAR-1, a proteolytically activated G protein coupled receptor expressed in endothelial cells, is activated by MMP-1 and can induce the expression of many proangiogenic genes through the MAPK pathway [60]. We have recently reported that activated endothelial cells have upregulated ADAM12 expression, which leads to increased cell migration and invasion *in vitro* [61]. In addition, ADAM12 may potentiate bFGF-mediated angiogenesis *in vivo* [61]. Recent studies have implicated MMPs in regulating endothelial progenitor cells (EPCs) functions. In mouse models of hindlimb and cerebral ischemia, MMP-9 is required for the formation of vascular network by EPCs. MMP-9 deficiency leads to severely impaired ischemia-induced neovascularization [62, 63]. MMPs have also been shown to serve as key regulators of the angiogenic switch, one of the earliest and rate-limiting steps in tumor progression. In particular, MMP-2 has been shown to regulate the transition from the dormant to the angiogenic phenotype in an animal model of the angiogenic switch [64]. MMP-9 has been shown to regulate the angiogenic switch in a pancreatic tumor model [33]. Importantly, although most commonly recognized as proangiogenic factors, MMPs can also play anti-angiogenic roles under certain circumstances. For example, by cleaving proteins in the plasma or ECM such as plasminogen and collagen XVIII, MMPs can generate endogenous angiogenesis inhibitors such as angiostatin [65, 66] and endostatin [67]. Certain MMPs such as MT1-MMP can shed endoglin from the cell surface, resulting in a soluble form of endoglin, which functions as an inhibitor of tumor angiogenesis [68]. These studies reveal the important and complicated role of MMPs in the regulation of angiogenesis.

10.3 MMPs as diagnostic and prognostic biomarkers of cancer

As a function of the important roles that MMPs/ADAMs have been shown to play in the development and progression of human cancers, these proteases have also been shown to be useful as potential diagnostic and prognostic cancer biomarkers. In this section, we review the potential of MMPs as tools for cancer detection and prognosis and as monitors of disease progression and therapeutic efficacy. We have specifically included those relevant biomarkers for which validation studies with statistically significant outcomes have accompanied the preliminary reports (Table 10.1).

Table 10.1 Candidate MMP and ADAM biomarkers of cancer.

Cancer type	MMPs/ADAMs/ TIMPs	Detected in (tissue/body fluid)	Method of analysis	References
Breast	MMP-9	Urine, serum, plasma, tissue	Gelatin zymography, IHC, ELISA	[69–72]
	MMP-2	Serum	ELISA	[73]
	ADAM12	Urine, tissue	Immunoblot, IHC	[74–76]
	ADAM17	Tissue	RT-PCR, IHC	[77]
	ADAM9	Tissue	RT-PCR	[78]
	MMP-9	Exocrine pancreatic secretions, serum	ELISA, Immunoblot	[79, 80]
Pancreas	MMP-2, Timp-1	Urine	ELISA	[81]
	MMP-7	Plasma, serum	IHC, RT-PCR, ELISA	[82, 83]
	ADAM9	Tissue	IHC	[84]
	MMP-9	Plasma	ELISA, gene analysis	[85]
Lung	MMP-2	Tissue	Gelatin zymography	[86]
	ADAM12	Tissue, serum, urine	RT-PCR, ELISA	[87, 88]
	ADAM28	Tissue, serum	IHC, ELISA	[89]
	MMP-7	Serum	ELISA	[90]
Ovarian	MMP-9	Tissue, serum	IHC, ELISA	[91, 92]
	MMP-2, MMP-9	Urine	ELISA	[93]
	MMP-2, MMP-14	Tissue	IHC	[94]
	ADAM17	Tissue	RT-PCR, IHC	[95]
	MMP-1, MMP-2, MMP-9	Tissue	IHC	[96]
Prostate	MMP-2, MMP-9, MMP-13	Plasma	ELISA	[97]
	MMP-9	Urine	Gelatin zymography	[98]
	MMP-7	Serum	ELISA	[99]
	ADAM9	Tissue	RT-PCR, IHC	[100]

Note: For all the studies listed above $n \geq 50$ with the exception of pancreatic tumor biomarker studies where samples sizes were smaller.

10.3.1 Breast cancer

In breast cancer, distinct members of the MMP family can be useful as potential biomarkers for diagnosis and prognosis, early detection, and risk assessment, as well as for therapeutic efficacy. Significantly elevated levels of MMP-1, MMP-9, and MMP-13 in blood, urine, and tumor tissues from breast cancer patients have been previously reported [69, 101, 70, 102, 71, 103, 72, 104, 105]. We have previously reported that urinary MMP-9 and MMP-9/NGAL complex are significantly upregulated in breast cancer patients [69, 70]. Several MMPs may add to the currently established breast tumor prognostic factors such as tumor grade and size, Ki-67, hormone receptor status, HER2 expression, and lymph node status. A recent meta-analysis of 2,344 patients from 15 studies suggests that elevated MMP-9 expression correlates with a higher risk of relapse and worse survival in breast cancer patients [106, 107]. Interestingly, although circulating (blood) MMP-2 levels do not appear to be prognostically useful, strong stromal expression of MMP-2 [108] and MMP-9 [109, 110] in tumor tissues has been associated with poor breast cancer patient prognosis. Elevated MMP-9 serum levels were found to be associated with reduced disease-free survival for breast cancer patients [111], whereas preoperative serum MMP-2 levels may be selectively associated with survival for patients with ER-negative breast tumors [73]. MMP-1 expression in breast tumor and stromal cells is associated with tumor progression and poor prognosis [112]. Strong MMP-13 expression in tumor but not stromal fibroblasts inversely correlated with overall survival of breast cancer patients [103].

ADAM9 mRNA levels were found to be higher in breast carcinoma tissues compared to normal breast, but were also upregulated in benign breast disease [78]. Interestingly, expression of the mature form of ADAM9 (~84 kDa) was higher in node-positive primary cancer tissues compared to benign disease, whereas the approximately 124 kDa precursor form of the enzyme was more prevalent in the latter [78]. We have previously reported that urinary ADAM12 detection in patients with breast cancer may be predictive of disease status and stage and that ADAM12 levels increase in urine during disease progression [74]. In addition, combined urinary MMP-9 and ADAM12 analysis provide important clinical information in the identification of women at increased risk of developing breast cancer [75]. ADAM17 and ADAM12 transcripts are upregulated in primary breast cancers compared to normal breast tissues [77, 113]. In particular, both ADAM17 and ADAM12 are expressed at significantly higher levels in triple negative breast cancer (TNBC) compared to non-TNBC tumor tissues [114, 76]. In TNBC, the membrane-associated isoform, ADAM12-L, was reported to be the primary protease responsible for the activation of EGFR in early stage, lymph node-negative disease, and correlated with decreased distant metastasis-free survival times for patients [76].

10.3.2 Prostate cancer

Several MMPs, including MMP-1, MMP-2, MMP-7, MMP-9, and MMP-13, have been reported to be upregulated in prostate cancer tissues or in blood from prostate cancer patients. We have previously reported that increased levels of urinary MMP-9 can distinguish between prostate and bladder cancer [98]. Immunohistochemistry (IHC) studies of paraffin-embedded specimens from 62 prostate cancer (PCA) and 15 benign prostatic hyperplasia (BPH) cases indicated that MMP-1, MMP-2, and MMP-9 expression was significantly higher in PCA compared to BPH tissues [96]. In this study, CD147 and MMP-2 expression correlated with TMN grade and Gleason score and in addition, patients with concurrent expression of CD147 and MMP-2 had the lowest survival rates [96]. Plasma levels of MMP-2, MMP-9, and MMP-13 are higher in PCA patients with metastasis as compared to those with organ-confined disease or BPH or healthy controls [97]. Similarly, serum MMP-7 levels were reported to be significantly elevated in PCA patients with metastatic disease, suggesting that circulating MMP-7 levels may serve as an independent risk factor for PCA-related deaths [99]. MMP-13 expression in prostate tumor tissues and MMP-9 expression in tumor-associated stromal cells were found to be independent factors for predicting biochemical recurrence (defined as increased serum PSA levels that may indicate development of distant metastasis) [115, 116]. It has been shown that the use of a combination of serum and urinary biomarkers can significantly improve diagnosis compared to the serum PSA test alone [117]. Measurement of serum alpha-methylacyl-CoA racemase (AMACR) and MMP-2 when combined with GSTP1/RASSF1A methylation status in urine sediments resulted in AUC of 0.788 compared to that for PSA alone of 0.476 [117]. In contrast, in a tissue microarray study including 278 patients undergoing radical prostatectomy for localized PCA, a higher MMP-9 expression in tumor cells was found to be associated with longer recurrence-free and disease-specific survival, whereas MMP-2, MMP-3, MMP-13, and MMP-19 were not [118].

To date, very few reports have described a role for ADAMs as potential biomarkers for PCA. Higher ADAM9 mRNA and protein expression in PCA tissues were shown to be an independent prognostic marker of PSA relapse-free survival following radical prostatectomy [100].

10.3.3 Lung cancer

Increased levels of several MMPs, such as MMP-1, MMP-2, MMP-9, and MMP-14, have been found in tumor tissues of lung cancer patients [119, 85]. MMP-1 levels are considerably higher in tumor tissues and plasma from lung cancer patients compared to healthy controls, and have been associated with advanced-stage cancer and significantly lower overall survival rates [120]. MMP-2 enzymatic activity was significantly increased in lung cancer tissues compared to normal lungs in stage I non-small lung cancer (NSCLC) patients [86]. A meta-analysis based on 11 published articles, including 1,439 patients, that analyzed the relationship between MMP-2 expression and overall patient survival, indicated that strong MMP-2 staining in tumor tissues predicted poorer patient survival [121]. Similarly, a meta-analysis of 17 studies including 2029 patients indicated a strong association (pooled hazard ratio (HR) of 1.84 (95% CI: 1.62–2.09) between MMP-9 overexpression and a poor prognosis for lung cancer patients [122]. Plasma MMP-9 levels were found to be increased in patients with lung cancer [85] and in the same study an MMP-9 C1562T polymorphism was found to be more frequent in the patient group compared to the controls. The efficacy of using serum MMP-9 levels as a longitudinal marker for response to chemotherapy has also been tested in patients with advanced NSCLC being treated with cisplatin-based standard chemotherapy. Pre-chemotherapy MMP-9 levels were significantly higher in patients that had at least a partial response to the treatment regimen compared to those with stable or progressive disease [123]. MMP-9 expression in tumor tissues along with other prognostic factors such as vessel invasion, and primary tumor (pT) stage may serve as independent prognostic factors to predict the prognosis of patients with pathologic stage IA NSCLC [124]. Concomitant analysis indicated that NSCLC tissues exhibited a lower expression of semaphorin 3A and a higher expression of MMP-14 compared to control lung tissues, which were associated with a poorer disease prognosis [125].

High mRNA levels of the membrane-associated isoform, ADAM12-L, in resected p-stage I lung carcinoma tissues correlated with a less differentiated tumor subtype and a significantly poorer postoperative prognosis for patients [87]. ADAM12 protein expression was upregulated in small cell lung cancer (SCLC) tissues, in a study including 70 patients and 40 normal controls; serum and urinary levels of ADAM12 were significantly higher in SCLC patients than in controls and in patients with extensive disease versus those with limited disease, suggesting that circulating ADAM12 levels may serve as an independent prognostic factor and/or diagnostic marker for SCLC [88]. ADAM28 may also serve as a serological and histochemical marker for NSCLC. Serum levels of ADAM28 were found to be approximately five-fold higher in lung cancer patients than in healthy controls and increased with disease progression, carcinoma recurrence or in patients with lymph node metastasis [89]. In a study including 122 advanced NSCLC cases, (37 patients with benign disease and 40 healthy controls) serum ADAM28 levels appeared to be a reliable surrogate marker to predict tumor response to chemotherapy and overall survival in patients [126]. ADAM28 protein expression in NSCLC tissues was almost 40-fold higher than in normal lung tissues [89]. ADAM28 gene expression was also significantly upregulated in asbestos-related lung cancer tissues, which constitute approximately 4–12% of lung cancers worldwide [127].

10.3.4 Pancreatic cancer

For the detection and prognosis of pancreatic cancer, several studies have evaluated the efficacy of using blood, serum and exocrine pancreatic secretion-based MMPs as biomarkers. Given that pancreatic cancer has a relatively low prevalence in the general population, some of the biomarker studies presented below may include $n \leq 50$ patient samples. Plasma MMP-7 levels are significantly higher in pancreatic ductal adenocarcinoma (PDAC) patients and can be used to differentiate pancreatic cancer from chronic pancreatitis [82]. MMP-7 levels in exocrine pancreatic secretions were higher in PDAC patients compared to those with chronic pancreatitis or benign disease, although the differences were not statistically significant [82]. More recently, a panel comprised of serum MMP-7, cathepsin D, and CA19-9 were reported to have increased diagnostic sensitivity (88%) and AUC (0.900) versus CA19-9 when used as a single marker (74%; 0.835) [83]. Proteomics analysis of exocrine pancreatic secretions from PDAC patients identified MMP-9 as a potential biomarker [79]. Subsequently, serum MMP-9 and TIMP-1 levels were reported to be significantly higher in PDAC patients compared to those with chronic pancreatitis and healthy controls [80]. We have previously reported that urinary MMP-2 and TIMP-1 levels may be significant independent predictors for distinguishing PDAC patients from healthy controls [81]. In addition, urinary MMP-2 may predict the presence of pancreatic neuroendocrine (pNET) tumors, whereas TIMP-1 levels may differentiate between PDAC and pNET patient groups [81]. Recently, the fingerprint of specific MMP-generated collagen fragments has been examined to differentiate PDAC patients from healthy controls. MMP-mediated degradation of collagen type I, type III, and type IV were assessed in serum, using a competitive ELISA approach [128] and the MMP-generated collagen fragments were found to be significantly higher in PDAC patients compared to controls [128].

Of the ADAMs, only ADAM9 has been analyzed in relation to pancreatic cancer to date [84]. IHC staining of tumor tissues indicated that positive ADAM9 expression was detected in a majority of the PDAC tissues (58/59; 98.3%) but very few of the acinar cell carcinomas (2/24; 8.3%) [84]. Interestingly, localization of ADAM9 expression appeared to be important in this study, such that strong cytoplasmic ADAM9 expression correlated with poor tumor differentiation and shorter overall survival for patients compared to cases with only apical membranous staining [84].

10.3.5 Ovarian cancer

The efficacy of MMP-2, MMP-7, MMP-9, and MMP-14 as potential biomarkers for ovarian cancer has now been widely assessed. A meta analysis of 30 individual studies concluded that increased MMP-9 expression was associated with poor prognosis in ovarian cancer patients [91]. In addition, MMP-9 was significantly associated with FIGO stage, grade of differentiation, and lymph node metastasis but not with histotype in ovarian cancer [91]. Interestingly, the cell type expressing the relevant MMP may be an important indicator. In a study of 292 primary ovarian tumors, high stromal MMP-9 but low tumor cell MMP-9 expression was shown to correlate with advanced stage of tumor as well as shorter disease-related survival [92]. In contrast, high tumor cell expression of EMMPRIN and MMP-2 were indicators of a more favorable prognosis [92]. We have recently reported that in patients with CA125 levels less than 35 U/ml (within normal range), for whom no diagnostics are available, urinary MMP-2 or MMP-9 or lipocalin can significantly discriminate between ovarian cancer patients and healthy controls. This finding demonstrates that urinary MMP-2 or MMP-9 analysis may be useful in the clinic in the diagnosis of advanced or recurrent ovarian cancer in patients with normal CA125 levels [93]. High levels of serum TIMP-1, but not TIMP-2, MMP-2, or MMP-9, were found to correlate with advanced stage of disease, aggressive tumor phenotype, and unfavorable prognosis in ovarian cancer [129]. Serum MMP-7 levels in combination with CA125, CCL18, and CCL11 have been shown to effectively detect early stages of ovarian cancer with high sensitivity [90]. MMP-14 and MMP-2 protein expression is higher in ovarian clear cell carcinoma tissues (>90%) relative to 30–55% in other ovarian tumor histotypes (serous, endometroid, mucinous), suggesting that these proteases may contribute to the progression of clear cell carcinomas [94]. In contrast, tumor tissue expression of MMP-14 in serous carcinomas was associated with lower progression and better prognosis [130].

ADAM17 expression has been found to be significantly increased in early and advanced ovarian tumor tissues in correlation with HB-EGF [95]. Interestingly, lysophosphatidic acid (LPA) has been found to induce an ADAM17-mediated HB-EGF cleavage and EGFR transactivation in ovarian cancer [131]. ADAM-mediated substrate shedding may also serve as prognostic marker(s) for ovarian cancer. Soluble CXCL16/CXCR6, in pre-treatment serum samples, can independently predict poor survival for cancer patients [132], suggesting that sCXCL16 (soluble CXCL16) may be a marker that may identify patients with highly metastatic tumors [132].

10.4 MMPs/ADAMs as diagnostic and prognostic biomarkers for non-neoplastic diseases

In addition to human cancers, MMPs and ADAMs have also been implicated in the development and progression of many non-neoplastic human diseases such as cardiovascular diseases, arthritis, and endometriosis. These MMPs and ADAMs are listed in Table 10.2. The role of MMPs as therapeutic and diagnostic targets has been investigated in these diseases as well as in the field of dental medicine [133–136]. Herein, we have included only those studies that have analyzed $n \geq 40$ patient samples for each study group, with the exception of some studies in endometriosis, where biomarker discovery is a new field and the sample sizes are significantly smaller (Table 10.2).

10.4.1 Cardiovascular diseases

In the cardiovascular system, certain MMPs are expressed by the vascular endothelial cells, smooth muscle cells, macrophages, and cells in the myocardium. Due to their versatile functions such as degrading the ECM and stimulating smooth muscle cell migration, these MMPs have been shown to play important roles during the development of cardiovascular diseases such as affecting plaque formation and instability in atherosclerosis, restenosis, and left ventricular remodeling after myocardial infarction [159, 160]. Therefore, numerous studies aiming to identify biomarkers for these cardiovascular diseases have focused on MMPs and TIMPs. In a study of patients with coronary artery disease, the baseline MMP-9 level in plasma was identified as a prognostic marker of cardiovascular mortality in these patients [137]. Baseline MMP-9 levels can also predict adverse effects including death and myocardial infarction after coronary revascularization treatments of patients with coronary artery disease [138]. In patients with acute myocardial infarction, increased MMP-9 levels in plasma are associated with a greater impairment of the left ventricular function and higher degree of left ventricular remodeling [139]. Serums levels of MMP-9 and its endogenous inhibitor, TIMP-1, are associated with decreased survival in patients with dilated cardiomyopathy [141]. In addition to MMP-9, MMP-2 has also demonstrated potential to serve as a biomarker of cardiovascular diseases. In patients with ST-elevation myocardial infarction, plasma MMP-2 levels measured at baseline and early after percutaneous coronary intervention (PCI) strongly correlated with infarct size and left ventricular intervention, indicating that MMP-2 is a potential biomarker of reperfusion injury after PCI [142].

10.4.2 Endometriosis

Endometriosis is a gynecological disease characterized by the growth of endometrial tissue outside of the uterus, usually in the peritoneal cavity. Although it is a non-malignant disease, the development of endometriosis shares several hallmarks with cancer, such as invasion of surrounding tissue via ECM degradation, and stimulation of angiogenesis in order to provide nutrients and oxygen for the lesion [162]. Therefore, MMPs that play essential roles during these processes often exhibit changes in their mRNA/protein levels and proteolytic activities in patients with endometriosis, and can therefore serve as biomarkers for this disease. For example, compared to endometrial tissues from healthy controls, eutopic endometrial tissues from patients with endometriosis exhibit increased MMP-9 proteolytic activity [163], increased MMP-9/TIMP-1 ratio [143], as well as increased levels of MMP-3 [147, 148] and MT5-MMP [150]. Increased levels of MMP-1 [164]

Table 10.2 Candidate MMP and ADAM biomarkers of non-malignant diseases.

Disease type		MMPs/ADAMs/TIMPs	Detected in (tissue/body fluid)	Method of analysis	References
Cardio-vascular diseases	Coronary artery disease	MMP-9	Plasma, Serum	ELISA	[137, 138]
	Acute myocardial infarction	MMP-2	Plasma	ELISA	[139, 140]
		MMP-9			[140]
		TIMP-1			
	Dilated cardiomyopathy	MMP-9	Serum	ELISA	[141]
		TIMP-1			
	ST-elevation myocardial infarction	MMP-2	Plasma	ELISA	[142]
Endometriosis		MMP-9	Eutopic endometrial tissue	Zymography, ELISA	[143]
			Peritoneal fluid	ELISA	[144]
			Urine	Zymography	[145]
			Peripheral blood	Genotyping (SNP)	[146]
		MMP-9/NGAL complex	Urine	Zymography	[145]
		MMP-9/TIMP-1 ratio	Eutopic endometrial tissue	ELISA, RT-PCR	[143]
		MMP-3	Eutopic endometrial tissue	ELISA, RT-PCR	[147, 148]
			Peripheral blood	RT-PCR	[149]
		MT5-MMP	Eutopic endometrial tissue	RT-PCR	[150]
			Eutopic endometrial tissue	RT-PCR	[150]
		MMP-2	Urine	Zymography	[145]
			Peripheral blood	Genotyping (SNP)	[146]
Preclampsia		MMP-9	Serum	RT-PCR	[151]
		ADAM12	Serum	ELISA	[152, 153]
Arthritis	Osteoarthritis	MMP-3	Serum	ELISA	[154, 155]
	Ankylosing spondylitis	MMP-3	Serum	ELISA	[156]
	Rheumatoid Arthritis	ADAMTS-4, ADAMTS-5	Synovial fluids, serum	ELISA	[157, 158]

Note: For all the studies listed above $n \geq 40$ with the exception of most biomarker studies for endometriosis where samples sizes were smaller.

and MT5-MMP [150] levels have also been observed in ectopic endometriosis tissues compared to eutopic endometrial tissues. Peritoneal fluid from patients with endometriosis exhibits higher levels of MMP-9 and lower levels of TIMP-1 [144], whereas another study found decreased levels of MMP-13 and MT1-MMP in the peritoneal fluid of these patients [165]. In peripheral blood, increased mRNA levels of MMP-3 can be detected in endometriosis patients [149]. Serum levels of MMP-2 have been shown to correlate with the severity of the disease in a pilot study. However, most of these studies were conducted using a limited number of patient samples. Very few studies, such as Becker et al. [145]. , have used large cohorts of patient samples and have identified MMP-2, MMP-9, and MMP-9/NGAL complex as non-invasive urinary biomarkers for endometriosis. In addition to the expression levels and proteolytic activities of these enzymes, the genetic variations of these enzymes have been investigated as potential biomarkers for endometriosis. For example, certain single-nucleotide polymorphisms (SNPs) in promoter regions of MMP-2 and MMP-9 have been shown to be associated with elevated risk of endometriosis [146]. These SNPs alter the binding affinity of transcriptional activators and repressors of the promoters which, in turn, may lead to changes in the expression levels of these MMPs and facilitate the development of the disease.

10.4.3 Preeclampsia

Preeclampsia (PE) is the most common complication of pregnancy worldwide, affecting approximately 3–10% of all pregnancies. If left untreated, it can be a leading cause of maternal and perinatal mortality. PE is characterized by a failure of the placenta to efficiently implant in the uterus resulting in reduced blood flow to the fetus. This leads to placental hypoxic stress, dysfunctional maternal endothelium, and systemic inflammatory responses ultimately manifesting in clinical signs such as maternal hypertension, proteinuria, and systemic vascular dysfunction [166–169]. Early detection, monitoring, and clinical care can prevent some of the adverse outcomes of PE [170]. Although the mechanism of PE development is poorly understood, MMP and ADAM proteases are known to contribute to a variety of normal placental functions, including trophoblast invasion and implantation and angiogenesis [171–173]. In a study analyzing serum samples from 160 patients with mild or severe PE and 112 normal pregnant controls, a significantly higher frequency of MMP-9 polymorphism (-1562 C:T) was observed compared to controls while the risk of severe PE and early-onset PE increased 2.7-fold in the carriers of this MMP-9 polymorphism [151]. Therefore, this MMP-9 genotypic variant could be a potential biomarker of susceptibility to early-onset and severe PE, although it is not yet clear what the outcome of this polymorphism would be with respect to MMP-9 function in PE. Interestingly, a recent study showed that MMP-9 deficiency in mice can cause physiological and placental abnormalities, intrauterine growth restriction, or embryonic death, features very similar to human PE [174]. MMP-9 or MMP-2 levels in the serum of PE patients are not significantly different as compared to normotensive pregnant women. Higher proMMP-9 levels and proMMP-9/TIMP-1 ratios were observed in women with gestational hypertension, but not in those with PE, compared to normotensive pregnant women [175–177]. These findings suggest that MMP-9 may play a role in the development of gestational hypertension, a disorder that can progress to PE. A different mechanism may be involved in PE-related hypertensive disorders.

ADAM12 has been extensively studied in relation to PE detection. Serum levels of the secreted isoform, ADAM12-S, increase markedly during pregnancy [178–180], such that serum ADAM12-S has been investigated in several pregnancy-related disorders. Reduced levels of ADAM12 are associated with pregnancies with fetal trisomy 21 and trisomy 18 [181, 152], with other aneuploidies and in patients with low gestational birth weight pregnancies [182]. In a study of first trimester serum samples from 160 pregnant women who later developed PE and 324 normal pregnant women, ADAM12-S levels detected using a semiautomated, time-resolved, immunofluorometric assay were found to be significantly downregulated during the first trimester [183]. These original findings have been corroborated by a number of subsequent reports [184, 153, 185, 186]. However, other studies have reported that ADAM12-S measurement may not be useful for early trimester antenatal screening for PE prediction [187–190]. Therefore, in light of these somewhat conflicting findings, future large scale studies must be undertaken in order to determine whether early trimester serum ADAM12 levels (in combination with other pregnancy-related markers) could be useful in the prediction of PE.

10.4.4 Arthritis

Arthritis is a joint disorder that involves inflammation of one or more joints. There are approximately 100 different forms of arthritis, and some of the common forms include osteoarthritis (OA), rheumatoid arthritis (RA), ankylosing spondylitis, and psoriatic arthritis. OA, a chronic degenerative joint disease that affects millions of people worldwide [191], is characterized by structural changes in load-bearing joints as a result of repetitive use, injury, infection, or obesity that leads to degradation of the joint surface articular cartilage, inflammation of the synovium, and changes to the subchondral bone [192]. Useful biomarkers for arthritis would provide tools for (i) assessing disease activity, (ii) predicting disease outcome, and (iii) stratification of patients with structural progression who are most in need of immediate treatment to maintain tissue integrity [193]. MMP expression and enzymatic function aid the remodeling of articular cartilage and have been implicated in the progression of OA and RA. In RA, MMP-3 was found to be significantly associated with disease activity and correlated with expression of inflammatory mediators and with cartilage breakdown; however, it did not reliably predict early stage RA [154, 155]. In a longitudinal observational cohort with an 8-year follow up, baseline serum MMP-3 levels were a strong independent predictor of radiographic disease outcome in RA patients [194]. Analysis of synovial fluids aspirated from joints for a panel of MMPs including MMP-1, MMP-2, MMP-7, MMP-8, MMP-9, MMP-12, and MMP-13 indicated that while MMP-2, MMP-8, and MMP-9 can differentiate between advanced OA and normal individuals, these MMP profiles did not distinguish early stage RA from healthy controls [195]. Serum MMP-3 was found to be elevated in patients with active ankylosing spondylitis [156].

Degradation of aggrecan in the cartilage is an early event in joint diseases such as OA and can be detected as elevated release of aggrecan from cartilage into the synovial fluids. ADAMTS-4 and ADAMTS-5 play a key role in aggrecan degradation. ADAMTS-4 levels were found to be increased in the synovial fluids from 11 knee surgery patients [157], indicating the potential of using this biomarker for early detection of cartilage-degrading joint diseases. Serum ADAMTS-5 mRNA levels were found to be significantly lower in RA patients that responded well to infliximab compared to those with a moderate response or nonresponders [158], suggesting that ADAMTS-5 levels may be useful for prediction of therapy response. Two SNPs have been identified in the ADAM12 gene that are strongly associated with early and late radiographic OA [196–198], suggesting that ADAM12 may be a candidate gene demonstrating susceptibility to OA. Serum ADAM12 levels have also been found to be elevated in OA patients and correlated with grades of disease [197].

Another strategy used for biomarker discovery in arthritis is based on the detection of the cleavage products of MMPs/ADAMs. For instance, aggrecan cleavage at the interglobular domain ^{392}Glu-^{393}Ala bond, an early event in arthritis, releases N-terminal ^{393}ARGS neoepitope fragments [199]. Serum ARGS neoepitope concentrations were found to be elevated in OA patients undergoing total knee replacement compared to non-surgical OA patients or healthy controls [200, 201]. Similarly, in a longitudinal 2-year study of 132 patients with early RA that were treated with nonbiologic therapies, serum MMP-3, C-telopeptide of type II collagen (CTX-II), cartilage oligomeric matrix protein (COMP), and TIMP-1 correlated significantly with radiographic progression of RA [202]. In particular, a model including MMP-3 and CTX-II provided the best prediction of radiographic progression at entry ($AUC = 0.76$) by multivariate analysis, and a combination of MMP-3, CTX-II, and swollen joint counts provided the best prediction for longitudinal progression ($AUC = 0.81$) [202] suggesting that MMPs and their breakdown products may serve as useful biochemical markers for the prediction of radiographic progression of RA. A serum protein called type I collagen degradation mediated by MMP-cleavage (C1M), can also serve as a biomarker for tissue destruction in RA. In the LITHE-biomarker study, a 1-year phase III, double blind, placebo-controlled, group study including 585 patients, baseline C1M serum levels correlated with worsening joint structure over 1 year, and C1M levels were also dose-dependently reduced in patients treated with a combination of tocilizumab and methotrexate [193]. Taken together, these findings suggest that analysis of MMP-cleavage products may prove useful in arthritis, not only to monitor active disease but to identify patients that are in most need of aggressive treatment.

10.5 MMPs as biomarkers of therapeutic efficacy

Recent studies have shown that MMPs can also be used as biomarkers to predict the therapeutic efficacy of a variety of drugs for a number of diseases. In patients with brain tumors, resection of the tumor is associated with clearance of urinary MMPs such as MMP-2, MMP-9, MMP-9/NGAL, and MMP-9 dimer. This absence of urinary MMPs is also correlated with radiographic imaging during follow-ups and can be used to monitor response to therapy [203]. In patients with chronic periodontitis, elevated MMP-8 levels in the gingival crevicular fluid is associated with poor response to scaling and root planning treatment [204].

Multiple studies have investigated the use of MMPs, either alone or in combination with cytokines, as biomarkers for the therapeutic efficacy of joint diseases. In a study comparing methotrexate (anti-proliferative and anti-inflammatory agent) alone and plus infliximab (TNF-α inhibitor), MMP-3 has been shown to correlate with clinical improvement in patients with RA [205]. Serum levels of MMP-1, MMP-3 and TIMP-1 can be used as biomarkers of RA disease activity and therapeutic response in a study comparing the therapeutic efficacy of anakinra, the IL-1 receptor antagonist, as monotherapy or in combination with the anti-TNF-α agent pegsunercept [206]. In RA patients who have failed methotrexate or TNF inhibitors and are receiving the IL-6 receptor antagonist tocilizumab, serum levels of MMP-3 correlated with clinical disease activity index [207]. Recently, researchers have developed a panel of 12 biomarkers, including MMP-1 and MMP-3, that has high potential to be used to measure clinical disease activity in RA patients and to detect their changes in response to treatment [208]. In patients with ankylosing spondylitis, serum MMP-3 levels have been shown to correlate with Bath Ankylosing Spondylitis Disease Activity Index (BASDAI) [209]. Blockade of TNF-α with infliximab or etanercept leads to decreased serum MMP-3 levels, indicating that MMP-3 may be useful in monitoring anti-TNF-α therapy in these patients [209, 210].

Due to the success in preclinical studies using MMPs as biomarkers for therapeutic efficacy, a number of clinical trials have measured levels of MMPs to monitor the response to therapy. These clinical trials are summarized in Table 10.3 and those relevant to the diseases described in this review are discussed below. Please note that not all clinical trials included in this Table have been completed to date or have resulted in a published report.

Many clinical studies focusing on malignant diseases have used MMPs as biomarkers for therapeutic efficacy. For example, a phase II clinical trial has been conducted to evaluate the use of MMP-2 as one of the biomarkers for therapeutic response to chemotherapy plus the Cox-2 inhibitor celecoxib in breast cancer patients (NCT00665457). A phase I clinical trial has also been initiated to determine whether MMP-9 levels in blood correlate with tumor remission and symptom improvement in metastatic prostate cancer patients treated with PCK3145, a drug that inhibits prostate cancer metastasis (NCT00695851). Both MMP-2 and MMP-9 levels were used as secondary outcome measures in a study evaluating the efficacy of green tea catechins in the treatment of prostate cancer patients (NCT00459407). In non-small cell lung cancer, serum levels of MMP-9 and MMP-9/NGAL complex have been under investigation to determine whether they can be used to predict response to erlotinib treatment (NCT01123460). In advanced or metastatic pancreatic cancer, MMP-7 levels in blood were used as an outcome measure for the efficacy of a chemotherapy drug, gemcitabine, in combination with a CBP/β-Catenin Inhibitor PRI-724 (NCT01764477). In a phase II study of cediranib, a pan-VEGFR2 inhibitor (NCT00305656), higher plasma levels of MMP-2 and higher urinary levels of MMP-9/NGAL after cediranib administration in recurrent glioblastoma patients have been shown to associate with poor progression-free survival or overall survival [211]. A phase II study has also been initiated to evaluate the changes in serum and plasma MMP-2 and MMP-9 levels in response to bevacizumab treatment in patients with renal cell carcinoma (NCT00113217).

Table 10.3 MMPs/TIMPs as biomarkers for therapeutic efficacy in clinical trials.

Clinical trial identifier	Phase	Recruitment status/Reference	MMPs or ADAMS	Condition	Treatment
NCT00665457	II	Completed; results unpublished	MMP-2	Breast cancer	Chemotherapy + Celecoxib
NCT00459407	I	Completed; results unpublished	MMP-2 MMP-9	Prostate cancer	Green tea extract
NCT00695851	I	Completed; results unpublished	MMP-9	Metastatic prostate cancer	PCK3145
NCT01123460	–	–	MMP-9 MMP-9/NGAL	Non-small cell lung cancer	Erlotinib
NCT01764477	I	Open	MMP-7	Advanced or metastatic pancreatic cancer	Gemcitabine + PRI-724
NCT00305656	II	Completed [211]	MMP-2 MMP-9/NGAL	Recurrent glioblastoma multiforme	Cediranib
NCT00113217	II	Completed; results unpublished	MMP-2 MMP-9	Renal cell carcinoma	Bevacizumab
NCT02135939	IV	Completed; results unpublished	MMP-9	Coronary artery disease	Behavioral: Vegan diet versus AHA diet
NCT01917721	II	Open	MMP-9 TIMPs	Kawasaki disease	Doxycycline
NCT00116792	II/III	–	MMP-9	Coronary artery disease and type 2 diabetes mellitus	Rosiglitazone
NCT00116831	III	Completed [212]	MMP-9	Progression of atherosclerosis in patients with heart disease and diabetes mellitus	Rosiglitazone versus Glipizide
NCT00048568	III	Completed [213]	MMP-3	Methotrexate-resistant active rheumatoid arthritis	Abatacept
NCT00048581	III	Completed [214]	MMP-3	Rheumatoid arthritis	Abatacept

(*continued*)

Table 10.3 *(Continued)*

Clinical trial identifier	Phase	Recruitment status/Reference	MMPs or ADAMS	Condition	Treatment
NCT00976599	II	Completed; results unpublished	MMP-3	Rheumatoid arthritis	Methotrexate + placebo versus Methotrexate + CP-690,550
NCT01163292	-	Completed; results unpublished	MMP-3	Rheumatoid arthritis	Adalimumab
NCT01374971	III	Completed; results unpublished	MMP-3 MMP-1	Rheumatoid arthritis	Certolizumab Pegol
NCT00195819	III	Complete [215]	MMP-3	Ankylosing Spondylitis	Adalimumab
NCT01034501	IV	Completed [216, 217]	MMP-8 MMP-2 TIMP-1	Chronic periodontitis	Photodynamic therapy
NCT01798225	IV	Completed; results unpublished	MMP-8	Peridontal disease Type 2 Diabetes Mellitus	Doxycycline
NCT00709111	-	Completed [218]	MMP-9	HIV infection	Maraviroc, antiretroviral therapy
NCT00796822	II	Completed [219]	MMP-9 TIMP-1	HIV-infected patients not receiving antiretroviral medications	Pentoxifylline
NCT00246324	IV	Completed [220]	MMP-9	Multiple sclerosis	Interferon-β-1α, doxycycline
NCT00770653	III	Completed; results unpublished	MMP-9	Diabetes Mellitus Dyslipidemia	Pioglitazone + Metformin versus Glimepiride + Metformin
NCT00562588	IV	Completed [221]	MMP-9	Cerebrovascular accident	Aggrenox versus acetylsalicylic acid

In clinical trials for cardiovascular diseases, MMP-9 is frequently used as a biomarker for response to therapy or behavioral changes. For example, in a phase IV clinical trial, MMP-9 was used as a biomarker to determine the effect of a whole-food plant-based vegan diet compared to the diet recommended by the American Heart Association in patients with coronary artery disease (NCT02135939). In children with Kawasaki disease, a disease defined by inflammation of the walls of arteries including the coronary arteries, serum levels of MMP-9 and TIMPs were used as biomarkers to determine the therapeutic effect of doxycycline in preventing coronary artery aneurysm formation and progression (NCT01917721). In a phase II/III clinical trial, MMP-9 levels were used to determine the effect of the PPAR-γ agonist rosiglitazone on patients with coronary artery disease and type 2 diabetes mellitus after percutaneous coronary intervention (NCT00116792).

In addition to cancer and cardiovascular diseases, clinical studies focusing on arthritis and periodontal diseases often use MMPs as biomarkers for therapeutic responses. In two phase III clinical trials evaluating the effect of abatacept (an inhibitor of T-cell activity) in RA patients resistant to anti-TNF-α treatment (NCT00048581, NCT00048568), serum levels of MMP-3 have been used as secondary outcome measures for disease activity [214, 213]. Similarly, in a phase III clinical trial studying the efficacy of the TNF-inhibiting anti-inflammatory agent adalimumab in patients with active ankylosing spondylitis (NCT00195819), serum levels of MMP-3 have been used to evaluate cartilage and bone degradation. For periodontal diseases, two clinical studies have used levels of MMPs or TIMPs in gingival crevicular fluid as measures of disease activity. One of these studies is a phase IV clinical trial to investigate the efficacy of photodynamic therapy on chronic periodontitis (NCT01034501). A panel of biomarkers including MMP-8, MMP-2, and TIMP-1 were used as secondary outcome measures. The other study is a phase IV trial on the efficacy of doxycycline, a broad spectrum MMP inhibitor, on patients with periodontal disease and type 2 diabetes mellitus, to investigate whether or not treatment of periodontal disease will result in better glycemic control in diabetic patients (NCT01798225). Changes in MMP-8 levels in gingival crevicular fluid were used as a secondary outcome measure in this study.

To date, there are over 100 clinical trials using MMP levels as primary or secondary outcome measures for the activity of diseases including those described above, as well as others such as HIV infection, multiple sclerosis, ischemic stroke, and diabetes mellitus. Due to page limitations, we are unable to discuss all of these studies in this chapter. Readers are encouraged to go to clinicaltrials.gov for a complete list of clinical trials using MMPs as biomarkers of therapeutic efficacy and/or to look for detailed information of a particular clinical trial using its clinicaltrial.gov identifier number.

10.6 MMP-specific molecular imaging for noninvasive disease detection

Since MMPs play a pivotal role in tissue remodeling and disease progression in a variety of pathological conditions, MMP-specific probes may be used efficiently for the noninvasive visualization and quantification of MMP expression and activity. Here we describe some of the novel probes being currently characterized using *in vitro* and *in vivo* approaches that may be potentially adapted as biomarkers for early diagnosis, staging, and/or therapeutic efficacy in the near future. MT1-hIC7L, a near-infrared (NIR) fluorescence probe that can be activated following interaction with MT1-MMP, was tested *in vivo* and could successfully visualize C6 glioma (high expression of MT1-MMP) xenografts whereas MCF-7 tumors that have low endogenous MT1-MMP expression showed no obvious fluorescence [222]. In a mouse model of colonic adenoma, screening the GI tract for colon cancer using MMP-9 and MMP-14 antibody-quantum dot (Ab-QD) conjugates demonstrated specific binding to the tumor, although a high rate of false positives indicated a need to increase the specificity for this method [223]. Another class of *in vivo* MMP targeting agents utilize an MMP cleavable linker to create activatable cell penetrating peptides (ACPPs) [224]. Absorption and uptake of the ACPP into cells is inhibited until the linker is proteolyzed by MMPs. MMP-2 and MMP-9-selective ACPPs have been shown to target both xenograft and transgenic breast tumors in mice and accumulation of ACPPs was found to be most concentrated at the tumor-stromal interface in both primary tumors and associated metastasis [224]. In addition, to improve sensitivity and specificity, an integrin $\alpha_v\beta_3$–binding domain was covalently linked to the ACPP, providing a co-targeting approach that relies on the interaction of MMP-2 with integrin $\alpha_v\beta_3$ [225]. This dual targeting approach greatly improved ACPP uptake in MDA-MB-231 and syngeneic Py230 murine tumor-bearing mice as well as greatly improved the efficacy of chemotherapeutic delivery (ACPP conjugated) to these tumors *in vivo* [225]. Similar molecular imaging strategies are currently also being explored for cardiovascular diseases as reviewed in [226]. Finally, in mouse models of OA and RA, biophotonic imaging via near-infrared fluorescent probes activatable by MMPs or cathepsins have been reported to discriminate between OA and RA [227], which are clinically distinct diseases.

10.7 Conclusions

The many years spent identifying and characterizing the key functional roles of MMPs, ADAMs, and their endogenous regulators, have provided the foundation for their current clinical relevance in a variety of human diseases. For example, the use of MMPs and ADAMs as biomarkers of disease status, stage, and therapeutic efficacy is a new and exciting area of Biomarker Medicine. In light of the studies discussed above, there is now ample evidence that MMP and ADAM family members have begun to make their way into the clinic, providing important and useful diagnostic and prognostic information to clinicians and their patients. As their clinical utilities are validated by larger cohort studies, we anticipate their eventual use in a variety of settings, including clinical laboratories, point-of-care, and perhaps, eventually, at home.

Acknowledgments

This work was supported by NIH 1RO1CA185530-01, The Ellison Foundation and The Breast Cancer Research Foundation. Due to space constraints, we have, in some cases, been forced to cite review articles as opposed to the original papers.

References

1. Roy, R., Yang, J. & Moses, M.A. (2009) Matrix metalloproteinases as novel biomarkers and potential therapeutic targets in human cancer. *Journal of Clinical Oncology : Official Journal of the American Society of Clinical Oncology*, 27, 5287–5297.
2. Westermarck, J. & Kahari, V.M. (1999) Regulation of matrix metalloproteinase expression in Tumor invasion. *FASEB journal : Official Publication of the Federation of American Societies for Experimental Biology*, 13, 781–792.
3. Piperi, C. & Papavassiliou, A.G. (2012) Molecular mechanisms regulating matrix metalloproteinases. *Current Topics in Medicinal Chemistry*, 12, 1095–1112.
4. Loffek, S., Schilling, O. & Franzke, C.W. (2011) Series "matrix metalloproteinases in lung health and disease": Biological role of matrix metalloproteinases: a critical balance. *The European Respiratory Journal*, 38, 191–208.
5. Roy, R., Zhang, B. & Moses, M.A. (2006) Making the cut: protease-mediated regulation of angiogenesis. *Experimental Cell Research*, 312, 608–622.
6. Nagase, H., Visse, R. & Murphy, G. (2006) Structure and function of matrix metalloproteinases and TIMPs. *Cardiovascular Research*, 69, 562–573.
7. Cruz-Munoz, W. & Khokha, R. (2008) The role of tissue inhibitors of metalloproteinases in tumorigenesis and metastasis. *Critical Reviews in Clinical Laboratory Sciences*, 45, 291–338.
8. Stetler-Stevenson, W.G., Krutzsch, H.C. & Liotta, L.A. (1989) Tissue inhibitor of metalloproteinase (TIMP-2). A new member of the metalloproteinase inhibitor family. *The Journal of Biological Chemistry*, 264, 17374–17378.
9. Stetler-Stevenson, W.G. (2008) Tissue inhibitors of metalloproteinases in cell signaling: metalloproteinase-independent biological activities. *Science Signaling*, 1, re6.
10. Murphy, G. & Nagase, H. (2008) Progress in matrix metalloproteinase research. *Molecular Aspects of Medicine*, 29, 290–308.
11. Harper, J. & Moses, M.A. (2006) Molecular regulation of tumor angiogenesis: mechanisms and therapeutic implications. *EXS*, 223–268.
12. van Hinsbergh, V.W. & Koolwijk, P. (2008) Endothelial sprouting and angiogenesis: matrix metalloproteinases in the lead. *Cardiovascular Research*, 78, 203–212.
13. Hadler-Olsen, E., Fadnes, B., Sylte, I., Uhlin-Hansen, L. & Winberg, J.O. (2011) Regulation of matrix metalloproteinase activity in health and disease. *The FEBS Journal*, 278, 28–45.
14. Gialeli, C., Theocharis, A.D. & Karamanos, N.K. (2011) Roles of matrix metalloproteinases in cancer progression and their pharmacological targeting. *The FEBS Journal*, 278, 16–27.
15. Stone, A.L., Kroeger, M. & Sang, Q.X. (1999) Structure-function analysis of the ADAM family of disintegrin-like and metalloproteinase-containing proteins (review). *Journal of Protein Chemistry*, 18, 447–465.
16. Murphy, G. (2008) The ADAMs: signalling scissors in the tumour microenvironment. *Nature Reviews Cancer*, 8, 929–941.
17. Duffy, M.J., McKiernan, E., O'Donovan, N. & McGowan, P.M. (2009) Role of ADAMs in cancer formation and progression. *Clinical Cancer Research*, 15, 1140–1144.
18. Klein, T. & Bischoff, R. (2011) Active metalloproteases of the A Disintegrin and Metalloprotease (ADAM) family: biological function and structure. *Journal of Proteome Research*, 10, 17–33.
19. Apte, S.S. (2004) A disintegrin-like and metalloprotease (reprolysin type) with thrombospondin type 1 motifs: the ADAMTS family. *The International Journal of Biochemistry & Cell Biology*, 36, 981–985.
20. Rocks, N., Paulissen, G., El Hour, M. *et al.* (2008) Emerging roles of ADAM and ADAMTS metalloproteinases in cancer. *Biochimie*, 90, 369–379.
21. Frantz, C., Stewart, K.M. & Weaver, V.M. (2010) The extracellular matrix at a glance. *Journal of Cell Science*, 123, 4195–4200.

22. Hadler-Olsen, E., Winberg, J.O. & Uhlin-Hansen, L. (2013) Matrix metalloproteinases in cancer: their value as diagnostic and prognostic markers and therapeutic targets. *Tumour Biology : the Journal of the International Society for Oncodevelopmental Biology and Medicine*, 34, 2041–2051.
23. Suzuki, M., Raab, G., Moses, M.A., Fernandez, C.A. & Klagsbrun, M. (1997) Matrix metalloproteinase-3 releases active heparin-binding EGF-like growth factor by cleavage at a specific juxtamembrane site. *The Journal of Biological Chemistry*, 272, 31730–31737.
24. Sahin, U., Weskamp, G., Kelly, K. *et al.* (2004) Distinct roles for ADAM10 and ADAM17 in ectodomain shedding of six EGFR ligands. *The Journal of Cell Biology*, 164, 769–779.
25. Mori, S., Tanaka, M., Nanba, D. *et al.* (2003) PACSIN3 binds ADAM12/meltrin alpha and up-regulates ectodomain shedding of heparin-binding epidermal growth factor-like growth factor. *The Journal of Biological Chemistry*, 278, 46029–46034.
26. Kivisaari, A.K., Kallajoki, M., Ala-aho, R. *et al.* (2010) Matrix metalloproteinase-7 activates heparin-binding epidermal growth factor-like growth factor in cutaneous squamous cell carcinoma. *The British Journal of Dermatology*, 163, 726–735.
27. Fowlkes, J.L., Enghild, J.J., Suzuki, K. & Nagase, H. (1994) Matrix metalloproteinases degrade insulin-like growth factor-binding protein-3 in dermal fibroblast cultures. *The Journal of Biological Chemistry*, 269, 25742–25746.
28. Loechel, F., Fox, J.W., Murphy, G., Albrechtsen, R. & Wewer, U.M. (2000) ADAM 12-S cleaves IGFBP-3 and IGFBP-5 and is inhibited by TIMP-3. *Biochemical and Biophysical Research Communications*, 278, 511–515.
29. Mochizuki, S., Shimoda, M., Shiomi, T., Fujii, Y. & Okada, Y. (2004) ADAM28 is activated by MMP-7 (matrilysin-1) and cleaves insulin-like growth factor binding protein-3. *Biochemical and Biophysical Research Communications*, 315, 79–84.
30. Manes, S., Mira, E., Barbacid, M.M. *et al.* (1997) Identification of insulin-like growth factor-binding protein-1 as a potential physiological substrate for human stromelysin-3. *The Journal of Biological Chemistry*, 272, 25706–25712.
31. Sadowski, T., Dietrich, S., Koschinsky, F. & Sedlacek, R. (2003) Matrix metalloproteinase 19 regulates insulin-like growth factor-mediated proliferation, migration, and adhesion in human keratinocytes through proteolysis of insulin-like growth factor binding protein-3. *Molecular Biology of the Cell*, 14, 4569–4580.
32. Houck, K.A., Leung, D.W., Rowland, A.M., Winer, J. & Ferrara, N. (1992) Dual regulation of vascular endothelial growth factor bioavailability by genetic and proteolytic mechanisms. *The Journal of Biological Chemistry*, 267, 26031–26037.
33. Bergers, G., Brekken, R., McMahon, G. *et al.* (2000) Matrix metalloproteinase-9 triggers the angiogenic switch during carcinogenesis. *Nature Cell Biology*, 2, 737–744.
34. Whitelock, J.M., Murdoch, A.D., Iozzo, R.V. & Underwood, P.A. (1996) The degradation of human endothelial cell-derived perlecan and release of bound basic fibroblast growth factor by stromelysin, collagenase, plasmin, and heparanases. *The Journal of Biological Chemistry*, 271, 10079–10086.
35. Jiao, Y., Feng, X., Zhan, Y. *et al.* (2012) Matrix metalloproteinase-2 promotes alphavbeta3 integrin-mediated adhesion and migration of human melanoma cells by cleaving fibronectin. *PLoS ONE*, 7, e41591.
36. Xu, J., Rodriguez, D., Petitclerc, E. *et al.* (2001) Proteolytic exposure of a cryptic site within collagen type IV is required for angiogenesis and tumor growth in vivo. *The Journal of Cell Biology*, 154, 1069–1079.
37. Koshikawa, N., Giannelli, G., Cirulli, V., Miyazaki, K. & Quaranta, V. (2000) Role of cell surface metalloprotease MT1-MMP in epithelial cell migration over laminin-5. *The Journal of Cell Biology*, 148, 615–624.
38. Cho, S.Y., Xu, M., Roboz, J., Lu, M., Mascarenhas, J. & Hoffman, R. (2010) The effect of CXCL12 processing on CD34+ cell migration in myeloproliferative neoplasms. *Cancer Research*, 70, 3402–3410.
39. Sugiyama, N., Gucciardo, E., Tatti, O. *et al.* (2013) EphA2 cleavage by MT1-MMP triggers single cancer cell invasion via homotypic cell repulsion. *The Journal of Cell Biology*, 201, 467–484.
40. Noe, V., Fingleton, B., Jacobs, K. *et al.* (2001) Release of an invasion promoter E-cadherin fragment by matrilysin and stromelysin-1. *Journal of Cell Science*, 114, 111–118.
41. Ratnikov, B.I., Rozanov, D.V., Postnova, T.I. *et al.* (2002) An alternative processing of integrin alpha(v) subunit in tumor cells by membrane type-1 matrix metalloproteinase. *The Journal of Biological Chemistry*, 277, 7377–7385.
42. Baciu, P.C., Suleiman, E.A., Deryugina, E.I. & Strongin, A.Y. (2003) Membrane type-1 matrix metalloproteinase (MT1-MMP) processing of pro-alphav integrin regulates cross-talk between alphavbeta3 and alpha2beta1 integrins in breast carcinoma cells. *Experimental Cell Research*, 291, 167–175.

43. Chetty, C., Vanamala, S.K., Gondi, C.S., Dinh, D.H., Gujrati, M. & Rao, J.S. (2012) MMP-9 induces CD44 cleavage and CD44 mediated cell migration in glioblastoma xenograft cells. *Cellular Signalling*, 24, 549–559.
44. Kung, C.I., Chen, C.Y., Yang, C.C., Lin, C.Y., Chen, T.H. & Wang, H.S. (2012) Enhanced membrane-type 1 matrix metalloproteinase expression by hyaluronan oligosaccharides in breast cancer cells facilitates CD44 cleavage and tumor cell migration. *Oncology Reports*, 28, 1808–1814.
45. Kuo, Y.C., Su, C.H., Liu, C.Y., Chen, T.H., Chen, C.P. & Wang, H.S. (2009) Transforming growth factor-beta induces CD44 cleavage that promotes migration of MDA-MB-435s cells through the up-regulation of membrane type 1-matrix metalloproteinase. *International Journal of Cancer*, 124, 2568–2576.
46. Maretzky, T., Reiss, K., Ludwig, A. *et al.* (2005) ADAM10 mediates E-cadherin shedding and regulates epithelial cell-cell adhesion, migration, and beta-catenin translocation. *Proceedings of the National Academy of Sciences of the United States of America*, 102, 9182–9187.
47. Roy, R., Rodig, S., Bielenberg, D., Zurakowski, D. & Moses, M.A. (2012) ADAM12 transmembrane and secreted isoforms promote breast tumor growth: a distinct role for ADAM12-S protein in tumor metastasis. *The Journal of Biological Chemistry*, 286, 20758–20768.
48. Hara, T., Mimura, K., Seiki, M. & Sakamoto, T. (2011) Genetic dissection of proteolytic and non-proteolytic contributions of MT1-MMP to macrophage invasion. *Biochemical and Biophysical Research Communications*, 413, 277–281.
49. Lynch, C.C., Vargo-Gogola, T., Matrisian, L.M. & Fingleton, B. (2010) Cleavage of E-Cadherin by Matrix Metalloproteinase-7 Promotes Cellular Proliferation in Nontransformed Cell Lines via Activation of RhoA. *Journal of Oncology*, 2010, 530745.
50. Dwivedi, A., Slater, S.C. & George, S.J. (2009) MMP-9 and -12 cause N-cadherin shedding and thereby beta-catenin signalling and vascular smooth muscle cell proliferation. *Cardiovascular Research*, 81, 178–186.
51. Su, G., Blaine, S.A., Qiao, D. & Friedl, A. (2007) Shedding of syndecan-1 by stromal fibroblasts stimulates human breast cancer cell proliferation via FGF2 activation. *The Journal of Biological Chemistry*, 282, 14906–14915.
52. Su, G., Blaine, S.A., Qiao, D. & Friedl, A. (2008) Membrane type 1 matrix metalloproteinase-mediated stromal syndecan-1 shedding stimulates breast carcinoma cell proliferation. *Cancer Research*, 68, 9558–9565.
53. Hanahan, D. & Weinberg, R.A. (2011) Hallmarks of cancer: the next generation. *Cell*, 144, 646–674.
54. Illman, S.A., Lehti, K., Keski-Oja, J. & Lohi, J. (2006) Epilysin (MMP-28) induces TGF-beta mediated epithelial to mesenchymal transition in lung carcinoma cells. *Journal of Cell Science*, 119, 3856–3865.
55. Lochter, A., Galosy, S., Muschler, J., Freedman, N., Werb, Z. & Bissell, M.J. (1997) Matrix metalloproteinase stromelysin-1 triggers a cascade of molecular alterations that leads to stable epithelial-to-mesenchymal conversion and a premalignant phenotype in mammary epithelial cells. *The Journal of Cell Biology*, 139, 1861–1872.
56. Radisky, D.C., Levy, D.D., Littlepage, L.E. *et al.* (2005) Rac1b and reactive oxygen species mediate MMP-3-induced EMT and genomic instability. *Nature*, 436, 123–127.
57. Matrisian, L.M., Wright, J., Newell, K. & Witty, J.P. (1994) Matrix-degrading metalloproteinases in tumor progression. *Princess Takamatsu Symposia*, 24, 152–161.
58. Birkedal-Hansen, H. (1995) Proteolytic remodeling of extracellular matrix. *Current Opinion in Cell Biology*, 7, 728–735.
59. Alfranca, A., Lopez-Oliva, J.M., Genis, L. *et al.* (2008) PGE2 induces angiogenesis via MT1-MMP-mediated activation of the TGFbeta/Alk5 signaling pathway. *Blood*, 112, 1120–1128.
60. Blackburn, J.S. & Brinckerhoff, C.E. (2008) Matrix metalloproteinase-1 and thrombin differentially activate gene expression in endothelial cells via PAR-1 and promote angiogenesis. *The American Journal of Pathology*, 173, 1736–1746.
61. Roy, R., and Moses, M. A. (2012) ADAM12: A novel mediator of tumor angiogenesis. in Proceedings of the 103rd Annual Meeting of the American Association for Cancer Research; 2012 Mar 31-Apr 4 Chicago, Illinois. AACR. Abstract No: 5285
62. Morancho, A., Hernandez-Guillamon, M., Boada, C. *et al.* (2013) Cerebral ischaemia and matrix metalloproteinase-9 modulate the angiogenic function of early and late outgrowth endothelial progenitor cells. *Journal of Cellular and Molecular Medicine*, 12, 1548–1553.
63. Huang, P.H., Chen, Y.H., Wang, C.H. *et al.* (2009) Matrix metalloproteinase-9 is essential for ischemia-induced neovascularization by modulating bone marrow-derived endothelial progenitor cells. *Arteriosclerosis, Thrombosis, and Vascular Biology*, 29, 1179–1184.

64. Fang, J., Shing, Y., Wiederschain, D. *et al.* (2000) Matrix metalloproteinase-2 is required for the switch to the angiogenic phenotype in a tumor model. *Proceedings of the National Academy of Sciences of the United States of America*, 97, 3884–3889.
65. O'Reilly, M.S., Wiederschain, D., Stetler-Stevenson, W.G., Folkman, J. & Moses, M.A. (1999) Regulation of angiostatin production by matrix metalloproteinase-2 in a model of concomitant resistance. *The Journal of Biological Chemistry*, 274, 29568–29571.
66. Dong, Z., Kumar, R., Yang, X. & Fidler, I.J. (1997) Macrophage-derived metalloelastase is responsible for the generation of angiostatin in Lewis lung carcinoma. *Cell*, 88, 801–810.
67. Wen, W., Moses, M.A., Wiederschain, D., Arbiser, J.L. & Folkman, J. (1999) The generation of endostatin is mediated by elastase. *Cancer Research*, 59, 6052–6056.
68. Hawinkels, L.J., Kuiper, P., Wiercinska, E. *et al.* (2013) Matrix metalloproteinase-14 (MT1-MMP)-mediated endoglin shedding inhibits tumor angiogenesis. *Cancer Research*, 70, 4141–4150.
69. Moses, M.A., Wiederschain, D., Loughlin, K.R., Zurakowski, D., Lamb, C.C. & Freeman, M.R. (1998) Increased incidence of matrix metalloproteinases in urine of cancer patients. *Cancer Research*, 58, 1395–1399.
70. Fernandez, C.A., Yan, L., Louis, G., Yang, J., Kutok, J.L. & Moses, M.A. (2005) The matrix metalloproteinase-9/neutrophil gelatinase-associated lipocalin complex plays a role in breast tumor growth and is present in the urine of breast cancer patients. *Clinical Cancer Research*, 11, 5390–5395.
71. Somiari, S.B., Somiari, R.I., Heckman, C.M. *et al.* (2006) Circulating MMP2 and MMP9 in breast cancer – potential role in classification of patients into low risk, high risk, benign disease and breast cancer categories. *International Journal of Cancer*, 119, 1403–1411.
72. Rahko, E., Kauppila, S., Paakko, P. *et al.* (2009) Immunohistochemical study of matrix metalloproteinase 9 and tissue inhibitor of matrix metalloproteinase 1 in benign and malignant breast tissue–strong expression in intraductal carcinomas of the breast. *Tumour Biology : the Journal of the International Society for Oncodevelopmental Biology and Medicine*, 30, 257–264.
73. Song, N., Sung, H., Choi, J.Y. *et al.* (2012) Preoperative serum levels of matrix metalloproteinase-2 (MMP-2) and survival of breast cancer among Korean women. *Cancer Epidemiology, Biomarkers & Prevention : a Publication of the American Association for Cancer Research, Cosponsored by the American Society of Preventive Oncology*, 21, 1371–1380.
74. Roy, R., Wewer, U.M., Zurakowski, D., Pories, S.E. & Moses, M.A. (2004) ADAM 12 cleaves extracellular matrix proteins and correlates with cancer status and stage. *The Journal of Biological Chemistry*, 279, 51323–51330.
75. Pories, S.E., Zurakowski, D., Roy, R. *et al.* (2008) Urinary metalloproteinases: noninvasive biomarkers for breast cancer risk assessment. *Cancer Epidemiology, Biomarkers & Prevention : a Publication of the American Association for Cancer Research, Cosponsored by the American Society of Preventive Oncology*, 17, 1034–1042.
76. Li, H., Duhachek-Muggy, S., Qi, Y., Hong, Y., Behbod, F. & Zolkiewska, A. (2012) An essential role of metalloprotease-disintegrin ADAM12 in triple-negative breast cancer. *Breast Cancer Research and Treatment*, 135, 759–769.
77. Narita, D., Seclaman, E., Ursoniu, S. & Anghel, A. (2012) Increased expression of ADAM12 and ADAM17 genes in laser-capture microdissected breast cancers and correlations with clinical and pathological characteristics. *Acta Histochemica*, 114, 131–139.
78. O'Shea, C., McKie, N., Buggy, Y. *et al.* (2003) Expression of ADAM-9 mRNA and protein in human breast cancer. *International Journal of Cancer*, 105, 754–761.
79. Tian, M., Cui, Y.Z., Song, G.H. *et al.* (2008) Proteomic analysis identifies MMP-9, DJ-1 and A1BG as overexpressed proteins in pancreatic juice from pancreatic ductal adenocarcinoma patients. *BMC Cancer*, 8, 241.
80. Mroczko, B., Lukaszewicz-Zajac, M., Wereszczynska-Siemiatkowska, U. *et al.* (2009) Clinical significance of the measurements of serum matrix metalloproteinase-9 and its inhibitor (tissue inhibitor of metalloproteinase-1) in patients with pancreatic cancer: metalloproteinase-9 as an independent prognostic factor. *Pancreas*, 38, 613–618.
81. Roy, R., Zurakowski, D., Wischhusen, J., Fraunhoffer, C., Hooshmand, S., Kulke, M., and Moses, M. (2014) Urinary TIMP-1 and MMP-2 levels detect the presence of pancreatic malignancies. *Br J Cancer*, 111, 1772–1779.
82. Kuhlmann, K.F., van Till, J.W., Boermeester, M.A. *et al.* (2007) Evaluation of matrix metalloproteinase 7 in plasma and pancreatic juice as a biomarker for pancreatic cancer. *Cancer Epidemiology, Biomarkers & Prevention: a Publication of the American Association for Cancer Research, Cosponsored by the American Society of Preventive Oncology*, 16, 886–891.

83. Park, H.D., Kang, E.S., Kim, J.W. *et al.* (2012) Serum CA19-9, cathepsin D, and matrix metalloproteinase-7 as a diagnostic panel for pancreatic ductal adenocarcinoma. *Proteomics*, 12, 3590–3597.
84. Grutzmann, R., Luttges, J., Sipos, B. *et al.* (2004) ADAM9 expression in pancreatic cancer is associated with tumour type and is a prognostic factor in ductal adenocarcinoma. *British Journal of Cancer*, 90, 1053–1058.
85. Bayramoglu, A., Gunes, H.V., Metintas, M., Degirmenci, I., Mutlu, F. & Alatas, F. (2009) The association of MMP-9 enzyme activity, MMP-9 C1562T polymorphism, and MMP-2 and -9 and TIMP-1, -2, -3, and -4 gene expression in lung cancer. *Genetic Testing and Molecular Biomarkers*, 13, 671–678.
86. Weng, Y., Cai, M., Zhu, J. *et al.* (2013) Matrix metalloproteinase activity in early-stage lung cancer. *Onkologie*, 36, 256–259.
87. Mino, N., Miyahara, R., Nakayama, E. *et al.* (2009) A disintegrin and metalloprotease 12 (ADAM12) is a prognostic factor in resected pathological stage I lung adenocarcinoma. *Journal of Surgical Oncology*, 100, 267–272.
88. Shao, S., Li, Z., Gao, W., Yu, G., Liu, D. & Pan, F. (2014) ADAM-12 as a diagnostic marker for the proliferation, migration and invasion in patients with small cell lung cancer. *PLoS ONE*, 9, e85936.
89. Kuroda, H., Mochizuki, S., Shimoda, M. *et al.* (2010) ADAM28 is a serological and histochemical marker for non-small-cell lung cancers. *International Journal of Cancer*, 127, 1844–1856.
90. Zohny, S.F. & Fayed, S.T. (2010) Clinical utility of circulating matrix metalloproteinase-7 (MMP-7), CC chemokine ligand 18 (CCL18) and CC chemokine ligand 11 (CCL11) as markers for diagnosis of epithelial ovarian cancer. *Medical Oncology*, 27, 1246–1253.
91. Li, L.N., Zhou, X., Gu, Y. & Yan, J. (2013) Prognostic value of MMP-9 in ovarian cancer: a meta-analysis. *Asian Pacific journal of Cancer Prevention : APJCP*, 14, 4107–4113.
92. Sillanpaa, S., Anttila, M., Suhonen, K. *et al.* (2007) Prognostic significance of extracellular matrix metalloproteinase inducer and matrix metalloproteinase 2 in epithelial ovarian cancer. *Tumour Biology : the Journal of the International Society for Oncodevelopmental Biology and Medicine*, 28, 280–289.
93. Coticchia, C.M., Curatolo, A.S., Zurakowski, D. *et al.* (2011) Urinary MMP-2 and MMP-9 predict the presence of ovarian cancer in women with normal CA125 levels. *Gynecologic Oncology*, 123, 295–300.
94. Adley, B.P., Gleason, K.J., Yang, X.J. & Stack, M.S. (2009) Expression of membrane type 1 matrix metalloproteinase (MMP-14) in epithelial ovarian cancer: high level expression in clear cell carcinoma. *Gynecologic Oncology*, 112, 319–324.
95. Tanaka, Y., Miyamoto, S., Suzuki, S.O. *et al.* (2005) Clinical significance of heparin-binding epidermal growth factor-like growth factor and a disintegrin and metalloprotease 17 expression in human ovarian cancer. *Clinical Cancer Research*, 11, 4783–4792.
96. Zhong, W.D., Han, Z.D., He, H.C. *et al.* (2008) CD147, MMP-1, MMP-2 and MMP-9 protein expression as significant prognostic factors in human prostate cancer. *Oncology*, 75, 230–236.
97. Morgia, G., Falsaperla, M., Malaponte, G. *et al.* (2005) Matrix metalloproteinases as diagnostic (MMP-13) and prognostic (MMP-2, MMP-9) markers of prostate cancer. *Urological Research*, 33, 44–50.
98. Roy, R., Louis, G., Loughlin, K.R. *et al.* (2008) Tumor-specific urinary matrix metalloproteinase fingerprinting: identification of high molecular weight urinary matrix metalloproteinase species. *Clinical Cancer Research*, 14, 6610–6617.
99. Szarvas, T., Becker, M., Vom Dorp, F. *et al.* (2011) Elevated serum matrix metalloproteinase 7 levels predict poor prognosis after radical prostatectomy. *International Journal of Cancer*, 128, 1486–1492.
100. Fritzsche, F.R., Jung, M., Tolle, A. *et al.* (2008) ADAM9 expression is a significant and independent prognostic marker of PSA relapse in prostate cancer. *European Urology*, 54, 1097–1106.
101. Zucker, S., Lysik, R.M., Zarrabi, H.M. *et al.* (1994) Plasma assay of matrix metalloproteinases (MMPs) and MMP-inhibitor complexes in cancer. Potential use in predicting metastasis and monitoring treatment. *Annals of the New York Academy of Sciences*, 732, 248–262.
102. Wu, Z.S., Wu, Q., Yang, J.H. *et al.* (2008) Prognostic significance of MMP-9 and TIMP-1 serum and tissue expression in breast cancer. *International Journal of Cancer*, 122, 2050–2056.
103. Zhang, B., Cao, X., Liu, Y. *et al.* (2008) Tumor-derived matrix metalloproteinase-13 (MMP-13) correlates with poor prognoses of invasive breast cancer. *BMC Cancer*, 8, 83.
104. Patel, S., Sumitra, G., Koner, B.C. & Saxena, A. (2011) Role of serum matrix metalloproteinase-2 and -9 to predict breast cancer progression. *Clinical Biochemistry*, 44, 869–872.
105. Roy, R., Zurakowski, D., Pories, S., Moss, M.L. & Moses, M.A. (2011) Potential of fluorescent metalloproteinase substrates for cancer detection. *Clinical Biochemistry*, 44, 1434–1439.
106. Song, J., Su, H., Zhou, Y.Y. & Guo, L.L. (2013) Prognostic value of matrix metalloproteinase 9 expression in breast cancer patients: a meta-analysis. *Asian Pacific Journal of Cancer Prevention: APJCP*, 14, 1615–1621.

107. Schveigert, D., Cicenas, S., Bruzas, S., Samalavicius, N.E., Gudleviciene, Z. & Didziapetriene, J. (2013) The value of MMP-9 for breast and non-small cell lung cancer patients' survival. *Advances in Medical Sciences*, 58, 73–82.
108. Niemiec, J., Adamczyk, A., Malecki, K., Ambicka, A. & Rys, J. (2013) Tumor grade and matrix metalloproteinase 2 expression in stromal fibroblasts help to stratify the high-risk group of patients with early breast cancer identified on the basis of st Gallen recommendations. *Clinical Breast Cancer*, 13, 119–128.
109. Del Casar, J.M., Gonzalez, L.O., Alvarez, E. *et al.* (2009) Comparative analysis and clinical value of the expression of metalloproteases and their inhibitors by intratumor stromal fibroblasts and those at the invasive front of breast carcinomas. *Breast Cancer Research and Treatment*, 116, 39–52.
110. Pellikainen, J.M., Ropponen, K.M., Kataja, V.V., Kellokoski, J.K., Eskelinen, M.J. & Kosma, V.M. (2004) Expression of matrix metalloproteinase (MMP)-2 and MMP-9 in breast cancer with a special reference to activator protein-2, HER2, and prognosis. *Clinical Cancer Research*, 10, 7621–7628.
111. Sung, H., Choi, J.Y., Lee, S.A. *et al.* (2012) The association between the preoperative serum levels of lipocalin-2 and matrix metalloproteinase-9 (MMP-9) and prognosis of breast cancer. *BMC Cancer*, 12, 193.
112. Bostrom, P., Soderstrom, M., Vahlberg, T. *et al.* (2011) MMP-1 expression has an independent prognostic value in breast cancer. *BMC Cancer*, 11, 348.
113. Narita, D., Seclaman, E., Ilina, R., Cireap, N., Ursoniu, S. & Anghel, A. (2011) ADAM12 and ADAM17 gene expression in laser-capture microdissected and non-microdissected breast tumors. *Pathology Oncology Research : POR*, 17, 375–385.
114. McGowan, P.M., Mullooly, M., Caiazza, F. *et al.* (2013) ADAM-17: a novel therapeutic target for triple negative breast cancer. *Annals of Oncology : Official Journal of the European Society for Medical Oncology/ESMO*, 24, 362–369.
115. Escaff, S., Fernandez, J.M., Gonzalez, L.O. *et al.* (2010) Study of matrix metalloproteinases and their inhibitors in prostate cancer. *British Journal of Cancer*, 102, 922–929.
116. Escaff, S., Fernandez, J.M., Gonzalez, L.O. *et al.* (2011) Collagenase-3 expression by tumor cells and gelatinase B expression by stromal fibroblast-like cells are associated with biochemical recurrence after radical prostatectomy in patients with prostate cancer. *World Journal of Urology*, 29, 657–663.
117. Prior, C., Guillen-Grima, F., Robles, J.E. *et al.* (2010) Use of a combination of biomarkers in serum and urine to improve detection of prostate cancer. *World Journal of Urology*, 28, 681–686.
118. Boxler, S., Djonov, V., Kessler, T.M. *et al.* (2010) Matrix metalloproteinases and angiogenic factors: predictors of survival after radical prostatectomy for clinically organ-confined prostate cancer? *The American Journal of Pathology*, 177, 2216–2224.
119. Sauter, W., Rosenberger, A., Beckmann, L. *et al.* (2008) Matrix metalloproteinase 1 (MMP1) is associated with early-onset lung cancer. *Cancer Epidemiology, Biomarkers & Prevention : a Publication of the American Association for Cancer Research, Cosponsored by the American Society of Preventive Oncology*, 17, 1127–1135.
120. Li, M., Xiao, T., Zhang, Y. *et al.* (2010) Prognostic significance of matrix metalloproteinase-1 levels in peripheral plasma and tumour tissues of lung cancer patients. *Lung Cancer*, 69, 341–347.
121. Qian, Q., Wang, Q., Zhan, P. *et al.* (2010) The role of matrix metalloproteinase 2 on the survival of patients with non-small cell lung cancer: a systematic review with meta-analysis. *Cancer Investigation*, 28, 661–669.
122. Peng, W.J., Zhang, J.Q., Wang, B.X., Pan, H.F., Lu, M.M. & Wang, J. (2012) Prognostic value of matrix metalloproteinase 9 expression in patients with non-small cell lung cancer. *Clinica Chimica Acta; International Journal of Clinical Chemistry*, 413, 1121–1126.
123. Ertan, E., Soydinc, H., Yazar, A., Ustuner, Z., Tas, F. & Yasasever, V. (2011) Matrix metalloproteinase-9 decreased after chemotherapy in patients with non-small cell lung cancer. *Tumori*, 97, 286–289.
124. Shao, W., Wang, W., Xiong, X.G. *et al.* (2011) Prognostic impact of MMP-2 and MMP-9 expression in pathologic stage IA non-small cell lung cancer. *Journal of Surgical Oncology*, 104, 841–846.
125. Zhou, H., Wu, A., Fu, W., Lv, Z. & Zhang, Z. (2014) Significance of semaphorin-3A and MMP-14 protein expression in non-small cell lung cancer. *Oncology Letters*, 7, 1395–1400.
126. Lv, Y.L., Yuan, D.M., Wang, Q.B. *et al.* (2012) Baseline and decline of serum ADAM28 during chemotherapy of advanced non-small cell lung cancer: a probable predictive and prognostic factor. *Medical Oncology*, 29, 2633–2639.
127. Wright, C.M., Larsen, J.E., Hayward, N.K. *et al.* (2010) ADAM28: a potential oncogene involved in asbestos-related lung adenocarcinomas. *Genes, Chromosomes & Cancer*, 49, 688–698.
128. Willumsen, N., Bager, C.L., Leeming, D.J. *et al.* (2013) Extracellular matrix specific protein fingerprints measured in serum can separate pancreatic cancer patients from healthy controls. *BMC Cancer*, 13, 554.

129. Rauvala, M., Puistola, U. & Turpeenniemi-Hujanen, T. (2005) Gelatinases and their tissue inhibitors in ovarian tumors; TIMP-1 is a predictive as well as a prognostic factor. *Gynecologic Oncology*, 99, 656–663.
130. Trudel, D., Desmeules, P., Turcotte, S. *et al.* (2014) Visual and automated assessment of matrix metalloproteinase-14 tissue expression for the evaluation of ovarian cancer prognosis. *Modern Pathology: an Official Journal of the United States and Canadian Academy of Pathology, Inc.*, 10, 1394–1404.
131. Braun, A.H. & Coffey, R.J. (2005) Lysophosphatidic acid, a disintegrin and metalloprotease-17 and heparin-binding epidermal growth factor-like growth factor in ovarian cancer: the first word, not the last. *Clinical Cancer Research*, 11, 4639–4643.
132. Gooden, M.J., Wiersma, V.R., Boerma, A. *et al.* (2014) Elevated serum CXCL16 is an independent predictor of poor survival in ovarian cancer and may reflect pro-metastatic ADAM protease activity. *British Journal of Cancer*, 110, 1535–1544.
133. Reynolds, J.J., Hembry, R.M. & Meikle, M.C. (1994) Connective tissue degradation in health and periodontal disease and the roles of matrix metalloproteinases and their natural inhibitors. *Advances in Dental Research*, 8, 312–319.
134. Sorsa, T., Tjaderhane, L., Konttinen, Y.T. *et al.* (2006) Matrix metalloproteinases: contribution to pathogenesis, diagnosis and treatment of periodontal inflammation. *Annals of Medicine*, 38, 306–321.
135. Sapna, G., Gokul, S. & Bagri-Manjrekar, K. (2013) Matrix metalloproteinases and periodontal diseases. *Oral Diseases*.
136. Garlet, G.P., Cardoso, C.R., Silva, T.A. *et al.* (2006) Cytokine pattern determines the progression of experimental periodontal disease induced by Actinobacillus actinomycetemcomitans through the modulation of MMPs, RANKL, and their physiological inhibitors. *Oral Microbiology and Immunology*, 21, 12–20.
137. Blankenberg, S., Rupprecht, H.J., Poirier, O. *et al.* (2003) Plasma concentrations and genetic variation of matrix metalloproteinase 9 and prognosis of patients with cardiovascular disease. *Circulation*, 107, 1579–1585.
138. Ye, Z.X., Leu, H.B., Wu, T.C., Lin, S.J. & Chen, J.W. (2008) Baseline serum matrix metalloproteinase-9 level predicts long-term prognosis after coronary revascularizations in stable coronary artery disease. *Clinical Biochemistry*, 41, 292–298.
139. Kelly, D., Khan, S.Q., Thompson, M. *et al.* (2008) Plasma tissue inhibitor of metalloproteinase-1 and matrix metalloproteinase-9: novel indicators of left ventricular remodelling and prognosis after acute myocardial infarction. *European Heart Journal*, 29, 2116–2124.
140. Kelly, D., Cockerill, G., Ng, L.L. *et al.* (2007) Plasma matrix metalloproteinase-9 and left ventricular remodelling after acute myocardial infarction in man: a prospective cohort study. *European Heart Journal*, 28, 711–718.
141. Franz, M., Berndt, A., Neri, D. *et al.* (2013) Matrix metalloproteinase-9, tissue inhibitor of metalloproteinase-1, B(+) tenascin-C and ED-A(+) fibronectin in dilated cardiomyopathy: potential impact on disease progression and patients' prognosis. *International Journal of Cardiology*, 168, 5344–5351.
142. Nilsson, L., Hallen, J., Atar, D., Jonasson, L. & Swahn, E. (2012) Early measurements of plasma matrix metalloproteinase-2 predict infarct size and ventricular dysfunction in ST-elevation myocardial infarction. *Heart*, 98, 31–36.
143. Collette, T., Maheux, R., Mailloux, J. & Akoum, A. (2006) Increased expression of matrix metalloproteinase-9 in the eutopic endometrial tissue of women with endometriosis. *Human Reproduction*, 21, 3059–3067.
144. Szamatowicz, J., Laudanski, P. & Tomaszewska, I. (2002) Matrix metalloproteinase-9 and tissue inhibitor of matrix metalloproteinase-1: a possible role in the pathogenesis of endometriosis. *Human Reproduction*, 17, 284–288.
145. Becker, C.M., Louis, G., Exarhopoulos, A. *et al.* (2010) Matrix metalloproteinases are elevated in the urine of patients with endometriosis. *Fertility and Sterility*, 94, 2343–2346.
146. Saare, M., Lamp, M., Kaart, T. *et al.* (2010) Polymorphisms in MMP-2 and MMP-9 promoter regions are associated with endometriosis. *Fertility and Sterility*, 94, 1560–1563.
147. Gilabert-Estelles, J., Estelles, A., Gilabert, J. *et al.* (2003) Expression of several components of the plasminogen activator and matrix metalloproteinase systems in endometriosis. *Human Reproduction*, 18, 1516–1522.
148. Ramon, L., Gilabert-Estelles, J., Castello, R. *et al.* (2005) mRNA analysis of several components of the plasminogen activator and matrix metalloproteinase systems in endometriosis using a real-time quantitative RT-PCR assay. *Human Reproduction*, 20, 272–278.

149. De Sanctis, P., Elmakky, A., Farina, A. *et al.* (2011) Matrix metalloproteinase-3 mRNA: a promising peripheral blood marker for diagnosis of endometriosis. *Gynecologic and Obstetric Investigation*, 71, 118–123.
150. Gaetje, R., Holtrich, U., Engels, K. *et al.* (2007) Expression of membrane-type 5 matrix metalloproteinase in human endometrium and endometriosis. *Gynecological Endocrinology*, 23, 567–573.
151. Rahimi, Z., Rahimi, Z., Shahsavandi, M.O., Bidoki, K. & Rezaei, M. (2013) MMP-9 (-1562 C:T) polymorphism as a biomarker of susceptibility to severe pre-eclampsia. *Biomarkers in Medicine*, 7, 93–98.
152. Laigaard, J., Christiansen, M., Frohlich, C., Pedersen, B.N., Ottesen, B. & Wewer, U.M. (2005) The level of ADAM12-S in maternal serum is an early first-trimester marker of fetal trisomy 18. *Prenatal Diagnosis*, 25, 45–46.
153. El-Sherbiny, W., Nasr, A. & Soliman, A. (2012) Metalloprotease (ADAM12-S) as a predictor of preeclampsia: correlation with severity, maternal complications, fetal outcome, and Doppler parameters. *Hypertension in Pregnancy* , 31, 442–450.
154. Ally, M.M., Hodkinson, B., Meyer, P.W., Musenge, E., Tikly, M. & Anderson, R. (2013) Serum matrix metalloproteinase-3 in comparison with acute phase proteins as a marker of disease activity and radiographic damage in early rheumatoid arthritis. *Mediators of Inflammation*, 2013, 183653.
155. Sun, S., Bay-Jensen, A.C., Karsdal, M.A. *et al.* (2014) The active form of MMP-3 is a marker of synovial inflammation and cartilage turnover in inflammatory joint diseases. *BMC Musculoskeletal Disorders*, 15, 93.
156. Soliman, E., Labib, W., el-Tantawi, G., Hamimy, A., Alhadidy, A. & Aldawoudy, A. (2012) Role of matrix metalloproteinase-3 (MMP-3) and magnetic resonance imaging of sacroiliitis in assessing disease activity in ankylosing spondylitis. *Rheumatology International*, 32, 1711–1720.
157. Peng, S., Zheng, Q., Zhang, X. *et al.* (2013) Detection of ADAMTS-4 activity using a fluorogenic peptide-conjugated Au nanoparticle probe in human knee synovial fluid. *ACS Applied Materials & Interfaces*, 5, 6089–6096.
158. Tsuzaka, K., Itami, Y., Takeuchi, T., Shinozaki, N. & Morishita, T. (2010) ADAMTS5 is a biomarker for prediction of response to infliximab in patients with rheumatoid arthritis. *The Journal of Rheumatology*, 37, 1454–1460.
159. Galis, Z.S. & Khatri, J.J. (2002) Matrix metalloproteinases in vascular remodeling and atherogenesis: the good, the bad, and the ugly. *Circulation Research*, 90, 251–262.
160. Halade, G.V., Jin, Y.F. & Lindsey, M.L. (2013) Matrix metalloproteinase (MMP)-9: a proximal biomarker for cardiac remodeling and a distal biomarker for inflammation. *Pharmacology & Therapeutics*, 139, 32–40.
161. Jordan, A., Roldan, V., Garcia, M. *et al.* (2007) Matrix metalloproteinase-1 and its inhibitor, TIMP-1, in systolic heart failure: relation to functional data and prognosis. *Journal of Internal Medicine*, 262, 385–392.
162. Burney, R.O. (2013) The genetics and biochemistry of endometriosis. *Current Opinion in Obstetrics & Gynecology*, 25, 280–286.
163. Collette, T., Bellehumeur, C., Kats, R. *et al.* (2004) Evidence for an increased release of proteolytic activity by the eutopic endometrial tissue in women with endometriosis and for involvement of matrix metalloproteinase-9. *Human Reproduction*, 19, 1257–1264.
164. Hudelist, G., Lass, H., Keckstein, J. *et al.* (2005) Interleukin 1alpha and tissue-lytic matrix metalloproteinase-1 are elevated in ectopic endometrium of patients with endometriosis. *Human Reproduction*, 20, 1695–1701.
165. Laudanski, P., Szamatowicz, J. & Ramel, P. (2005) Matrix metalloproteinase-13 and membrane type-1 matrix metalloproteinase in peritoneal fluid of women with endometriosis. *Gynecological Endocrinology*, 21, 106–110.
166. Huppertz, B. (2007) The feto-maternal interface: setting the stage for potential immune interactions. *Seminars in Immunopathology*, 29, 83–94.
167. Hung, T.H., Skepper, J.N., Charnock-Jones, D.S. & Burton, G.J. (2002) Hypoxia-reoxygenation: a potent inducer of apoptotic changes in the human placenta and possible etiological factor in preeclampsia. *Circulation Research*, 90, 1274–1281.
168. Maynard, S.E., Min, J.Y., Merchan, J. *et al.* (2003) Excess placental soluble fms-like tyrosine kinase 1 (sFlt1) may contribute to endothelial dysfunction, hypertension, and proteinuria in preeclampsia. *The Journal of Clinical Investigation*, 111, 649–658.
169. Steegers, E.A., von Dadelszen, P., Duvekot, J.J. & Pijnenborg, R. (2010) Pre-eclampsia. *Lancet*, 376, 631–644.
170. Hadker, N., Garg, S., Costanzo, C. *et al.* (2010) Financial impact of a novel pre-eclampsia diagnostic test versus standard practice: a decision-analytic modeling analysis from a UK healthcare payer perspective. *Journal of Medical Economics*, 13, 728–737.

171. de Jager, C.A., Linton, E.A., Spyropoulou, I., Sargent, I.L. & Redman, C.W. (2003) Matrix metalloprotease-9, placental syncytiotrophoblast and the endothelial dysfunction of pre-eclampsia. *Placenta*, 24, 84–91.
172. Huisman, M.A., Timmer, A., Zeinstra, M. *et al.* (2004) Matrix-metalloproteinase activity in first trimester placental bed biopsies in further complicated and uncomplicated pregnancies. *Placenta*, 25, 253–258.
173. Staun-Ram, E., Goldman, S., Gabarin, D. & Shalev, E. (2004) Expression and importance of matrix metalloproteinase 2 and 9 (MMP-2 and -9) in human trophoblast invasion. *Reproductive Biology and Endocrinology : RB&E*, 2, 59.
174. Plaks, V., Rinkenberger, J., Dai, J. *et al.* (2013) Matrix metalloproteinase-9 deficiency phenocopies features of preeclampsia and intrauterine growth restriction. *Proceedings of the National Academy of Sciences of the United States of America*, 110, 11109–11114.
175. Palei, A.C., Sandrim, V.C., Cavalli, R.C. & Tanus-Santos, J.E. (2008) Comparative assessment of matrix metalloproteinase (MMP)-2 and MMP-9, and their inhibitors, tissue inhibitors of metalloproteinase (TIMP)-1 and TIMP-2 in preeclampsia and gestational hypertension. *Clinical Biochemistry*, 41, 875–880.
176. Rahimi, Z., Rahimi, Z., Aghaei, A. & Vaisi-Raygani, A. (2014) AT2R -1332 G:A polymorphism and its interaction with AT1R 1166 A:C, ACE I/D and MMP-9 -1562 C:T polymorphisms: risk factors for susceptibility to preeclampsia. *Gene*, 538, 176–181.
177. Ab Hamid, J., Mohtarrudin, N., Osman, M., Andi Asri, A.A., Wan Hassan, W.H. & Aziz, R. (2012) Matrix metalloproteinase-9 and tissue inhibitors of metalloproteinases 1 and 2 as potential biomarkers for gestational hypertension. *Singapore Medical Journal*, 53, 681–683.
178. Makrydimas, G., Sotiriadis, A., Spencer, K., Cowans, N.J. & Nicolaides, K.H. (2006) ADAM12-s in coelomic fluid and maternal serum in early pregnancy. *Prenatal Diagnosis*, 26, 1197–1200.
179. Laigaard, J., Sorensen, T., Frohlich, C. *et al.* (2003) ADAM12: a novel first-trimester maternal serum marker for Down syndrome. *Prenatal Diagnosis*, 23, 1086–1091.
180. Valinen, Y., Peuhkurinen, S., Jarvela, I.Y., Laitinen, P. & Ryynanen, M. (2010) Maternal serum ADAM12 levels correlate with PAPP-A levels during the first trimester. *Gynecologic and Obstetric Investigation*, 70, 60–63.
181. Laigaard, J., Cuckle, H., Wewer, U.M. & Christiansen, M. (2006) Maternal serum ADAM12 levels in Down and Edwards' syndrome pregnancies at 9-12 weeks' gestation. *Prenatal Diagnosis*, 26, 689–691.
182. Spencer, K., Cowans, N.J. & Stamatopoulou, A. (2007) Maternal serum ADAM12s as a marker of rare aneuploidies in the first or second trimester of pregnancy. *Prenatal Diagnosis*, 27, 1233–1237.
183. Laigaard, J., Sorensen, T., Placing, S. *et al.* (2005) Reduction of the disintegrin and metalloprotease ADAM12 in preeclampsia. *Obstetrics and Gynecology*, 106, 144–149.
184. Spencer, K., Cowans, N.J. & Stamatopoulou, A. (2008) ADAM12s in maternal serum as a potential marker of pre-eclampsia. *Prenatal Diagnosis*, 28, 212–216.
185. Kuc, S., Koster, M.P., Franx, A., Schielen, P.C. & Visser, G.H. (2013) Maternal characteristics, mean arterial pressure and serum markers in early prediction of preeclampsia. *PLoS ONE*, 8, e63546.
186. Odibo, A.O., Patel, K.R., Spitalnik, A., Odibo, L. & Huettner, P. (2014) Placental pathology, first-trimester biomarkers and adverse pregnancy outcomes. *Journal of Perinatology : Official Journal of the California Perinatal Association*, 34, 186–191.
187. Bestwick, J.P., George, L.M., Wu, T., Morris, J.K. & Wald, N.J. (2012) The value of early second trimester PAPP-A and ADAM12 in screening for pre-eclampsia. *Journal of Medical Screening*, 19, 51–54.
188. Deurloo, K.L., Linskens, I.H., Heymans, M.W., Heijboer, A.C., Blankenstein, M.A. & van Vugt, J.M. (2013) ADAM12s and PP13 as first trimester screening markers for adverse pregnancy outcome. *Clinical Chemistry and Laboratory Medicine : CCLM/FESCC*, 51, 1279–1284.
189. Boucoiran, I., Suarthana, E., Rey, E., Delvin, E., Fraser, W.B. & Audibert, F. (2013) Repeated measures of placental growth factor, placental protein 13, and a disintegrin and metalloprotease 12 at first and second trimesters for preeclampsia screening. *American Journal of Perinatology*, 30, 681–688.
190. Goetzinger, K.R., Zhong, Y., Cahill, A.G., Odibo, L., Macones, G.A. & Odibo, A.O. (2013) Efficiency of first-trimester uterine artery Doppler, a-disintegrin and metalloprotease 12, pregnancy-associated plasma protein a, and maternal characteristics in the prediction of preeclampsia. *Journal of Ultrasound in Medicine : Official Journal of the American Institute of Ultrasound in Medicine*, 32, 1593–1600.
191. Wieland, H.A., Michaelis, M., Kirschbaum, B.J. & Rudolphi, K.A. (2005) Osteoarthritis – an untreatable disease? *Nature Reviews. Drug Discovery*, 4, 331–344.
192. Troeberg, L. & Nagase, H. (2012) Proteases involved in cartilage matrix degradation in osteoarthritis. *Biochimica et Biophysica Acta*, 1824, 133–145.
193. Siebuhr, A.S., Bay-Jensen, A.C., Leeming, D.J. *et al.* (2013) Serological identification of fast progressors of structural damage with rheumatoid arthritis. *Arthritis Research & Therapy*, 15, R86.

194. Houseman, M., Potter, C., Marshall, N. *et al.* (2012) Baseline serum MMP-3 levels in patients with Rheumatoid Arthritis are still independently predictive of radiographic progression in a longitudinal observational cohort at 8 years follow up. *Arthritis Research & Therapy*, 14, R30.
195. Heard, B.J., Martin, L., Rattner, J.B., Frank, C.B., Hart, D.A. & Krawetz, R. (2012) Matrix metalloproteinase protein expression profiles cannot distinguish between normal and early osteoarthritic synovial fluid. *BMC Musculoskeletal Disorders*, 13, 126.
196. Kerna, I., Kisand, K., Tamm, A.E., Kumm, J. & Tamm, A.O. (2013) Two single-nucleotide polymorphisms in ADAM12 gene are associated with early and late radiographic knee osteoarthritis in estonian population. *Arthritis*, 2013, 878126.
197. Kerna, I., Kisand, K., Laitinen, P. *et al.* (2012) Association of ADAM12-S protein with radiographic features of knee osteoarthritis and bone and cartilage markers. *Rheumatology International*, 32, 519–523.
198. Valdes, A.M., Van Oene, M., Hart, D.J. *et al.* (2006) Reproducible genetic associations between candidate genes and clinical knee osteoarthritis in men and women. *Arthritis and Rheumatism*, 54, 533–539.
199. Larsson, S., Lohmander, L.S. & Struglics, A. (2009) Synovial fluid level of aggrecan ARGS fragments is a more sensitive marker of joint disease than glycosaminoglycan or aggrecan levels: a cross-sectional study. *Arthritis Research & Therapy*, 11, R92.
200. Germaschewski, F.M., Matheny, C.J., Larkin, J. *et al.* (2014) Quantitation oF ARGS aggrecan fragments in synovial fluid, serum and urine from osteoarthritis patients. *Osteoarthritis and cartilage/OARS, Osteoarthritis Research Society*, 22, 690–697.
201. Dufield, D.R., Nemirovskiy, O.V., Jennings, M.G., Tortorella, M.D., Malfait, A.M. & Mathews, W.R. (2010) An immunoaffinity liquid chromatography-tandem mass spectrometry assay for detection of endogenous aggrecan fragments in biological fluids: Use as a biomarker for aggrecanase activity and cartilage degradation. *Analytical Biochemistry*, 406, 113–123.
202. Young-Min, S., Cawston, T., Marshall, N. *et al.* (2007) Biomarkers predict radiographic progression in early rheumatoid arthritis and perform well compared with traditional markers. *Arthritis and Rheumatism*, 56, 3236–3247.
203. Smith, E.R., Zurakowski, D., Saad, A., Scott, R.M. & Moses, M.A. (2008) Urinary biomarkers predict brain tumor presence and response to therapy. *Clinical Cancer Research*, 14, 2378–2386.
204. Mantyla, P., Stenman, M., Kinane, D. *et al.* (2006) Monitoring periodontal disease status in smokers and nonsmokers using a gingival crevicular fluid matrix metalloproteinase-8-specific chair-side test. *Journal of Periodontal Research*, 41, 503–512.
205. Visvanathan, S., Marini, J.C., Smolen, J.S. *et al.* (2007) Changes in biomarkers of inflammation and bone turnover and associations with clinical efficacy following infliximab plus methotrexate therapy in patients with early rheumatoid arthritis. *The Journal of Rheumatology*, 34, 1465–1474.
206. Rooney, T., Roux-Lombard, P., Veale, D.J., FitzGerald, O., Dayer, J.M. & Bresnihan, B. (2010) Synovial tissue and serum biomarkers of disease activity, therapeutic response and radiographic progression: analysis of a proof-of-concept randomised clinical trial of cytokine blockade. *Annals of the Rheumatic Diseases*, 69, 706–714.
207. Kaneko, A., Kida, D., Saito, K., Tsukamoto, M. & Sato, T. (2012) Clinical results for tocilizumab over one year in the clinical setting as assessed by CDAI (clinical disease activity index): CRP at week 12 and MMP-3 at week 24 are predictive factors for CDAI. *Rheumatology International*, 32, 3631–3637.
208. Bakker, M.F., Cavet, G., Jacobs, J.W. *et al.* (2012) Performance of a multi-biomarker score measuring rheumatoid arthritis disease activity in the CAMERA tight control study. *Annals of the Rheumatic Diseases*, 71, 1692–1697.
209. Yang, C., Gu, J., Rihl, M. *et al.* (2004) Serum levels of matrix metalloproteinase 3 and macrophage colony-stimulating factor 1 correlate with disease activity in ankylosing spondylitis. *Arthritis and Rheumatism*, 51, 691–699.
210. Pedersen, S.J., Hetland, M.L., Sorensen, I.J., Ostergaard, M., Nielsen, H.J. & Johansen, J.S. (2010) Circulating levels of interleukin-6, vascular endothelial growth factor, YKL-40, matrix metalloproteinase-3, and total aggrecan in spondyloarthritis patients during 3 years of treatment with TNFalpha inhibitors. *Clinical Rheumatology*, 29, 1301–1309.
211. Batchelor, T.T., Duda, D.G., di Tomaso, E. *et al.* (2010) Phase II study of cediranib, an oral pan-vascular endothelial growth factor receptor tyrosine kinase inhibitor, in patients with recurrent glioblastoma. *Journal of Clinical Oncology : Official Journal of the American Society of Clinical Oncology*, 28, 2817–2823.
212. Gerstein, H.C., Ratner, R.E., Cannon, C.P. *et al.* (2010) Effect of rosiglitazone on progression of coronary atherosclerosis in patients with type 2 diabetes mellitus and coronary artery disease: the assessment on the prevention of progression by rosiglitazone on atherosclerosis in diabetes patients with cardiovascular history trial. *Circulation*, 121, 1176–1187.

213. Kremer, J.M., Genant, H.K., Moreland, L.W. *et al.* (2006) Effects of abatacept in patients with methotrexate-resistant active rheumatoid arthritis: a randomized trial. *Annals of Internal Medicine*, 144, 865–876.
214. Hassett, A.L., Li, T., Buyske, S., Savage, S.V. & Gignac, M.A. (2008) The multi-faceted assessment of independence in patients with rheumatoid arthritis: preliminary validation from the ATTAIN study. *Current Medical Research and Opinion*, 24, 1443–1453.
215. Lambert, R.G., Salonen, D., Rahman, P. *et al.* (2007) Adalimumab significantly reduces both spinal and sacroiliac joint inflammation in patients with ankylosing spondylitis: a multicenter, randomized, double-blind, placebo-controlled study. *Arthritis and Rheumatism*, 56, 4005–4014.
216. Braun, A., Dehn, C., Krause, F. & Jepsen, S. (2008) Short-term clinical effects of adjunctive antimicrobial photodynamic therapy in periodontal treatment: a randomized clinical trial. *Journal of Clinical Periodontology*, 35, 877–884.
217. Christodoulides, N., Nikolidakis, D., Chondros, P. *et al.* (2008) Photodynamic therapy as an adjunct to non-surgical periodontal treatment: a randomized, controlled clinical trial. *The Journal of Periodontology*, 79, 1638–1644.
218. Wilkin, T.J., Lalama, C.M., McKinnon, J. *et al.* (2012) A pilot trial of adding maraviroc to suppressive antiretroviral therapy for suboptimal CD4(+) T-cell recovery despite sustained virologic suppression: ACTG A5256. *The Journal of Infectious Diseases*, 206, 534–542.
219. Gupta, S.K., Mi, D., Dube, M.P. *et al.* (2013) Pentoxifylline, inflammation, and endothelial function in HIV-infected persons: a randomized, placebo-controlled trial. *PLoS ONE*, 8, e60852.
220. Minagar, A., Alexander, J.S., Schwendimann, R.N. *et al.* (2008) Combination therapy with interferon beta-1a and doxycycline in multiple sclerosis: an open-label trial. *Archives of Neurology*, 65, 199–204.
221. Dengler, R., Diener, H.C., Schwartz, A. *et al.* (2010) Early treatment with aspirin plus extended-release dipyridamole for transient ischaemic attack or ischaemic stroke within 24 h of symptom onset (EARLY trial): a randomised, open-label, blinded-endpoint trial. *Lancet Neurology*, 9, 159–166.
222. Shimizu, Y., Temma, T., Hara, I. *et al.* (2014) In vivo imaging of membrane type-1 matrix metalloproteinase with a novel activatable near-infrared fluorescence probe. *Cancer Science*, 105, 1056–1062.
223. Oh, G., Yoo, S.W., Jung, Y. *et al.* (2014) Intravital imaging of mouse colonic adenoma using MMP-based molecular probes with multi-channel fluorescence endoscopy. *Biomedical Optics Express*, 5, 1677–1689.
224. Olson, E.S., Aguilera, T.A., Jiang, T. *et al.* (2009) In vivo characterization of activatable cell penetrating peptides for targeting protease activity in cancer. *Integrative Biology : Quantitative Biosciences from Nano to Macro*, 1, 382–393.
225. Crisp, J.L., Savariar, E.N., Glasgow, H.L., Ellies, L.G., Whitney, M.A. & Tsien, R.Y. (2014) Dual yargeting of integrin alphavbeta3 and matrix metalloproteinase-2 for optical imaging of tumors and chemotherapeutic delivery. *Molecular Cancer Therapeutics*, 13, 1514–1525.
226. Osborn, E.A. & Jaffer, F.A. (2013) The advancing clinical impact of molecular imaging in CVD. *JACC: Cardiovascular Imaging*, 6, 1327–1341.
227. Vermeij, E.A., Koenders, M.I., Blom, A.B. *et al.* (2014) In vivo molecular imaging of cathepsin and matrix metalloproteinase activity discriminates between arthritic and osteoarthritic processes in mice. *Molecular Imaging*, 13, 1–10.

Index

Matrix Metalloproteinase Biology, First Edition. Edited by Irit Sagi and Jean P. Gaffney.